KU-052-267

Low Back Disorders

Evidence-Based Prevention and Rehabilitation

Stuart McGill, PhD
University of Waterloo
Canada

Human Kinetics

Library of Congress Cataloging-in-Publication Data

McGill, Stuart, 1957-
Low back disorders : evidence-based prevention and rehabilitation /
Stuart McGill.
p. ; cm.
Includes bibliographical references and index.
ISBN 0-7360-4241-5 (hard cover)
1. Backache. 2. Evidence-based medicine.
[DNLM: 1. Back Injuries--prevention & control. 2. Back
Injuries--rehabilitation. 3. Evidence-Based Medicine. 4. Low Back
Pain--prevention & control. 5. Low Back Pain--rehabilitation. WE 755
M4778 2002] I. Title.
RD771.B217 M42 2002
617.5'64052--dc21 2002005695

ISBN: 0-7360-4241-5

Acquisitions Editor: Loarn D. Robertson, PhD; **Developmental Editor:** Elaine H. Mustain; **Assistant Editor:** Maggie Schwarzentraub; **Copyeditor:** Patsy Fortney; **Proofreader:** Erin Cler; **Indexer:** Craig Brown; **Permission Manager:** Dalene Reeder; **Graphic Designer:** Fred Starbird; **Graphic Artist:** Yvonne Griffith; **Photo Manager:** Leslie A. Woodrum; **Cover Designer:** Jack W. Davis; **Cover Image:** Primal Pictures; **Photographer (interior):** Leslie A. Woodrum, except where noted in Photo Credits (page 280) or on individual photos; **Art Manager:** Kelly Hendren; **Illustrator (Mac):** Tom Roberts; **Illustrator (medical):** Argosy; **Printer:** Sheridan Books

Printed in the United States of America 10 9 8 7 6 5 4 3 2

Human Kinetics
Web site: www.HumanKinetics.com

United States: Human Kinetics, P.O. Box 5076, Champaign, IL 61825-5076
800-747-4457
e-mail: humank@hkusa.com

Canada: Human Kinetics, 475 Devonshire Road, Unit 100, Windsor, ON N8Y 2L5
800-465-7301 (in Canada only)
e-mail: orders@hkcanada.com

Europe: Human Kinetics, 107 Bradford Road, Stanningley
Leeds LS28 6AT, United Kingdom
+44 (0) 113 255 5665
e-mail: hk@hkeurope.com

Australia: Human Kinetics, 57A Price Avenue, Lower Mitcham, South Australia 5062
08 8277 1555
e-mail: liahka@senet.com.au

New Zealand: Human Kinetics, P.O. Box 105-231, Auckland Central
09-523-3462
e-mail: hkp@ihug.co.nz

I dedicate this book to my teachers. My parents, John and Elizabeth, who believed in the therapy of hard work—it makes for better-tasting beer at the end of the day and a better tomorrow. The few high school teachers who were able to see past my impatience and belligerence, in particular Ralph Colucci, who withstood my antics in history class but then spent countless hours coaching on the track and football fields. He practiced the highest example of daily dedication and the drive to continue after others had given up. Those lessons I will never forget. My university professors in the tough early years, particularly Dr. Klambauer, who was able to transform mathematics from abstract magic into something very mechanical that I could suddenly feel in my hands. This was a turning point. Professor David Winter, who taught me, among many other things, how to read a scientific paper. Professor Robert Norman, my PhD supervisor and mentor, who is the embodiment of integrity and the master of viewing any issue from the highest of intellectual vantage points. My many research colleagues from around the world, too numerous to mention here, who have taught me the many perspectives needed to temper the arrogance that comes so naturally when one is the only person privy to new research results. And to my wonderful graduate students—Jacek Cholewicki, Vanessa Yingling, Jack Callaghan, Sylvain Grenier, Lina Santaguida, Crisanto Sutarno, John Peach, Craig Axler, Lisa Brereton, Greg Lehman, Jennifer Gunning, Richard Preuss, Joan Scannell, David Bereznick, Kim Ross, Doug Richards, Kelly Walker, and Natasa Kavcic. The old dog has learned a few of your new tricks. All of these people have contributed to my forging. Any remaining personal failings I can only attribute to the unfortunate combination of running Canadian software on Irish hardware.

Finally, to my wife and children, Kathryn, John, and Sarah, who have taught me to see the joy of the moment and enjoy the security of unconditional love. I'm just a chunk of coal, but I'm gonna be a diamond some day!

Contents

List of Tests and Exercises

Why and How You Should Read This Book

Given the myriad of stretch and strengthening programs available for the professional or layperson with a bad back, why write another book offering guidelines for better prevention and rehabilitation of low back troubles? The reason is that very few of the available books/manuals actually quantify the demands of the exercises, specifically, the resulting tissue loads and the muscle challenge. This has led to some programs recommending routines that replicate the mechanical causes of the damage! Such routines may actually increase the risk of developing back troubles or exacerbate existing ones. Other more specialized books promote spine stabilization exercises without quantifying stability, thus basing the recommended exercises on a good guess—albeit a guess based on years of clinical observation. A scientific foundation to justify the recommendations has been thin at best. *This book is not about perpetuating clinical myths—it is about challenging them and proposing valid and justifiable alternatives.*

How the Book Is Organized

The intention of this book is to provide the best available scientific evidence to optimize injury prevention and rehabilitation efforts. It is designed to assist those who work to prevent low back troubles and those who are charged with the responsibility of rehabilitating them. Part I begins with a scan of the landscape related to the low back question. Here the main issues will be introduced together with a brief introduction to the legislative and institutional constraints—some of which assist efforts to enhance back health while others act as direct barriers. This is followed with evidence as to how the low back structure and associated tissues work during conditions of good health and how they become injured. These chapters describe

- how the spine works,
- injury mechanisms involving individual tissues, and
- injury mechanisms involving the full lumbar mechanism.

Specific information—such as tissue loads, muscle activation levels, and measures of stability—is provided so that readers can meaningfully compare approaches when deciding how to prevent or rehabilitate low back injury. Thus, the first part provides the scientific foundation for the rest of the book.

Identifying and implementing the best programs result from considering the evidence. Thus, part II supports evidence-based practice by applying the scientific foundation in part I to the prevention of low back injury. Part III applies this same information to the rehabilitation of low back injury.

Focus and Features of the Text

This text will be of value for those involved in designing and delivering back trouble prevention and rehabilitation programs—physical therapists, physiatrists, chiropractors, kinesiologists, ergonomists, and orthopedic specialists.

Low Back Disorders features a number of elements besides the basic text to ensure that all segments of the very diverse readership that deals with low back issues will be able to absorb and use the material. Each chapter introduction includes learning objectives. Special elements in some chapters highlight issues of methodological concern, for example, or simply provide further elucidation of some concepts or terms that are not common across disciplines. Every section that has clinical relevance contains highlighted statements identified as such. In addition, a wide variety of occupational and athletic examples throughout the book can be generalized to all activities of daily living.

A major theme of this book is the benefits of blending rehabilitation efforts with ergonomics and prevention. Traditionally, ergonomists and industrial engineers have been responsible for prevention efforts mainly in the workplace, while the medical community has been responsible for rehabilitation. This natural division has caused an unfortunate compromise. Rehabilitation efforts will not be effective without preventing the cause of the tissue irritants, and prevention efforts will be retarded without an understanding of the variables obtained from medical investigations into the causes of tissue damage. *No clinician will be effective if the cause of the patient's troubles is not removed.* The expertise required for optimal prevention and rehabilitation can be found in the fields of ergonomics and rehabilitation medicine, which are naturally symbiotic. Many in the sport and business world have debated whether it is better to develop a stronger offense or defense. The obvious answer is that it is best to have both. I hope that this book will foster a better process resulting in more effective injury prevention and successful rehabilitation.

Some Encouraging Words

Some readers may be tempted to apply the material in parts II and III without digesting the foundation material contained in part I. This is a mistake. People who have attended my courses and were previously familiar with our published scientific papers often state that they were unable to fully synthesize the story and integrate the critical information. Once aware of the full story contained in the course and told in this book, they were much more effective at preventing low back disorders and prescribing exercises that help patients become better rather than worse.

A Note to Lay Readers

Although this book was written for clinicians, many lay people have bought copies from its initial print run. As a result, they have written me letters expressing their appreciation for their better understanding the science and their own back troubles, and many have reported substantial improvement. This leads me to add the thought that although successful self-treatment is gratifying, it removes some of the safeguards in the rehabilitation process. If you are an owner of a troubled back and are not a clinician, I remind you always to consult a physician at the outset to rule out any possible existence of a condition such as a tumor or a contraindication for exercise.

Acknowledgments

No one will know the effort involved in writing a book until they have done so themselves. Many people have been part of the team. I specifically want to thank and acknowledge my editors at Human Kinetics Publishers: Loarn Robertson for his early suggestions and, in particular, Elaine Mustain for her expertise in the English language, ability to grasp mechanical and medical concepts, and dogged patience during the rewriting stage. Also Sylvain Grenier, my PhD student and lab manager, for the preparation of many figures that appear in this book.

PART I

Scientific Foundation

The very best professionals with expertise in low back injury prevention and rehabilitation have two things in common: an insight and wisdom developed from experience and a strong scientific foundation to enable evidence-based practice. Each problematic back is part of a whole person, who in turn is the sum of influences that range from those at the cellular level to those at the societal level. Some refer to this as the biopsychosocial approach. No two problematic backs (or patients) are the same, suggesting that optimal intervention would be different for each one. The best approach for optimal rehabilitation and prevention of those factors that exacerbate bad backs will result from wise and scientifically based decisions. Part I of this book provides a foundation for back professionals to enable them to make better decisions.

CHAPTER 1

Introduction to the Issues

There is no shortage of manuals and books offering wisdom on low back health. Authors range from those with formal medical or rehabilitation training to laypeople who have found an approach to alleviate their own back troubles and become self-proclaimed low back health prophets. Their intentions are honorable, but their advice is rarely based on a sound scientific foundation. Too many of these authors offer inappropriate recommendations or even harmful suggestions. Years ago, as I began to develop scientific investigations into various aspects of low back problems, I would ask my graduate students to find the scientific foundation for many of the "commonsense" recommendations I was hearing both in the clinic and in industrial settings. To my surprise they often reported that the literature yielded no, or very thin, evidence (note that I choose my students carefully and that they are very competent and reliable). Examples of such thinly supported "commonsense" recommendations include the following:

- Bend the knees to perform a sit-up.
- Bend the knees and keep the back straight to perform a lift.
- Reduce the load being handled in order to reduce the risk of back troubles.

In fact, each of these recommendations may be appropriate in some situations but, as will be shown, not in all.

The famous economist John Kenneth Galbraith was well-known for demonstrating that actions based on common wisdom, at least in economic terms, were often doomed to fail. He stated that common wisdom is generally neither common nor wise. Galbraith eloquently expressed exactly what I had experienced with "clinical wisdom" pertaining to the low back. Many attempts at preventing low back troubles and rehabilitating symptomatic ones have failed simply because they relied on ill-conceived clinical wisdom. This history of failed attempts is particularly unfortunate because it has lent credence to the assertions of a number of increasingly well-known "authorities" that low back injury prevention and rehabilitation programs are a waste of resources. These authorities claim that the majority of low back problems are not organic at all—that, for example, most of these difficulties have materialized because workers are paid too much for injury compensation, have been subject to psychosocial influences, or crave sympathy. These dismissals of back injury are not justified. Back injury prevention and rehabilitation programs with strong scientific foundations, executions, and follow-up can be effective.

Having stated this, I must acknowledge that justifying improved practice on scientific evidence is a dynamic process. With new evidence, the foundation will change. To account for such inevitable shifts, I have developed a balanced approach in these pages, reviewing the assets and liabilities and opposing views of an argument where appropriate. But fair warning! As you read this book, be prepared to challenge current thoughts and rethink currently accepted practices of injury prevention and approaches to rehabilitation.

This chapter will familiarize you with some of the debatable issues regarding low back function, together with some opinions that rehabilitation professionals hold about patients, diagnosis, compensation, and disability. It will also explore the circumstances that lead to back injury and discuss the need to apply this knowledge to improve low back injury prevention and rehabilitation.

Legislative Landscape: The Unfortunate Adverse Impact on Bad Backs

Although most legislation and legal activity involving bad backs are enacted with good intentions, much is counterproductive. A good example is the issue of spine range of motion (ROM). The American Medical Association (AMA) guidelines (1990) for quantifying the degree of back disability are based mostly on loss of spine ROM. Lawyers and compensation boards who need numbers for the purpose of defining disability and rewarding compensation have latched onto spine ROM as an objective and easily measured factor. In the legal arena, therapy is considered successful when the ROM has been restored or at least improved.

Scientific evidence suggests, however, that after back injury, many people do not do well with an emphasis on enhancing spine mobility. In some cases, back problems are actually exacerbated by this approach. In fact, evidence shows that many back injuries improve with stabilizing approaches—motor control training, enhancement of muscle endurance, and training with the spine in a neutral position (Saal and Saal, 1989, may be considered the classic work). Our most recent work has shown that three-dimensional low back range of motion has no correlation to functional test scores or even the ability to perform occupational work (Parks et al., in press). In the best practice, spine flexibility may not be emphasized until the very late stages of rehabilitation, if ever.

How, then, did this idea of flexibility as the best measure of successful rehabilitation become so entrenched? The current metric for determining disability appears to have been chosen for legal convenience rather than for a positive impact on low back troubles. The current landscape creates a reward system for therapy that arguably hinders optimal rehabilitation. Perhaps the criteria for determining disability need to be reassessed and justified with scientific evidence.

Another example illustrates the perverse impact of well-intended equity legislation. We all have equal rights under the law, but we are not physical equals. Although individual variance is present in every group, different populations within society demonstrate quite different capabilities. For example, the data of Jager and colleagues (1991), compiled from many studies, clearly showed that young men can tolerate more compressive load down their lumbar spines than can older men, and similarly, men can tolerate about an additional third more load than can women when matched for age. Yet human rights legislation, which is designed to create fairness and equity by discouraging distinctions among groups, actually puts older females at greater risk than younger men. By not allowing a 64-year-old osteoporotic woman to be treated (and protected) differently from a 20-year-old, fit, 90-kg (200-lb) male in terms of tolerating spine load, the legislation presents a major barrier for intelligently implementing tolerance-based guidelines for protecting workers.

Deficiencies in Current Low Back Disorder Diagnostic Practices

It is currently popular for many authorities to suggest that back trouble is not a medical condition. They assert that physical loading has little to do with low back injury compensation claims; rather, they believe workers complain of back problems in order to benefit from overly generous compensation packages or to convince physicians they are sick. According to this view, any biomechanically based injury prevention or rehabilitation program is useless. Variables within the **psychosocial** sphere dominate any biological or mechanical variable. If this is true, then this book is of no value—it should be about psychosocial intervention.

Those who contend that psychosocial factors dominate low back issues are well-published scientists and physicians. For example, Professor Richard Deyo (1998) summarized the view of many: "Consider the following paradox. The American economy is increasingly postindustrial, with less heavy labor, more automation and more robotics, and medicine has consistently improved diagnostic imaging of the spine and developed new forms of surgical and nonsurgical therapy. But work disability caused by back pain has steadily risen." This line of reasoning assumes that modern work (i.e., more repetitive, more sedentary) is healthier for the back than the predominantly physical labor of past generations. The evidence suggests, however, that the repetitive motions required by some specialized modern work or the sedentary nature characterizing others, produce damaging biomechanical stressors. In fact, the variety of work performed by our great-grandparents may have been far healthier. Deyo also seems to assume that nonsurgical therapy has been appropriately chosen for each individual, whereas I suggest that inappropriate therapy prescriptions remain quite common.

Professor Alf Nachemson (1992) wrote that "most case control studies of cross-sectional design that have addressed the mechanical and psychosocial factors influencing LBP (low back pain), including job satisfaction, have concluded that the latter play a more important role than the extensively studied mechanical factors." Yet none of the several references cited to support this opinion made reasonable quantification of the physical job demands. Generally, these studies found that psychosocial variables were related to low back troubles, but in the absence of measuring mechanical loading, they had no chance to evaluate a loading relationship.

Finally, Dr. Nortin Hadler (2001) has been rather outspoken, stating, for example, that "it is unclear whether there is any meaningful association between task content and disabling regional musculoskeletal disorders for a wide range of physical disorders" and that "on the other hand, nearly all multivariate cross sectional and longitudinal studies designed to probe for associations beyond the physical demand of tasks, detect associations with the psychosocial context of working."

Recent evidence clearly shows that, although psychosocial factors can be important in modulating patient behavior, biomechanical components are important in leading to low back disorders and in their prevention. The position that biomechanics plays no role in back health and activity tolerance can be held only by those who have never performed physical labor and have not experienced firsthand the work methods that must be employed to avoid disabling injury. While the scientific evidence is absolutely necessary, it will only confirm the obvious to those who have this experience. I find it perversely satisfying when physicians tell me that they are now, after missing work as a result of a nasty back episode related to physical work, able to relate to their patients. Perhaps experience with a variety of heavy work and with disabling pain should be required for some medics!

It is, then, essential to investigate and understand the links among loading, tissue damage or irritation, psychosocial factors, and performance to provide clues for the design and implementation of better prevention and rehabilitation strategies for low back troubles. In the following sections I will address several commonly held beliefs about back injury.

Is It True That 85% of Back Troubles Are of Unknown Etiology?

Low back injury reports often mention the statistic that 85% of low back troubles are of unknown **etiology**. This has led to the popular belief that disabling back troubles are inevitable and just happen, a statement that defies the plethora of literature linking specific mechanical scenarios to specific tissue damage. Some have argued that this statement is simply the product of poor diagnosis—or of clinicians reaching the end of their expertise (e.g., Finch, 1999). In fairness I must point out that diagnosis often depends on the profession of the diagnostician. Each group attempts to identify the primary dysfunction according to its particular type of treatment. For example, a physical therapist will attempt diagnosis to guide decisions regarding manual therapy approaches, while a surgeon may find a diagnosis directed toward making surgical decisions more helpful. Some clinicians (surgeons, for example) seek a specific tissue as a pain candidate. From this perspective, nerve block procedures have shown conclusive pain source diagnosis in well over 50% of cases (e.g., Bogduk et al., 1996; Lord et al., 1996; Finch, 1999). This has prompted research into which tissues are innervated and are candidates as pain generators. Biomechanists often argue that this may be irrelevant since a spine with altered biomechanics has altered tissue stresses. Thus a damaged tissue may cause overload on another tissue causing pain whether the damaged tissue is innervated or not. This is why other clinicians employ skilled provocative mechanical loading of specific tissues to reveal those that hurt or at least to reveal loading patterns or motion patterns that cause pain. These types of functional diagnoses are helpful in designing therapy and in developing less painful motion patterns, but the process of functional diagnoses will be hindered with a poor understanding of spine biomechanics. Furthermore, those with a thorough understanding of the biomechanics of tissue damage can be guided to a general diagnosis by reconstructing the instigating mechanical scenario. An additional benefit of this approach is that once the cause is understood, it can be removed or reduced. Unfortunately, many patients continue to have troubles simply because they continue to engage in the mechanical cause. Familiarity with spine mechanics will dispel this myth of undiagnosable back trouble and reduce the percentage of those with back troubles of no known cause.

Even with a tissue-based diagnosis, the practice of treating all patients with a specific diagnosis with a singular therapy has not proven productive (Rose, 1989). For example, success rates with many cancer therapies greatly improved with the combination of chemotherapy and radiotherapy. Optimal back rehabilitation requires removal of the cause and the addition perhaps of stability, manual soft tissue therapy, or something else depending on the patient. Few patients fall into a "complete fit" for functional diagnosis where a singular approach will yield optimal results. Both for interpretation of the literature and for clinical decision making, it would appear prudent to question the diagnostic criteria needed before a given diagnosis is assigned.

Limitations in tissue-based diagnosis should not be used to suggest that determining the cause of back troubles is irrelevant nor that the manual/medical treatment in some cases is fruitless, leaving psychosocial approaches to prevail by default. Even given the current limitations, the diagnostic approach is productive for guiding prevention and rehabilitation approaches.

Is It True That Most Chronic Back Complaints Are Rooted in Psychological Factors?

While there is no doubt that many chronic back cases have psychological overlays, the significance of psychology for back problems is often greatly exaggerated. Dr. Ellen Thompson (1997) coined the phrase "bankrupt expertise" when referring to spine docs who are unable to guide improvement in their patients and default to blaming the patients and their psychoses. These physicians either dismiss mechanical causation or assume that mechanical causation has been adequately addressed.

At our university clinic I see patients who have been referred by physicians for consult. These are either elite performers or the very difficult chronic bad backs who have failed with all other approaches. In spite of the fact that these people have received very thorough attention, I am continually heartbroken to hear about the minimal notice paid to ongoing back stressors and of the exercises that these "basket case backs" have been prescribed that have only exacerbated their condition. The day before I wrote this section, I saw a classic example.

A woman had suffered for five years on disability and had seen no fewer than 12 specialists from a variety of disciplines. Although several had acknowledged that she had physical concerns, her troubles were largely attributed to mental depression. She consistently reported being unable to tolerate specific activities while being able to tolerate others. Some provocative testing confirmed her report together with uncovering a previously undiagnosed arthritic hip. For years she had been faithfully following the instructions of her health care providers to perform pelvic tilts, knees-to-chest stretches first thing in the morning, and sit-ups; to take her large dog for walks; and so on. All of these ill-chosen suggestions had prevented her posterior disc (with sciatica)–based troubles from improving. As we will see later, these types of troubles typically do not recover with flexion-based approaches—particularly first thing in the morning. Moreover, the lead-imposed torsional loads that she experienced every time she walked her dog exceeded her tolerance. Although she reported vacuuming as a major exacerbator of her troubles, her health care providers had never shown her how to vacuum her home in a way to spare her back. I suggested that removing these daily activities and replacing the flexion stretches with neutral spine position awareness training and isometric torso challenges would likely start a slow, progressive recovery pattern. I believed that her psychological concerns would probably disappear with her back symptoms if she fell into the typical pattern. This patient, with this typical story, has a reasonably good chance to enjoy life once again. (Note: this patient was back to work and off her antidepressant medication at the time of proofing this manuscript.)

None of the "experts" this woman had seen—including physical therapists, chiropractors, psychologists, physiatrists, neurologists, and orthopods—addressed mechanical concerns. This is not to condemn these professions but rather to suggest that sharing experiences and approaches will help us to be more successful in helping bad backs. Perhaps these professionals were unaware of the principles of spine function, the types of loads that are imposed on the spine tissues during certain activities, and how these activities and spine postures can be changed to greatly reduce the loads—in other words, the biomechanical components.

This book is an attempt to heighten the awareness and potential of this mechanical approach. While it sounds very harsh, I have found relatively few experts who appear willing to adequately address the causes of back troubles while working to find the most appropriate therapy. My years of laboratory-based work combined with collaboration in recent years with my clinical colleagues have provided me with unique insight. As a result, I am not so quick to blame the chronic patient.

Does Pain Cause Activity Intolerance?

Evidence that mechanical tissue overload causes damage is conclusive. But does the damage cause pain, and does the chronic pain cause work intolerance? Several, but limited numbers of, studies have mechanically or chemically stimulated tissues to reproduce clinical pain patterns. (The absence of definitive, large-scale studies is due to the ethical issues of performing invasive procedures and probably not because such studies are without scientific merit.) For example, the pioneering work of Hirsch and colleagues (1963-64) documented pain from the injection of hypertonic saline into specific spine tissues thought to be candidates for damage. Subsequent to this work, several other studies suggested the link between mechanical stimulation and pain, for example, the work of Hsu and colleagues (1988), documenting pain in damaged discs.

There is irrefutable evidence that vertebral disc end-plate fractures are very common and only result from mechanical overload (Brinckmann, Bigemann, and Hilweg, 1989; Gunning, Callaghan, McGill, 2001). That these fractures are also found in necropsy specimens that were subjected to whiplash (Taylor, Twomey, and Corker, 1990) also strengthens another facet of this relationship. Hsu performed discograms (injections of radio contrast) into 692 discs of which 14 demonstrated leakage into the vertebral body, confirming an end-plate fracture. Four of these discs (28%) produced severe pain, nine (64%) produced fully concordant pain, and one produced mildly discordant pain. In contrast, only 11% of the remaining 678 discs with no end-plate disruption produced severe pain, 31% with concordant pain, 17% with mild pain, and 41% with no pain. This evidence provides strong support for the notion that loading causes damage and damage causes pain.

Even though pain can limit function and activity in other areas of the body, some still suggest that these are not linked when a bad back is at issue. Teasell (1997) provided an interesting perspective when he argued that in sports medicine, as opposed to occupational medicine, it is well accepted that some injuries require months of therapy or can even cause retirement from the activity. He noted that athletes receiving specialized sports medicine care are an interesting group to consider since many are highly motivated, are in top physical condition, are well paid, have access to good medical care, and are fully compensated even while injured. Their injuries and pain can cause absence from play for substantial amounts of time and can even end their lucrative careers. Teasell reminded us that not all long-term chronic pain is an entirely psychosocial concern, as implied by some clinicians. These clinicians' dismissal of the usefulness of physical approaches simply because they have not been successful in reducing long-term troubles is a disservice to the patient.

Inadequacies in Current Care and Prevention of Low Back Disorders

Many back patients can testify that the care they have received for their troubles is not satisfactory. What are some of the factors that contribute to the inadequacy of their experience? Certainly one of them is the fact that the epidemiological evidence on which many professionals base their treatment recommendations can be quite confusing. Following are some examples of the issues that cause confusion.

- **Plethora of studies on "backache."** Nonspecific "backache" is nearly impossible to quantify and, even if it could be quantified, offers no guidance for intervention. As such, any study of treatment interventions on nonspecific backache is of little use. Some backs suffer with discogenic problems, for example, and will respond quite differently from those with ligamentous damage or facet-based problems. Efficacy studies that do not subclassify bad backs end up with nonspecific "average" responses. This has led to the belief that nothing works—or that everything does, but to a limited degree. More studies on nonspecific backache treatment will not be helpful, nor will the large epidemiological reviews of these studies offer real insight.

- **U-shaped function of loading and resulting injury risk.** Like many health-related phenomena, the relationship of low back tissue loading to injury risk appears to form a U-shaped function—not a monotonically rising line. For example, virtually every nutrient will cause poisoning with excessive dosage levels, but health suffers in their absence; thus, there is a moderate optimum. In the case with low back loading, evidence suggests that two regions in the U-shaped relationship are problematic—too much and too little. Porter (1987) suggested that heavy work is good for the back—but how does one define *heavy*? Porter was probably referring to work of sufficient challenge and variability to reach the bottom of the U and hence lower symptoms. From a biological perspective, sufficient loading is necessary to cause strengthening and toughening of tissues, but excessive levels will result in weakening. This lack of consensus in the literature regarding the measurement of exposure has been problematic (Marras et al., 2000). A more advanced understanding is required.

- **Relationship of intensity, duration of loading, and rest periods.** As Ferguson and Marras (1997) pointed out, some studies suggest that a certain type of loading is not related to pain, injury, or disability, while others suggest it is, depending on how the exposure was measured and where the moderate optimum for tissue health resides for the experimental population. The subjectivity of such studies is further underscored when we consider the question, Is there a clinical difference between tissue irritation and tissue damage? Loading experiments on human and animal tissues to produce damage reveal the "ultimate tolerable load" beyond which injuries cause biomechanical changes, pain, and gross failure to structures. In real life, any of us could irritate tissues to produce tremendous pain at loading levels well below the cadavericly determined **tolerance** by repeated and prolonged loading. In fact, evidence presented by Videman and colleagues (1995) suggests that the progressive development of conditions such as spinal stenosis results from years of specific **subfailure** activity. The fundamental question is, Could such conditions be avoided by evidence-based prevention strategies that include optimal loading, rest periods, and controlling the duration of exposure?

Ill-Advised Rehabilitation Recommendations

These failures to frame research and its results appropriately have resulted in many oversimplifications about low back treatment, which have in turn led to some inadequate treatment practices and recommendations. A few of the most common recommendations for back health are discussed here.

- **Strengthen muscles in the torso to protect the back.** Despite the clinical emphasis on increasing back muscle strength, several studies have shown that muscle strength cannot predict who will have future back troubles (Biering-Sorenson, 1984). On the other hand, Luoto and colleagues (1995) have shown that muscular endurance (as opposed to strength) is protective. Why, then, do many therapeutic programs continue to emphasize strength and neglect endurance? Perhaps it is a holdover influence from the athletic world in which the goal of training is to enhance performance. Perhaps it is an influence from the pervasive use of bodybuilding approaches in rehabilitation. As will be shown, optimal exercise therapy occurs when the emphasis shifts away from the enhancement of performance and toward the establishment of improved health. In many cases the two are mutually exclusive!

- **Bend the knees when performing sit-ups.** Clinicians widely recommend bending the knees during a sit-up, but on what evidence? A frustrating literature search suggests that this perception may be the result of "clinical wisdom." Interestingly, Axler and McGill (1997) demonstrated that there is little advantage to one knee position over the other, and in fact the issue is probably moot because there are far better ways to challenge the abdominal musculature and impose lower lumbar spine loads (traditional sit-ups cause spine loading conditions that greatly elevate the risk of injury). This issue is one of many that will be challenged in this book.
- **Performing sit-ups will increase back health.** Is this a true statement or an artifact of experimental methodology? Despite what many would like to believe, there is only mild literature support for the belief that people who are fit have less back trouble (although positive evidence is increasing, for example, Stevenson et al., 2001). Interestingly, many of the studies attempting to evaluate the role of increased fitness in back health actually included exercises that have been known to cause back troubles in many people. For example, many have attempted to enhance abdominal strength with sit-ups. After examining the lumbar compression that results from performing sit-ups with full flexion of the lumbar spine, it is clear that enough sit-ups will cause damage in most people. Each sit-up produces low back compression levels close to the National Institute for Occupational Safety and Health (**NIOSH**) action limit, and repeatedly compressing the spine to levels higher than the NIOSH action limit has been shown to increase the risk of back disorders (Axler and McGill, 1997). Thus, reaching a conclusion over the role of fitness from the published literature has been obscured by ill-chosen exercises. Increased fitness does have support, but the way in which fitness is increased appears to be critical.
- **To avoid back injury when lifting, bend the knees, not the back.** Probably the most common advice given by the clinician to the patient who must lift is to bend the knees and keep the back straight. In addition, this forms the foundation for virtually every set of ergonomic guidelines provided to reduce the risk of work-related injury. Very few jobs can be performed this way. Further, despite the research that has compared stooping and squatting styles of lifting, no conclusion as to which is better has been reached. The issue of whether to stoop or squat during a lift depends on the dimensions and properties of the load, the characteristics of the lifter, the number of times the lift is to be repeated, and so forth, and there may in fact be safer techniques altogether. Much more justifiable guidelines will be developed later in the text.
- **A single exercise or back stability program is adequate for all cases.** It is currently popular to promote the training of single muscles to enhance spine stability. While the original research was motivated by the intention to reeducate perturbed motor patterns that were documented to be the result of injury, others have misinterpreted the data and are promoting exercises to train muscles they believe are the most important stabilizers of the spine. Unfortunately, they did not quantify stability. The process of quantifying the contribution of the anatomical components to stability reveals that virtually all muscles can be important, but their importance continually changes with the demands of the activity and task. It is true that damage to any of the spinal tissues from mechanical overload results in unstable joint behavior. Because of

biomechanical changes to the joint, however, the perturbed tissue is rarely linked to the symptomatology in a simple way. More likely other tissues become involved, and which ones are involved will result in different accompanying motor disturbances. This variety in possible etiologies means that a single, simple rehabilitation approach often will not work. Is mobility to be restored at the expense of normal joint stability? Or is stability to be established first with enhanced mobility as a secondary delayed rehabilitation goal? Or is the clinical picture complex, for example, in a situation in which spine stability is needed but tonic psoas activity causing chronic hip flexure necessitates hip mobilization? This example, one of many that could have been chosen, illustrates the challenge of ensuring sufficient stability for the spinal tissues. No simple, or single, approach will produce the best results in all cases. The description and data presented in this book will help guide the formulation of exercises that ensure spine stability.

Can Back Rehabilitation Be Completed in 6 to 12 Weeks?

Some have suggested that damaged tissues should heal within 6 to 12 weeks. In fact, many have used this argument to support the notion that work intolerance exceeding this period has no pathoanatomical basis (e.g., Fordyce, 1995) but stems from psychosocial issues. Further, some have suggested that patient recovery would be better served by redirecting rehabilitative efforts away from physical approaches. This position can be refuted by data and indicates a misunderstanding of the complexities of spine pathomechanics. The concept that tissues heal within 6 to 12 weeks appears to be originally based on animal studies (reviewed in Spitzer, 1993). However, not all human patients get better so quickly (Mendelson, 1982), and the follow-up studies from some defined disorders such as whiplash are compelling in the support of lingering tissue disruption (e.g., Radanov et al., 1994).

Evidence will be presented in later chapters of both mechanical and neurological changes that linger for years subsequent to injury. This includes loss of various motor control parameters together with documented asymmetric muscle atrophy and other disorders. This suggests that postinjury changes are not a simple matter of gross damage "healing." Following are only a few of the types of damage that can be long term indeed:

- Specific tissues such as ligaments, for example, have been shown to take years to recover from relatively minor insult (Woo, Gomez, and Akeson, 1985).
- The intervertebral motion units form a complex mechanism involving intricate interplay among the parts such that damage to one part changes the biomechanics and loading on another part. From the perspective of **pathomechanics**, many reports have documented the cascade of biomechanical change associated with initial disc damage and subsequent joint instability and secondary arthritis, which may take years to progress (e.g., Brinckmann, 1985; Kirkaldy-Willis, 1998).
- Videman and colleagues (1995) documented that vertebral osteophytes were most highly associated with end-plate irregularities and disc bulging. Osteophytes are generally accepted to be secondary to disc and end-plate trauma but take years to develop.

Thus, to suggest that back troubles are not mechanically based if they linger longer than a few months only demonstrates a limited expertise.

Another question is, Can these back troubles linger for a lifetime? In this connection, it is interesting that elderly people appear to complain about bad backs less than younger people. Valkenburg and Haanen (1982) showed that back troubles are more frequent during the younger years. Weber (1983) provided further insight by reporting on patients who were 10 years post disc herniations (some of them had had surgery while others had not) and were engaged in strenuous daily activity—yet all were still receiving total disability benefits! It would appear that the cascade of changes resulting from some forms of tissue damage can take years but generally not longer than 10 years. Although the bad news is that the affected joints stiffen during the cascade of change, the good news is that eventually the pain is gone.

To summarize, the expectation that damaged low back tissues should heal within a matter of weeks has no foundation. In fact, longer-term troubles do have a substantial biomechanical/pathoanatomical basis. On the other hand, troublesome backs are generally not a life sentence.

Should the Primary Goal of Rehabilitation Be Restoring the Range of Motion?

Research has shown that an increased range of motion in the spine can increase the risk of future back troubles (e.g., Battie et al., 1990; Biering-Sorenson, 1984; Burton, Tillotson, and Troup, 1989). Why then does increasing the range of motion remain a rehabilitation objective? The first reason, discussed earlier, is the need to quantify reduced disability as defined by the AMA. Second, there is a holdover philosophy from the athletic world that increased range of motion enhances performance. This may be true for some activities, but it is untrue for others. As will be shown, this philosophy may work for other joints, but it generally does not work for the back. In fact, successful rehabilitation for the back is generally retarded by following athletic principles.

Mechanical Loading and the Process of Injury: A Low Back Tissue Injury Primer

Any clinician completing a worker/patient compensation form is required to identify the event that caused the injury. Very few back injuries, however, result from a single event. This section documents the more common cumulative trauma pathway leading to the culminating event of a back injury. Because the culminating event is falsely presumed to be the cause, prevention efforts are then focused on that event. This misdirection of efforts fails to deal with the real cause of the cumulative trauma.

While a generic scenario for injury is presented here, chapter 4 offers a more in-depth discussion of injury from repeated and prolonged loading to specific tissue. The purpose of this section is to promote consideration of the many factors that modulate the risk of tissue failure and to encourage probing to generate appropriate hypotheses about injury etiology.

Injury, or failure of a tissue, occurs when the applied load exceeds the **failure tolerance** (or strength of the tissue). For the purposes of this discussion, injury will be defined as the full continuum from the most minor of tissue irritation (but microtrauma nonetheless) to the grossest of tissue failure, for example, vertebral fracture or ligament avulsion. We will proceed on the premise that such damage generates pain.

Obviously, a load that exceeds the failure tolerance of the tissue, applied once, produces injury (see figure 1.1b in which a Canadian snowmobiler airborne and about to experience an axial impact with the spine fully flexed is at risk of posterior disc herniation upon landing). This injury process is depicted in figure 1.1a, where a margin of safety is observed in the first cycle of subfailure load. In the second loading cycle, the applied load increases in magnitude, simultaneously decreasing the margin of safety to zero, at which point an injury occurs. While this description of low back injury is common, particularly among medical practitioners who are required to identify

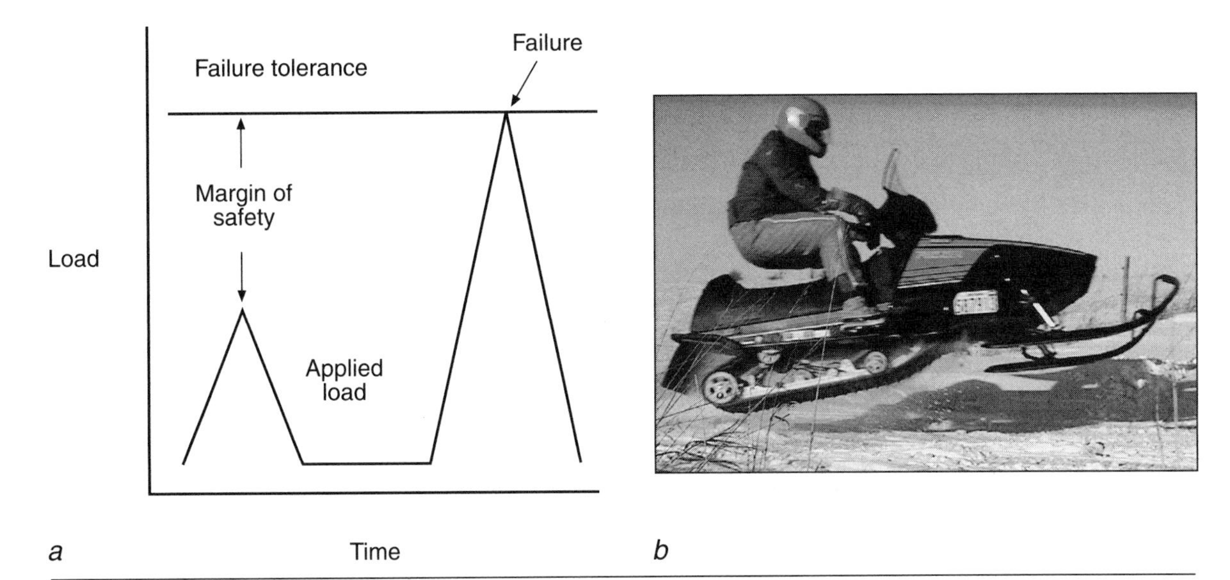

Figure 1.1 (*a*) A margin of safety is observed in the first cycle of subfailure load. In the second loading cycle, the applied load increases in magnitude, simultaneously decreasing the margin of safety to zero, at which point an injury occurs. (*b*) The Canadian snowmobile driver (the author in this case who should know better) is about to experience an axial compressive impact load to a fully flexed spine. A one-time application of load can reduce the margin of safety to zero as the applied load exceeds the strength or failure tolerance of the supporting tissues.

Reprinted from *Journal of Biomechanics,* 30 (5), S.M. McGill, "Invited paper: Biomechanics of low back injury: Implications on current practice and the clinic," 456-475, 1997, with permission from Elsevier Science.

an injury-causing event when completing forms for workers' compensation reports, my experience suggests that relatively few low back injuries occur in this manner.

More commonly, injury during occupational and athletic endeavors involves cumulative trauma from repetitive subfailure magnitude loads. In such cases, injury is the result of accumulated trauma produced by either the repeated application of relatively low load or the application of a sustained load for a long duration (as in a sitting task, for example). An individual loading boxes on a pallet who is repeatedly loading the tissues of the low back (several tissues could be at risk) to a subfailure level (see figure 1.2, a-b)

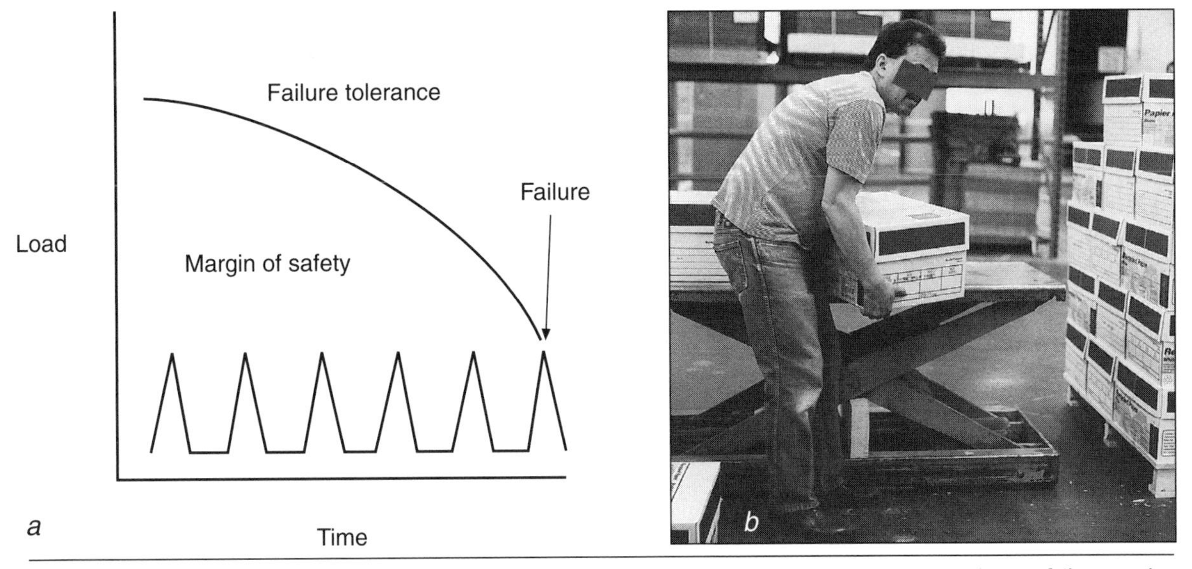

Figure 1.2 (*a*) Repeated subfailure loads lead to tissue fatigue, reducing the failure tolerance, leading to failure on the *N*th repetition of load, or box lift in this example (*b*).

Reprinted from *Journal of Biomechanics,* 30 (5), S.M. McGill, "Invited paper: Biomechanics of low back injury: Implications on current practice and the clinic," 456-475, 1997, with permission from Elsevier Science.

experiences a slow degradation of failure tolerance (e.g., vertebrae, Adams and Hutton, 1985; Brinckmann, Biggemann, and Hilweg, 1989). As tissues fatigue with each cycle of load and correspondingly the failure tolerance lowers, the margin of safety eventually approaches zero, at which point this individual will experience low back injury. Obviously, the accumulation of trauma is more rapid with higher loads. Carter and Hayes (1977) noted that, at least with bone, fatigue failure occurs with fewer repetitions when the applied load is closer to the yield strength.

Yet another way to produce injury with a subfailure load is to sustain stresses constantly over a period of time. The rodmen shown in figure 1.3b, with their spines fully flexed for a prolonged period of time, are loading the posterior passive tissues and are initiating time-dependent changes in disc mechanics (figure 1.3a). Under sustained loads these **viscoelastic** tissues slowly deform and creep. The sustained load and resultant creep causes a progressive reduction in the tissue strength. Correspondingly, the margin of safety also declines until injury occurs at a specific percentage of tissue **strain** (i.e., at the **breaking strain** of that particular tissue). Note that these workers are not lifting a heavy load; simply staying in this posture long enough will eventually ensure injurious damage. The injury may involve a single tissue, or a complex picture may emerge in which several tissues become involved. For example, the prolonged stooped posture imposes loads on the posterior ligaments of the spine and posterior fibers of the intervertebral disc. The associated creep deformation that ultimately produces microfailure (e.g., Adams, Hutton, and Stott, 1980; McGill and Brown, 1992) may initiate another chain of events. Stretched ligaments increase joint laxity, which can lead to hyperflexion injury (to the disc) and to the following sequence of events:

1. Local instability
2. Injury of unisegmental structures
3. Ever-increasing shearing and bending loads on the neural arch

This laxity remains for a substantial period after the prolonged stoop.

a

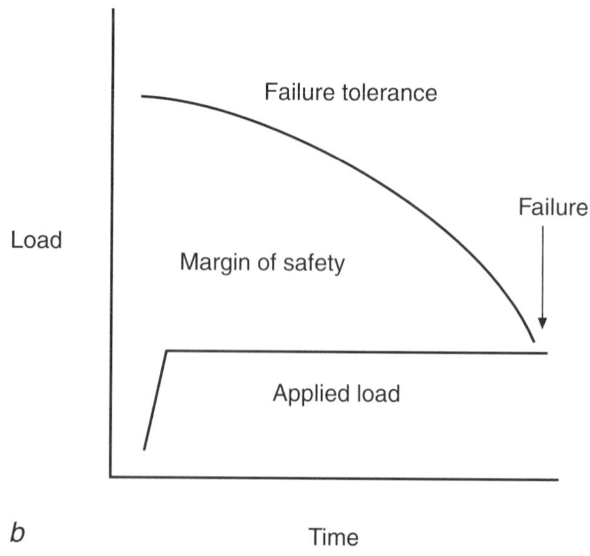

b

Figure 1.3 (*a*) These rodmen with fully flexed lumbar spines are loading posterior passive tissues for a long duration, (*b*) reducing the failure tolerance leading to failure at the *N*th% of tissue strain.

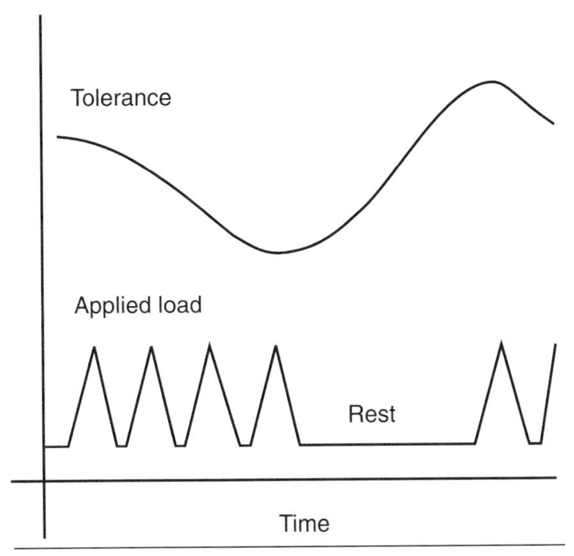

Figure 1.4 Loading is necessary for optimal tissue health. When loading and the subsequent degradation of tolerance are followed by a period of rest, an adaptive tissue response increases tolerance. Tissue "training" results from the optimal blend of art and science in medicine and tissue biomechanics.

Understanding the process of tissue damage in this way emphasizes why simple injury prevention approaches often fail. More effective injury intervention strategies recognize and address the complexities of tissue overload.

Avoidance of loading altogether is undesirable. The objective of injury prevention strategies is to ensure that tissue adaptation stimulated from exposure to load keeps pace with, and ideally exceeds, the accumulated tissue damage. Thus, exposure to load is necessary, but in the process of accumulating microtrauma, the applied loads must be removed (with rest) to allow the healing/adaptation process to gradually increase the failure tolerance to a higher level. We have already seen how tissue loading and injury risk form an optimal U-shaped relationship of not too much and not too little load. Determining the optimal load for health encompasses both the art and science of medicine and tissue biomechanics. Figure 1.4 presents a final load-time history to demonstrate the links among loading, rest, and adaptive tissue tolerance.

In summary, the injury process may be associated either with very high loads or with relatively low loads that are repeated or sustained. This either/or causation necessitates rigorous examination of injury and tissue loading history for substantial periods of time prior to the culminating injury event. It is important to recognize that simply focusing on a single variable such as one-time load magnitude may not result in a successful index of risk of injury, particularly across a wide variety of activities.

A Final Note

The selected controversies introduced in this chapter illustrate the need for the evidence presented in the rest of the text and the relevance of the discussions that follow. Resist the urge to assume that conventional wisdom is correct: first consider the evidence and then form your own opinions.

References

Adams, M.A., and Hutton, W.C. (1985) Gradual disc prolapse. *Spine,* 10: 524.

Adams, M.A., Hutton, W.C., and Stott, J.R.R. (1980) The resistance to flexion of the lumbar intervertebral joint. *Spine* 5: 245.

American Medical Association. (1990) Guides to the evaluation of permanent impairment (3rd edition).

Axler, C.T., and McGill, S.M. (1997) Choosing the best abdominal exercises based on knowledge of tissue loads. *Medicine and Science in Sports and Exercise,* 29: 804-811.

Battie, M.C., Bigos, S.J., Fisher, L.D., Spengler, D.M., Hansson, T.H., Nachemson, A.L., and Wortley, M.D. (1990) The role of spinal flexibility in back pain complaints within industry: A prospective study. *Spine,* 15: 768-773.

Biering-Sorensen, F. (1984) Physical measurements as risk indicators for low-back trouble over a one-year period. *Spine,* 9: 106-119.

Bogduk, N., Derby, R., April, C., Louis, S., and Schwarzer, R. (1996) Precision diagnosis in spinal pain. In: Campbell, J. (Ed.), *Pain 1996—An updated review* (pp. 313-323). Seattle: IASP Press.

Brinckmann, P. (1985) Pathology of the vertebral column. *Ergonomics,* 28: 235-244.

Brinckmann, P., Biggemann, M., and Hilweg, D. (1989) Prediction of the compressive strength of human lumbar vertebrae. *Clinical Biomechanics*, 4 (Suppl. 2): S1-S27.

Burton, A.L., Tillotson, K.M., and Troup, J.D.G. (1989) Prediction of low back trouble frequency in a working population. *Spine,* 14: 939-946.

Carter, D.R., and Hayes, W.C. (1977) The comprehensive behaviour of bone as a two phase porous structure. *Journal of Bone and Joint Surgery,* 58A: 954.

Deyo, R.A. (1998) Low back pain. *Scientific American*, August, 49-53.

Ferguson, S.A., and Marras, W.S. (1997) A literature review of low back disorder surveillance measures and risk factors. *Clinical Biomechanics*, 12 (4): 211-226.

Finch, P. (1999, November 11-14) Spinal pain—An Australian perspective. In the proceedings of the 13th World Congress of the International Federation of Physical Medicine and Rehabilitation, Washington DC, 243-246.

Fordyce, W.E. (Ed.). (1995) *Back pain in the workplace.* Seattle: IASP Press.

Gunning, J.L., Callaghan, J.P., and McGill, S.M. (2001) The role of prior loading history and spinal posture on the compressive tolerance and type of failure in the spine using a porcine trauma model. *Clinical Biomechanics,* 16 (6): 471-480.

Hadler, N. (2001) Editorial—The bane of the aging worker. *Spine*, 26 (12): 1309-1310.

Hirsch, C., Ingelmark, B.E., and Miller, M. (1963-64) The anatomical basis for low back pain. *Acta Orthopaedica Scandinavica,* 1 (33).

Hsu, K., Zucherman, J.F., Derby, R., White, A.H., Goldthwaite, N., and Wynne, G, (1988) Painful lumbar end plate disruptions. A significant discographic finding. *Spine,* 13 (1): 76-78.

Jager, M., Luttman, A., and Laurig, W. (1991) Lumbar load during one-handed bricklaying. *International Journal of Industrial Ergonomics*, 8: 261-277.

Kirkaldy-Willis, W.H. (1998) The three phases of the spectrum of degenerative disease. In: *Managing low back pain* (2nd ed.). New York: Churchill-Livingston.

Lord, S.M., Barnsley, L., Wallis, B.J., McDonald, G.J., and Bogduk, N. (1996) Percutaneous radio frequency neurotomy for chronic cervical zygapophyseal joint pain. *New England Journal of Medicine*, 335: 1721-1726.

Luoto, S., Heliovaara, M., Hurri, H., and Alarenta, M. (1995) Static back endurance and the risk of low back pain. *Clinical Biomechanics,* 10: 323-324.

Marras, W.S., Davis, K.G., Heaney, C.A., Maronitis, A.B., and Allread, W.G. (2000) The influence of psychosocial stress, gender, and personality on mechanical loading of the lumbar spine. *Spine,* 25: 3045-3054.

McGill, S.M., and Brown, S. (1992) Creep response of the lumbar spine to prolonged lumbar flexion. *Clinical Biomechanics,* 7: 43.

Mendelson, G. (1982) Not cured by a verdict: Effect of a level settlement on compensation claimants. *Medical Journal of Australia*, 2: 219-230.

Nachemson, A.L. (1992) Newest knowledge of low back pain: A critical look. *Clinical Orthopaedics and Related Research,* 279: 8-20.

Parks, K.A., Crichton, K.S., Goldford, R.J., McGill, S.M. (in press) A comparison of lumbar range of motion with functional ability scores on low back pain patients: Assessment of the validity of range of motion.

Porter, R.W. (1987) Does hard work prevent disc protrusion? *Clinical Biomechanics*, 2: 196-198.

Radanov, B.P., Sturzenegger, M., DeStefano, G., and Schinrig, A. (1994) Relationship between early somatic, radiological, cognitive and psychosocial findings and outcome during a one-year follow up in 117 patients suffering from common whiplash. *British Journal of Rheumatology*, 33: 442-448.

Rose, S.J. (1989) Physical therapy diagnosis: Role and function. *Physical Therapy,* 69: 535-537.

Saal, J.A., and Saal, J.S. (1989) Nonoperative treatment of herniated lumbar intervertebral disc with radiculopathy: An outcome study. *Journal of Biomechanics*, 14: 431-437.

Spitzer, W.O. (1993) Editorial: Low back pain in the workplace: Attainable benefits not attained. *British Journal of Industrial Medicine,* 50: 385-388.

Stevenson, J.M., Weber, C.L., Smith, J.T., Dumas, G.A., and Albert, W.J. (2001) A longitudinal study of the development of low back pain in an industrial population. *Spine,* 26 (12): 1370-1377.

Taylor, J.R., Twomey, L.T., and Corker, M. (1990) Bone and soft tissue injuries in post-mortem lumbar spines. *Paraplegia,* 28: 119-129.

Teasell, R.W. (1997) The denial of chronic pain. *Journal of Pain Research Management,* 2: 89-91.

Thompson, E.N. (1997) Back pain: Bankrupt expertise and new directions. *Journal of Pain Research Management,* 2: 195-196.

Valkenburg, H.A., and Haanen, H.C.M. (1982) The epidemiology of low back pain. In: White, A.A., and Gordon, S.L. (Eds.), *Symposium on idiopathic low back pain.* St. Louis: Mosby.

Videman, T., Battie, M.C., Gill, K., Manninen, H., Gibbons, L.E., and Fisher, L.D. (1995). Magnetic resonance imaging findings and their relationships in the thoracic and lumbar spine: Insights into the etiopathogenesis of spinal degeneration. *Spine,* 20 (8): 928-935.

Weber, H. (1983) Lumbar disk herniation: A controlled prospective study with ten years of observation. *Spine,* 8: 131.

Woo, S.L.-Y., Gomez, M.A., and Akeson, W.H. (1985) Mechanical behaviors of soft tissues: Measurements, modifications, injuries, and treatment. In: Nahum, H.M., and Melvin, J. (Eds.), *Biomechanics of trauma* (pp. 109-133). Norwalk, CT: Appleton Century Crofts.

Scientific Approach Unique to This Book

This book contains many nontraditional viewpoints on how the spine functions and becomes damaged. Most of these perspectives have emerged from a unique biomechanically based methodological approach. This chapter will familiarize you with the unique general approach taken for obtaining much of the data in this book and will help you begin to understand both the limitations and the unique insights provided by such an approach.

As spine biomechanists, our methods of inquiry are similar to those used by mechanical or civil engineers. For example, a civil engineer charged with the task of building a bridge needs three types of information:

- Traffic to be accommodated, or the design load
- Structure to be used (e.g., space truss or roman arch), as each architecture possesses specific mechanical traits and features
- Characteristics of proposed materials that will affect strength, endurance, stability, resistance to structural fatigue, and so on.

Our approach to investigating spine function is similar to that of our engineering colleagues. We begin with the following relationship to predict the risk of tissue damage:

Applied load > tissue strength = tissue failure (injury)

Recall from the tissue injury primer at the end of chapter 1 that tissue strength is reduced by repeated and prolonged loading but is increased with subsequent rest and adaptation. Analyzing tissue failure in this way requires two distinct methodological approaches. This is why we developed two quite distinct laboratories, which led to much of the progress documented in this book. (Please note that the "we" used in this chapter includes my research team of graduate students, visiting scholars, and technicians.) Our first lab is equipped for in-vitro testing of spines, in which we purposefully try to create herniated discs, damaged end plates, and other tissue-specific injuries. The second lab is the in-vivo lab where living people (both normals and patients) are tested for their response to stress and loading. Individual tissue loads are obtained from sophisticated modeling procedures.

In-Vitro Lab

The in-vitro lab is equipped with loading machines, an acceleration rack, tissue sectioning equipment, and an X-ray suite to document progressive tissue damage. For example, by performing discograms with radio opaque contrast liquids, we can document the mechanics of progressive disc herniation. We investigate any other injury mechanisms in the same way—that is, by applying physiological loads and motion patterns and then documenting the damage with appropriate technology.

Since many technique issues can affect the experimental results, the decision to use one over another is governed by the specific research question. For example, since a matched set of human spines to run a controlled failure test cannot be obtained, animal models must be used. Here, control is exercised over genetic homogeneity, diet, physical activity, and so forth to contrast an experimental cohort with a matched set of control spines.

Of course, the results must be validated and interpreted for them to be relevant to humans. In addition, identifying the limitations for relevance in interpretation is critical. Some hypotheses that demand the use of human material are compromised by the lack of available young, healthy, undegenerated specimens. Having healthy specimens is critical because biomechanics and injury mechanisms radically change with age. Other major methodological issues include the way in which biological tissues are loaded, perhaps at specific load rates or at specific rates of displacement. The researcher must decide which has the most relevance to the issue at hand. Devising these experiments is not a trivial task.

In-Vivo Lab

The in-vivo lab is unique in its approach in attempting to document the loads on the many lumbar tissues in vivo. This knowledge lends powerful insight into spine mechanics, both of normal functioning and of failure mechanics. Since transducers cannot be routinely implanted in the tissues to measure force, noninvasive methods are necessary. The intention of the basic approach is to create a virtual spine. This virtual model must accurately represent the anatomy that responds dynamically to the three-dimensional motion patterns of each test subject/patient and must mimic the muscle activation patterns chosen by the individual. In so doing, it enables us to evaluate subjects' unique motor patterns and the consequences of their choices and skill.

How the Virtual Spine Works

While two groups (the Marras group [e.g., Granata and Marras, 1993] and the McGill group) have devoted much effort to the development of biologically driven models, the McGill model will be described here given its familiarity to the author. The model—a dynamic, three-dimensional, anatomically complex, biologically driven approach to predicting individual lumbar tissue loads—is composed of two distinct parts: a linked-segment model and a highly detailed spine model that determines tissue loads and spine stability.

- The first part of the McGill model is a three-dimensional linked-segment representation of the body using a dynamic load in the hands as input. Two or more video cameras at 30 Hz record joint displacements to reconstruct the joints and body segments in three dimensions. Working through the arm and trunk linkage using linked-segment mechanics, reaction forces and **moments** are computed about a joint in the

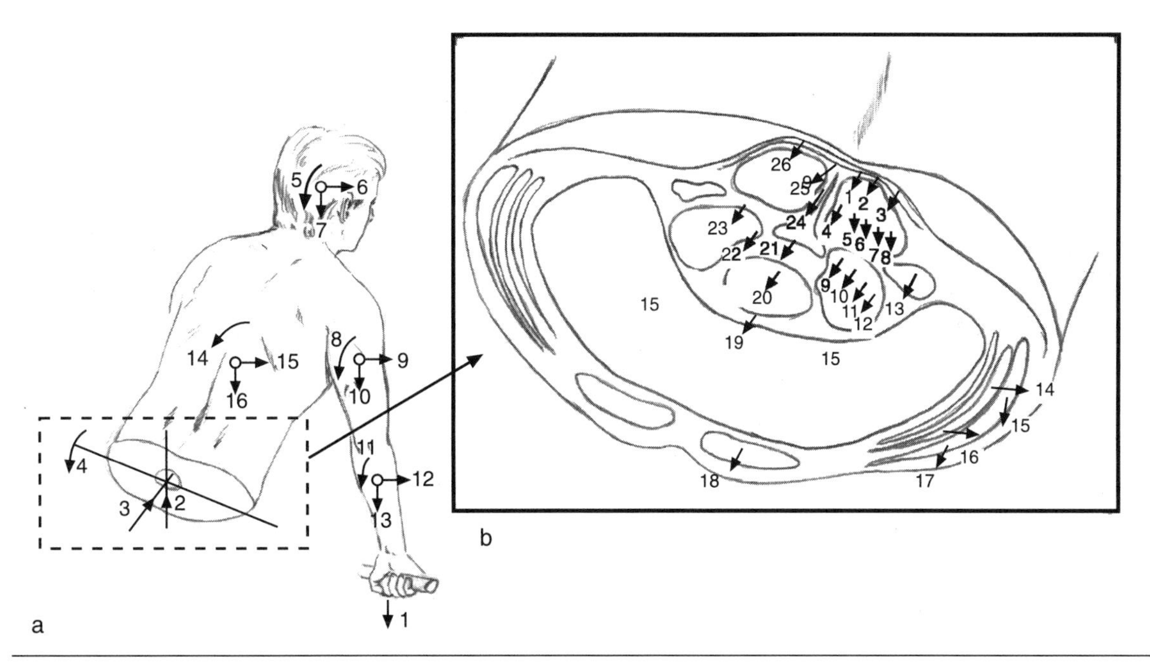

Figure 2.1 The tissue load prediction approach requires two models. The first *(a)* is a dynamic three-dimensional linked-segment model to obtain the three reaction moments about the low back. The second model *(b)* partitions the moments into tissue forces (muscle forces 1-18; ligaments 19-26; and moment contributions from deformed disc, gut, and skin in bending).

low back (usually L4-L5) (previously described in McGill and Norman, 1985) (see figure 2.1a). Using pelvic and spine markers, the three reaction moments are converted into moments about the three orthopedic axes of the low back (flexion/extension, lateral bend, and axial twist).

- The second part of the McGill model enables the partitioning of the reaction moments obtained from the linked-segment model into the substantial restorative moment components (supporting tissues) using an anatomically detailed three-dimensional representation of the skeleton, muscles, ligaments, nonlinear elastic intervertebral discs, and so on (see figure 2.1b). This part of the model was first described in McGill and Norman (1986), with full three-dimensional methods described in McGill (1992), and the most recent update provided by Cholewicki and McGill (1996). In total, 90 low back and torso muscles are represented. Very briefly, first the passive tissue forces are predicted by assuming stress–strain or load deformation relationships for the individual passive tissues. Deformations are modeled from the three-dimensional lumbar kinematics measured from the subject, which drive the vertebral kinematics of the model. Passive tissue stresses are calibrated for the differences in flexibility of each subject by normalizing the stress–strain curves to the passive range of motion of the subject. Electromagnetic instrumentation, which monitors the relative lumbar angles in three dimensions, detects the isolated lumbar motion. The remaining moment is then partitioned among the many fascicles of muscle based on their activation profiles (measured from EMG) and their physiological cross-sectional area. The moment is then modulated with known relationships for instantaneous muscle length of either shortening or lengthening velocity. Sutarno and McGill (1995) described the most recent improvements of the force–velocity relationship. In this way, the modeled spine moves according to the movements of the subject's spine, and the virtual muscles are activated according to the activation measured directly from the subject (see figures 2.2, a-b and 2.3).

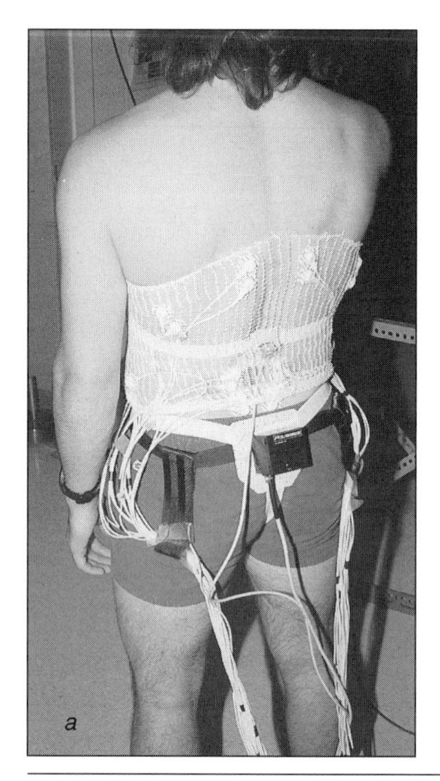

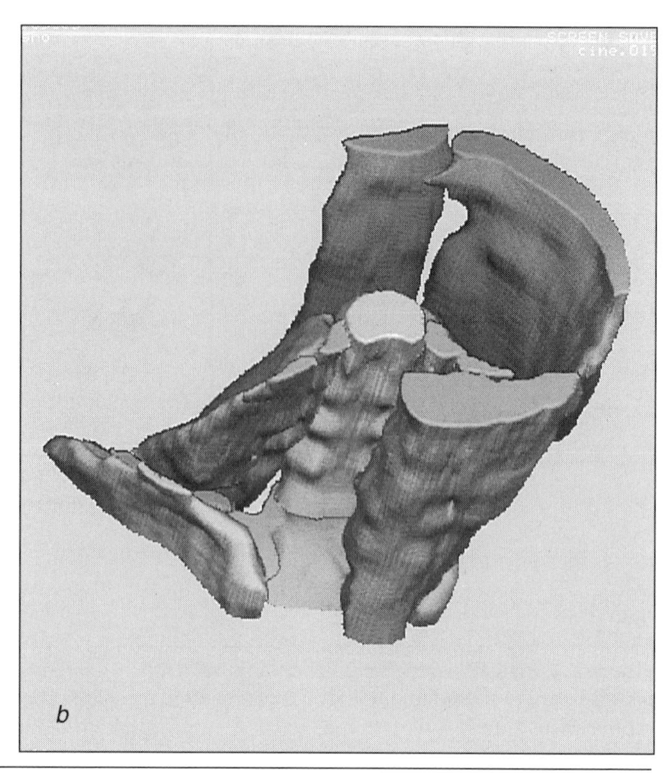

Figure 2.2 *(a)* Subject monitored with EMG electrodes and electromagnetic instrumentation to directly measure three-dimensional lumbar kinematics and muscle activity. *(b)* The modeled spine (partially reconstructed for illustration purposes, although for the purposes of analysis it remains in mathematical form) moves in accordance with the subject's spine. The virtual muscles are activated by the EMG signals recorded from the subject's muscles.

Using biological signals in this fashion to solve the indeterminacy of multiple load-bearing tissues facilitates the assessment of the many ways that we choose to support loads. Such an assessment is necessary for evaluating injury mechanisms and formulating injury-avoidance initiatives. From a clinical perspective, this ability to mimic individual spine motions and muscle activation patterns enables us to evaluate the consequences of a chosen motor control strategy. For example, we can see that some people are able to stabilize their backs and spare their lumbar tissues from overload when performing specific tasks. Conversely, we are able to evaluate the consequences of poorly chosen motor strategies. In this way we can identify those individuals with perturbed motor patterns and devise specific therapies to regroove healthy motor patterns that ensure sufficient spine stability and spare their tissues from damaging load.

Our challenge has been to ensure sufficient biological fidelity so that estimations of tissue forces are valid and robust over a wide variety of activities. The three-dimensional anatomy is represented in computer memory (muscle areas are provided in appendix A.1). On occasion, if the expense is warranted, we create a virtual spine of an individual from a three-dimensional reconstruction of serial MRI slices from the hip trochanter to T4 (see figure 2.2b). This component of the modeling process is well documented for the interested reader in McGill and Norman (1986) and McGill (1992). A list of the large number of associated research papers pertaining to the many detailed aspects of the process is provided in a separate reference section at the end of this chapter. See figures 2.4 and 2.5 for a flowchart and example of the modeling process.

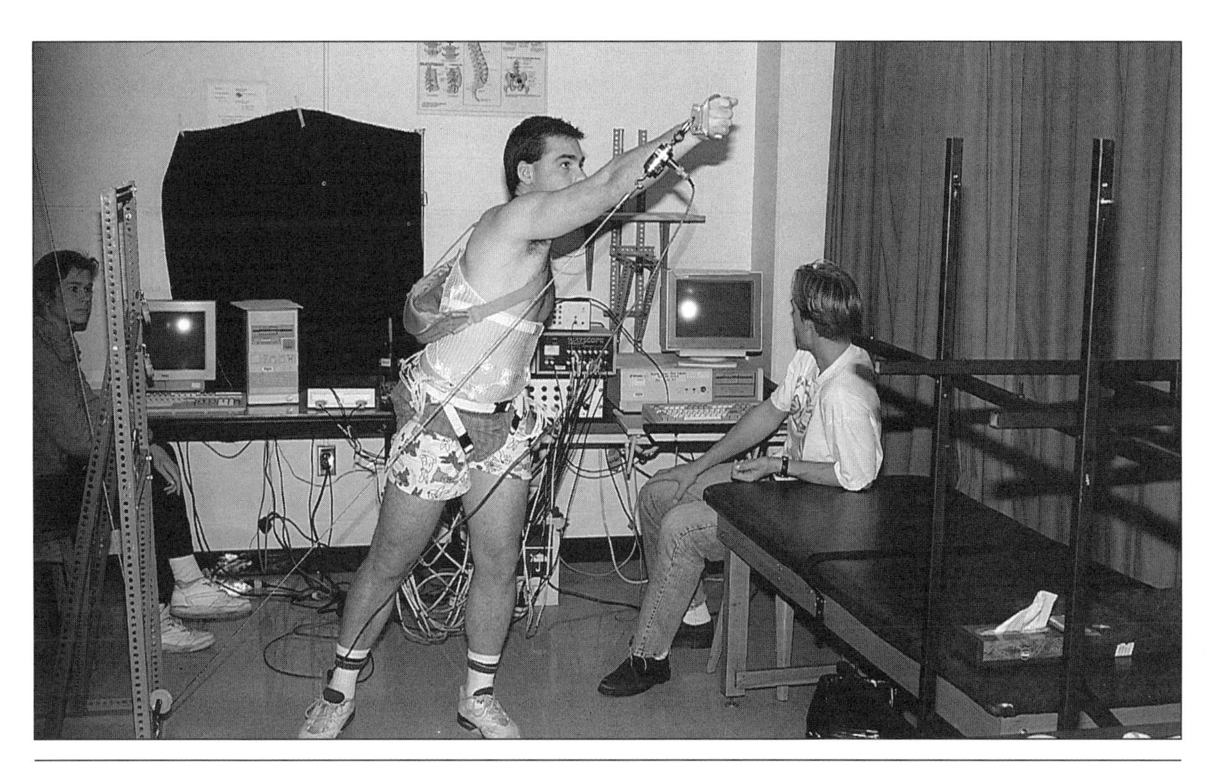

Figure 2.3 In this historical photo an instrumented subject simulating the complex three-dimensional task of tossing an object. The instrumentation includes three-dimensional video to capture body segment kinematics, a recording of the three-dimensional force vector applied to the hand, a 3-SPACE electromagnetic device to record isolated three-dimensional lumbar motion and assist in partitioning the passive tissue forces, and 16 channels of electromyographic electrodes to capture muscle activation patterns.

Development of the Virtual Spine

The development of the virtual spine approach has been an evolutionary process spanning 20 years. In that time we have had to confront several specific issues, namely, issues of validation and how to handle deep muscles that are inaccessible with surface EMG electrodes. Briefly, although we have tried intramuscular electrodes in highly selected conditions, this is a limiting invasive procedure. Generally, we estimate the deep muscle activation amplitudes from movement synergists (McGill, Juker, and Kropf, 1996). This method is limited, however, because it requires a prior knowledge of muscle patterns for a given moment combination in a specific task. We try to incorporate the highest level of content validity by using detailed representations of the anatomy and physiological cross-sectional areas, recording stress–strain relationships, and incorporating known modulators of muscle force such as length and velocity. One of our validation exercises is to compare the three measured reaction moments with the sum of individual tissue moments predicted by the virtual spine. A close match suggests that we have succeeded in accurately representing force.

Over the past five years this approach for predicting individual lumbar tissue loads has evolved to enable us to document spine stability. In this way, we can evaluate an individual's motor patterns and identify strategies that ensure safety and those that could result in injury. This is not a trivial task. It requires converting tissue forces to stiffness and using convergence algorithms to separate the forces needed to create the torques that sustain postures and movements from the additional forces needed to

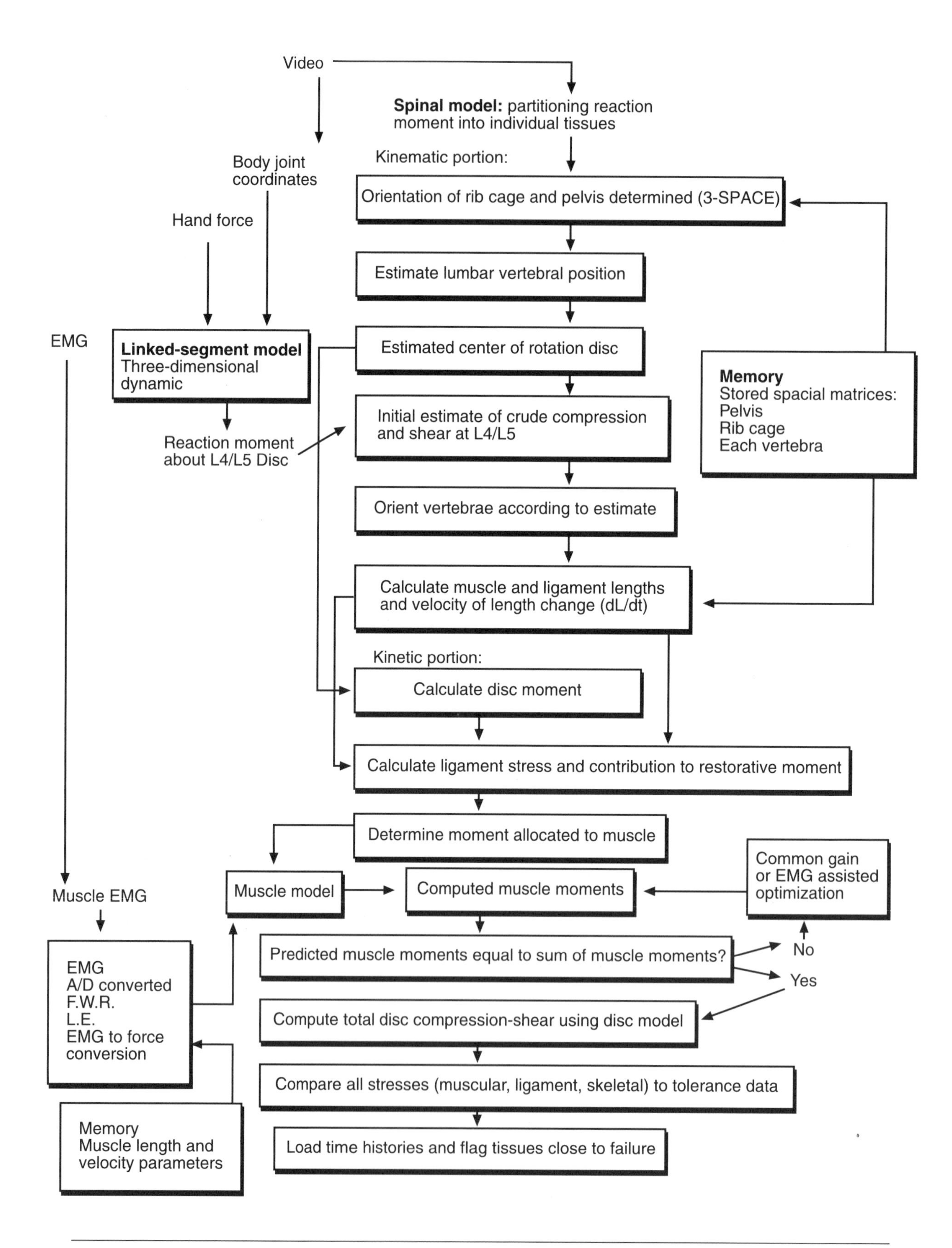

Figure 2.4 The model input and output are illustrated in this flowchart up to the point of calculating moments and tissue loads. Spine stability is calculated with an additional module.

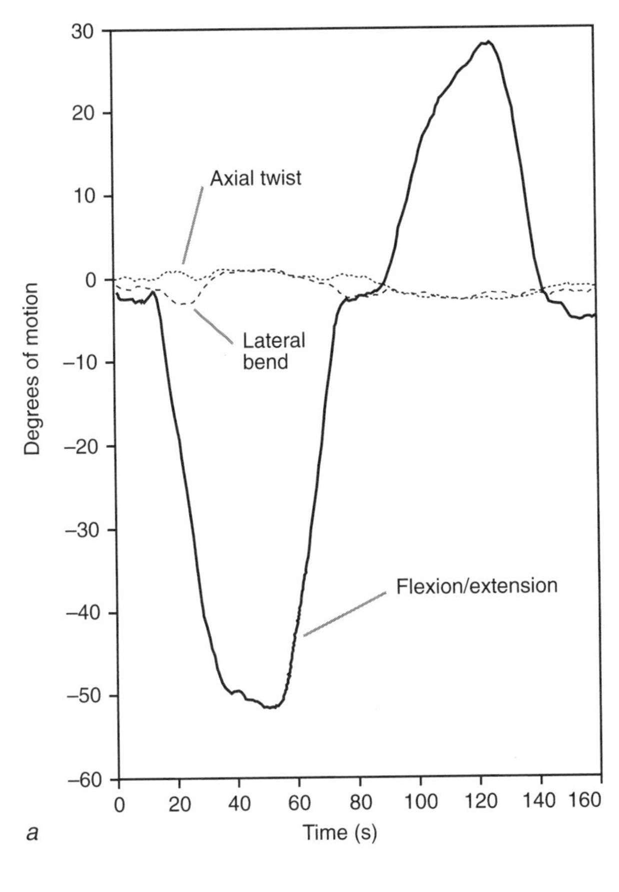

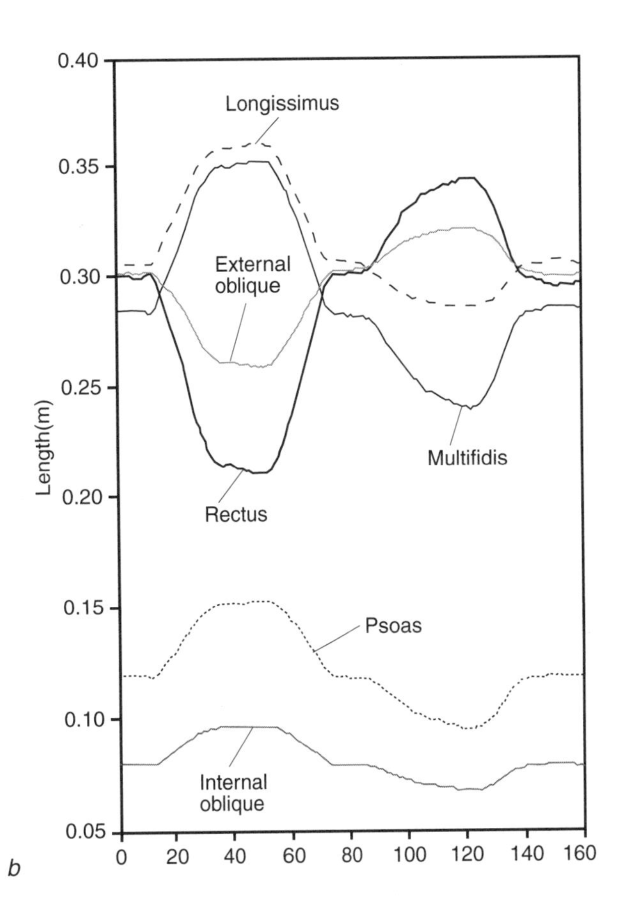

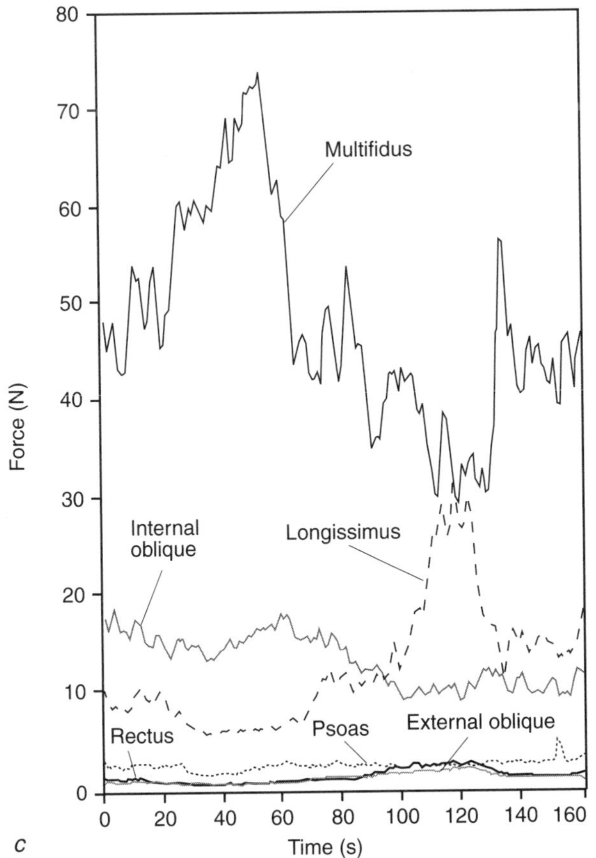

Figure 2.5 Various stages of model output in this example of a subject flexing, picking up a weight, and extending: lumbar motion about the three axes—flexion-extension, lateral bend, and axial twist *(a)*; lengths of a few selected muscles throughout the motion *(b)*; some muscle forces *(c)*; L4-L5 joint forces of compression and shear *(d)*; and stability index, where larger positive numbers indicate higher stability and a zero or negative number would suggest that unstable behavior is possible *(e)*.

(continued)

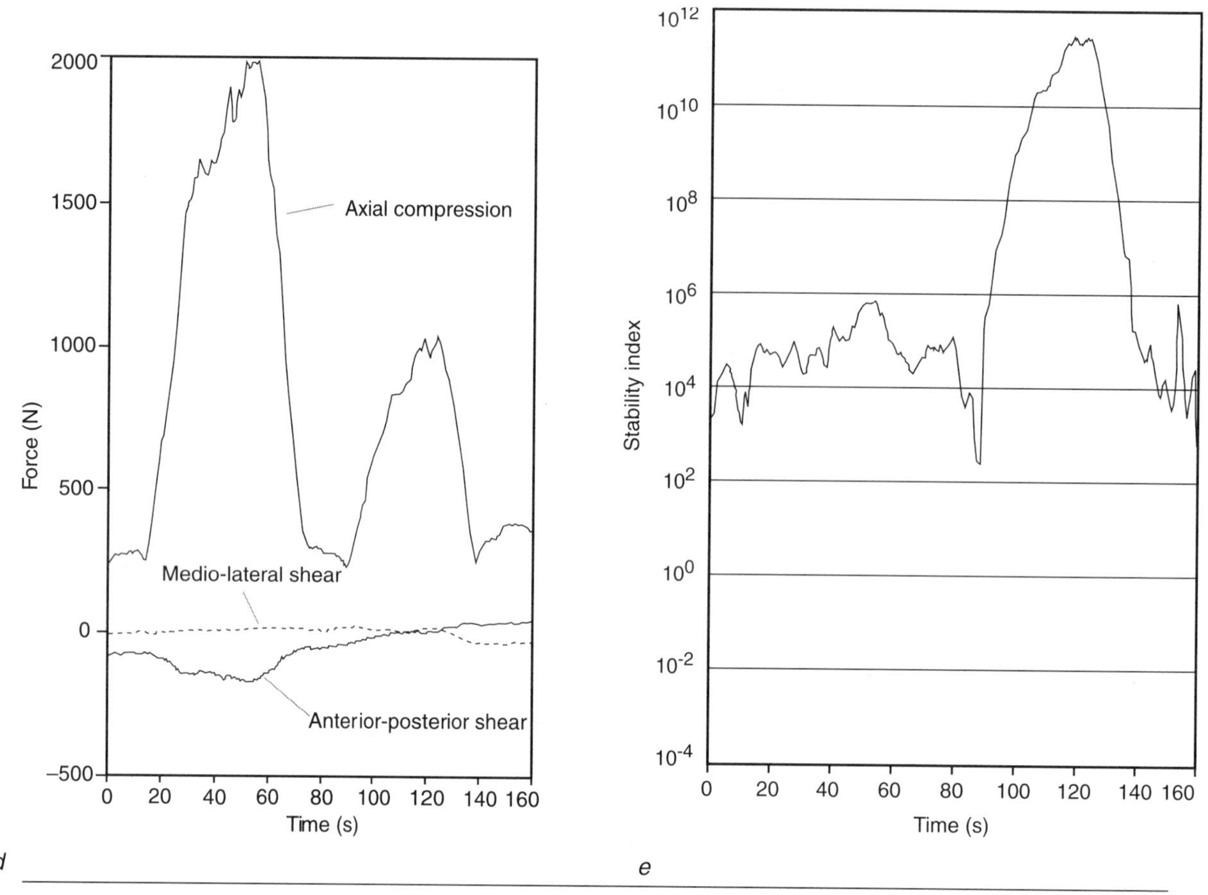

Figure 2.5 (continued).

ensure stability. Potential energy-based analyses are then employed to identify the "stability index" in each degree of freedom of the joint, which in turn reveal the joint's ability to survive a given loading scenario. This level of modeling represents the highest level of sophistication currently available. It has enabled us to challenge concepts pertaining to spine stability—more on this in chapters 6 and 13.

References

Cholewicki, J., and McGill, S.M. (1996) Mechanical stability of the in vivo lumbar spine: Implications for injury and chronic low back pain. *Clinical Biomechanics,* 11 (1): 1-15.

Granata, K.P., and Marras, W.S. (1993) An EMG-assisted model of loads on the lumbar spine during assymetric trunk extensions. *Journal of Biomechanics,* 26: 1429-1438.

McGill, S.M. (1992) A myoelectrically based dynamic 3-D model to predict loads on lumbar spine tissues during lateral bending. *Journal of Biomechanics,* 25 (4): 395-414.

McGill, S.M., and Norman, R.W. (1986) The Volvo Award for 1986: Partitioning of the L4/L5 dynamic moment into disc, ligamentous and muscular components during lifting. *Spine,* 11 (7): 666-678.

References and Bibliography Pertaining to the Finer Details of the Modeling Process

Cholewicki, J., and McGill, S.M. (1994) EMG Assisted Optimization: A hybrid approach for estimating muscle forces in an indeterminate biomechanical model. *Journal of Biomechanics,* 27 (10): 1287-1289.

Cholewicki, J., and McGill, S.M. (1995) Relationship between muscle force and stiffness in the whole mammalian muscle: A simulation study. *Journal of Biomechanical Engineering,* 117: 339-342.

Cholewicki, J., McGill, S.M., and Norman, R.W. (1995) Comparison of muscle forces and joint load from an optimization and EMG assisted lumbar spine model: Towards development of a hybrid approach. *Journal of Biomechanics,* 28 (3): 321-331.

McGill, S.M. (1988) Estimation of force and extensor moment contributions of the disc and ligaments at L4/L5. *Spine,* 12: 1395-1402.

McGill, S.M. (1996) A revised anatomical model of the abdominal musculature for torso flexion efforts. *Journal of Biomechanics,* 29 (7): 973-977.

McGill, S.M., Juker, D., and Axler, C. (1996) Correcting trunk muscle geometry obtained from MRI and CT scans of supine postures for use in standing postures. *Journal of Biomechanics,* 29 (5): 643-646.

McGill, S.M., Juker, D., and Kropf, P. (1996) Appropriately placed surface EMG electrodes reflect deep muscle activity (psoas, quadratus lumborum, abdominal wall) in the lumbar spine. *Journal of Biomechanics,* 29 (11): 1503-1507.

McGill, S.M., and Norman, R.W. (1985) Dynamically and statically determined low back moments during lifting. *Journal of Biomechanics,* 18 (12): 877-885.

McGill, S.M., and Norman, R.W. (1987a) Effects of an anatomically detailed erector spinae model on L4/L5 disc compression and shear. *Journal of Biomechanics*, 20 (6): 591-600.

McGill, S.M., and Norman, R.W. (1987b) An assessment of intra-abdominal pressure as a viable mechanism to reduce spinal compression. *Ergonomics,* 30 (11): 1565-1588.

McGill, S.M., and Norman, R.W. (1988) The potential of lumbodorsal fascia forces to generate back extension moments during squat lifts. *Journal of Biomedical Engineering,* 10: 312-318.

McGill, S.M., Patt, N., and Norman, R.W. (1988) Measurement of the trunk musculature of active males using CT scan radiography: Implications for force and moment generating capacity about the L4/L5 joint. *Journal of Biomechanics,* 21 (4): 329-341.

McGill, S.M., Santaguida, L., and Stevens, J. (1993) Measurement of the trunk musculature from T6 to L5 using MRI scans of 15 young males corrected for muscle fibre orientation. *Clinical Biomechanics,* 8: 171-178.

McGill, S.M., Seguin, J., and Bennett, G. (1994) Passive stiffness of the lumbar torso about the flexion-extension, lateral bend and axial twist axes: The effect of belt wearing and breath holding. *Spine*, 19 (6): 696-704.

McGill, S.M., Thorstensson, A., and Norman, R.W. (1989) Non-rigid response of the trunk to dynamic axial loading: An evaluation of current modelling assumptions. *Clinical Biomechanics,* 4: 45-50.

Santaguida, L., and McGill, S.M. (1995) The psoas major muscle: A three-dimensional mechanical modelling study with respect to the spine based on MRI measurement. *Journal of Biomechanics,* 28 (3): 339-345.

Sutarno, C., and McGill, S.M. (1995) Iso-velocity investigation of the lengthening behaviour of the erector spinae muscles. *European Journal of Applied Physiology and Occupational Physiology,* 70 (2): 146-153.

Epidemiological Studies on Low Back Disorders (LBDs)

Definitive experiments are rare in the fields of science and medicine. Instead, conclusions generally emerge from the integration and synthesis of evidence from a variety of sources. Using a similar approach, lawyers argue cases in which each piece of evidence is considered circumstantial in the hope that eventually the circumstantial evidence will become overwhelming. Like lawyers, scientific researchers gather and regather circumstantial evidence from several perspectives with the goal of understanding cause and effect. By studying the relationships among variables and investigating mechanisms, they are able to form perspectives that are robust and plausible. This type of research work, together with longitudinal studies, tests the causative factors identified in the mechanistic studies.

Other chapters in this book are dedicated to investigating the various mechanisms within the low back and their links with good health and disability. This chapter focuses on the study of associations of variables through various epidemiological approaches. Some readers will find this a boring chapter—so my students inform me. You may choose to skip this chapter. But those wishing a fuller understanding of the challenges that lie in building a strong foundation for the very best injury prevention and rehabilitation programs are encouraged to read on.

Several sections in this chapter will help you understand the risk factors for low back trouble—specifically, the changes in personal factors and whether they cause back troubles or are a consequence of having them. For the purpose of this review, purported disabling low back troubles and possibly related work intolerance will be referred to collectively as low back disorders (LBDs). Furthermore, the term *personal factors* can include anthropometric and fitness variables, as well as motor control ability, injury history, and so forth.

Upon completion of this chapter you will appreciate the epidemiological perspective, have a more complete appreciation for the positions taken in this book to reduce the economic impact of low back disorders (LBDs), and understand certain subsequent approaches for prevention and rehabilitation.

Multidimensional Links Among Biomechanical, Psychosocial, and Personal Variables

As noted in chapter 2, several prominent people have declared that psychosocial variables are the most significant factors in LBD. This is an important issue since effective intervention must address the real cause and consequence of back troubles. In this section we will see that virtually all studies that properly measured or calculated the physical demands of tasks found that people subjected to specific mechanical stressors are at higher risk of LBD but that there also appear to be some mitigating issues.

Three Important Studies

The following three studies are reviewed here because of their past influence and future implications.

- **Bigos et al.** In 1986 Bigos and colleagues performed a highly quoted study—one that has been very influential in shaping opinion regarding injury prevention and rehabilitation—at the Boeing plant in Washington State in the United States. This retrospective investigation analyzed 4645 injuries (of which 900 were to the low back) over a 15-month period in 1979-1980. They reported a correlation between the incidence of back injury and poor appraisal ratings of employees performed by their supervisors within six months prior to the reported injury. The authors considered the poor ratings to represent a psychosocial factor. In 1991 Bigos and colleagues conducted a longitudinal prospective study of 3020 employees at Boeing, during which there were 279 reported low back injuries. The researchers collected personality inventories as well as questionnaires regarding family and coworker support and job satisfaction. They also analyzed personal factors such as isometric strength, flexibility, aerobic capacity, height, and weight. The authors concluded that psychosocial measures—particularly those related to job enjoyment—had the strongest influence of all the variables analyzed. In fact, those who stated they did not enjoy their job were 1.85 times more likely to report a back injury (odds ratio = 1.85). Job satisfaction counted for less than 15% of the variance as an injury risk factor, meaning that more than 85% of the variance was unaccounted for. In other words, psychosocial factors failed to account for 85% of the causation of LBDs. The authors concluded "that the statistically significant, though clinically modest, predictive power of work perceptions and psychosocial factors for reports of acute back pain among industrial workers argues against the exclusive use of an injury model to explain such problems." This is a fair summary of the implications of their work. It does not mean that mechanical loading is unimportant. Yet, this study is often quoted to support the viewpoint that psychosocial factors are the most important causes of back disorders. Interestingly, Marras and colleagues (1993) also found similar odds ratios for job satisfaction (1.56) in a massive study of 400 repetitive industrial lifting jobs across 48 industries.

Odds Ratios

Perhaps the most lucid definition of an **odds ratio** can be achieved through an example. If smokers have three times the risk of developing lung cancer as nonsmokers do (perhaps 6 out of 10 smokers as opposed to 2 out of 10 in nonsmokers), they have an odds ratio of 3. Thus, an odds ratio greater than 1 suggests an increased risk from a specific factor.

Very few epidemiologically based studies have employed reasonably robust quantifications of biomechanical, psychosocial, and personal factors. Two important studies meet this requirement.

- **Marras et al.** The first important study was reported by Marras and colleagues (1995), who surveyed over 400 industrial lifting jobs across 48 different industries. They examined medical records in these industries to classify each type of job as being either low, medium, or high risk for causing LBD. They documented a variety of mechanical variables as well as reported job satisfaction. The most powerful single variable for predicting those jobs with LBD was maximum low back moment (odds ratio = 4.04) between low- and medium-risk groups and which produced an odds ratio of 3.32 between the low- and high-risk groups. Other single variables produced impressive odds ratios, for example, sagittal trunk velocity (odds ratio = 2.48) for the low- and high-risk comparison and 2.42 for maximum weight handled between the low- and high-risk groups. Job satisfaction produced an odds ratio of 1.48 between the low- and high-risk groups and 1.32 between the low- and medium-risk groups. The researchers entered the single variables into a multiple logistic regression model. The group of various measures selected by the model described the risk index well between the low- and high-risk groups and also between the low- and medium-risk groups. Suitably varying the five measures chosen by the regression process (maximum load moment, maximum lateral trunk angular velocity, average trunk twisting velocity, lifting frequency, and the maximum sagittal trunk angle) decreased the odds of being a member of the high-risk LBD group over 10 times (odds ratio = 10.6). This was an important study for specifying certain physical characteristics of job design that reduce the risk of LBD and linking epidemiological findings with quantitative biomechanical analysis and psychosocial factors across a large working population.
- **Norman et al.** The second important study to successfully integrate biomechanical, psychosocial, and personal factors was conducted by Norman and colleagues (1998), who examined injuries that occurred in an auto assembly plant that employed more than 10,000 hourly paid workers. During a two-year period of observation in the plant, the authors reported analyses on 104 cases and 130 randomly selected controls. Cases were people who reported low back pain (LBP) to a nursing station; controls were people randomly selected from company rosters who did not report pain. This is a notable study because the authors attempted to obtain good-quality psychosocial, personal, and **psychophysical** data on all participants from an interviewer-assisted questionnaire, as well as good-quality, directly measured biomechanical data on the physical demands of the jobs of all participants. (Note that the psychophysical approach is based on worker self-perceived stresses.) The study revealed that several independent and highly significant biomechanical, psychosocial, and psychophysical factors (identified as risk factors) existed in those who reported LBP. The personal risk factors that were included were much less important. After adjusting for personal risk factors, the statistically independent biomechanical risk factors that emerged were peak lumbar shear force (conservatively estimated odds ratio = 1.7), lumbar disc compression integrated over the work shift (odds ratio = 2.0), and peak force on the hands (odds ratio = 1.9). The odds ratios for the independent psychosocial risk factors, from among many studied, were worker perceptions of poorer workplace social environment (2.6), higher job satisfaction (not lower, as shown in the Boeing study) (1.7), higher coworker support (1.6), and perception of being more highly educated (2.2). Perceptions of higher physical exertion, a psychophysical factor, resulted in an odds ratio of 3.0, which is possibly related to the capacity of the worker relative to the job demands. Nearly 45% of the total variance was accounted for by these risk factors, with approximately 12% accounted for by the psychosocial factors and 31% by the

biomechanical factors. These results are very consistent with those reported by Marras and colleagues (1995) and by Punnett and colleagues (1991). Only a few of the personal factors were associated with reporting LBP: body mass index (odds ratio = 2.0) and prior compensation claim (odds ratio: 2.2). This case/control study is of high quality because it used a battery of many of the best measurement methods available for field use to assess many psychosocial, biomechanical, and personal factors on all participants in the data pool.

The evidence from the comprehensive studies suggests that both psychosocial and biomechanical variables are important risk factors for LBD. In particular, cumulative loading, joint moments, and spine shear forces are important. Those claiming that only psychosocial factors are important or that only physical loading factors are important cannot mount a creditable, data-based defense, as it appears that the data they quote fail to measure properly either (or both) physical or psychosocial variables.

Do Workers Experience LBDs Because They Are Paid to Act Disabled?

Some papers in the literature appear to dismiss the link between pain and disability. Most of these papers clearly state that this notion is restricted to "nonspecific back pain," noting that specific diagnoses do impair the ability of a worker to perform a demanding job. However, some based their argument on the concept that low back tissue injury heals in 6 to 12 weeks, while others base their arguments on a behavioral model of chronic pain that is not totally consistent with the findings of other scientific approaches. A short discussion of the issues and evidence related to physician diagnosis, compensation, tissue damage, and pain is necessary.

The position that chronic pain and disability are a function of compensation (and not mechanical factors) is contradicted by evidence that low back troubles continue after legal settlement of injury compensation (Mendelson, 1982). Hadler (1991) believes that the contest between the patient and the medical officer charged with determining the compensatory award causes the patient to act disabled and thwarts any incentive to act well. Teasell and Shapiro (1998) shared this opinion in a review of several different chronic pain disorders, as did Rainville and colleagues (1997) in another very nice study that specifically addressed chronic low back pain. Unfortunately, Hadler has taken the position that mechanical factors are, for the most part, of little importance in either causing or rehabilitating bad backs when compared with the psychosocial modulators (e.g., Hadler, 2001). His selective citation of the literature excludes evidence linking mechanical overload to tissue damage and ignores several important intervention studies. For example, Werneke and Hart (2001) showed that pain patterns upon patient presentation, specifically whether the pain "centralizes" or not, are much more powerful predictors of chronicity than the psychosocial variables they studied. Thus, although the topic of compensation is important, it is irrelevant in discussions of the links between loading and LBD. Compensation issues should not be used to argue against the existence of a mechanical link between injury and work tolerance or, worse yet, to suggest that the removal of compensation will eliminate the cause of tissue damage.

Does Pain Have an Organic Basis—Or Is It All in the Head?

Much has been written about the apparent absence of an organic basis for chronic low back disability (and other chronic pain syndromes). As noted in chapter 1, as high as 85% of disabling LBD cases are claimed to have no definitive pathoanatomical diagnosis (White and Gordon, 1982). Two conclusions have been proposed:

- Many LBD patients present with "nonorganic signs" suggesting psychological disturbance as the cause of their condition.
- Poor diagnostic techniques, either from inadequately trained doctors or from limitations of widely available diagnostic technology, have precluded the making of many solid diagnoses.

The first conclusion suggests that the correct course of action in these "undiagnosed" cases is to ignore physiological issues and address only psychological/psychosocial factors. The second conclusion suggests that one should not ignore the possibilities that more thorough diagnostic techniques could unearth physical causes and that rehabilitation based on that assumption could be more effective than rehabilitation for psychological disturbance alone. Let's examine each of these arguments.

Does Absence of Diagnosis Imply Psychological Cause?

Waddell and colleagues wrote many manuscripts and guidelines (e.g., 1980, 1984, 1987) on nonorganic signs in patients to support the notion of psychological disturbance overriding any pathoanatomical tissue damage (because none was diagnosed). Teasell and Shapiro (1998) wrote a nice summary of an extensive experimental literature suggesting that these pain symptoms may indeed have a physiological basis. They reviewed the recent science on the spread of neuron excitability and sensitization of adjacent neurons to explain the sensation of radiating pain in chronic conditions. Changes in neuroanatomy are coupled with biochemical changes with chronic pain. For example, in fibromyalgia patients numerous studies have shown levels of substance "P" in the cerebrospinal fluid elevated two to three times over controls (reported in a review by Teasell, 1997). While the nonorganic signs described in Waddell's papers are important considerations in many cases and are a contribution to clinical practice, strong evidence suggests that many nonorganic signs may not be exclusive of a pathoanatomical mechanism that has eluded diagnosis.

Some have argued that no link exists between pain and tissue damage and activity intolerance. In the absence of no direct evidence, some groups have simply assumed that **nociceptive** pain that is not surgically correctable or that has not improved within six weeks should not be regarded as disabling (e.g., Fordyce, 1995). In fact, the Fordyce monograph initiated much discussion, including several letters (e.g., Thompson/Merskey/Teasell/Fordyce, 1996). Fordyce's 1996 statements that "the course we are presently on threatens disaster" and that "we change or go broke" were particularly revealing. The high cost of treating chronic pain and disability appears to have motivated the elevation of the importance of the psychosocial factors so emphasized in this monograph. It is also interesting to observe the absence of any eminent biomechanical expert among the author list of the Fordyce report. The Canadian Pain Society stated that the Fordyce report literature review "is incomplete and does not reflect the contemporary understanding of chronic low back pain" (*Pain,* vol. 65, p. 114, 1996). Moreover, the Fordyce report largely ignored the evidence linking mechanical overload to measurable changes in spine biomechanics and spinal pain neuromechanical mechanisms. Another assumption of the report was that most spine LBDs are nonspecific, meaning simply that a diagnosis was not made.

Could Inadequate Diagnosis Be a Factor in Nonorganic LBP?

Bogduk and colleagues (1996) argued that pain arising from many spinal tissues can be attributed to a detectable painful lesion. For example, the facet joints will produce pain upon stimulation (McCall, Park, and O'Brien, 1979). Bogduk's point is certainly correct. However, because not all lesions are easily detectable, it cannot be used to argue that if a lesion is not detected there is no organic basis for pain. For example, fractures and meniscal tears that have been detected in postmortem studies have not shown radiologically on planar X ray (Jonsson et al., 1991; Taylor, Twomey, and Corker, 1990) or on CT (Schwarzer et al., 1995). Nor have freshly produced fractures and articular damage been outwardly detectable radiographically in animal models (Yingling and McGill, 2000). Yet these are the typical diagnostic procedures used.

Interestingly, Bogduk and colleagues (1996) and Lord and colleagues (1996) showed that injection of anesthetic (placebo-controlled diagnostic blocks) convincingly demonstrates that facet joints are often the site of pain origin. Further, disc studies examining the pain response of well over 1000 discs in over 400 people undergoing discography by Vanharanta and colleagues (1987) (subsequently reappraised by Moneta et al. in 1994 and reported by Bogduk et al. in 1996) showed a clear and statistically significant correlation between disc pain and grade 3 fissures of the annulus fibrosis. Most physicians would probably not detect these deep fissures of the annulus. In a rigorous and systematic study of diagnosis based on anesthetic blocks, Schwarzer and colleagues (1995) were able to diagnose over 60% of the LBD cases cited in the study by Bogduk and colleagues (1996) as being internal disc disruption (39%), facet joint pain (15%), and sacroiliac pain (12%). Bogduk and colleagues (1996) stated, "If inappropriate tests such as EMG and imaging are used, nothing will be found in the majority of cases, falsely justifying the impression that nothing can be found." Clearly we must question the statement that 85% of LBD cases are idiopathic or have no definitive pathoanatomical cause.

Helpful Strategies for Undiagnosed LBDs

Some physicians are clearly frustrated with the delayed improvement of undiagnosed chronic LBD patients. This frustration, together with concern for the financial health of the compensation system, may have motivated the Fordyce report to emphasize psychosocial modulators rather than organically based variables to explain intolerance to certain types of activity. But dealing with frustration on the basis of false assumptions will not help the situation. What, then, are more useful approaches to undiagnosed chronic cases of LBD?

As noted earlier, many clinicians do not have the expertise or tools to diagnose back troubles at a tissue-based level. Provocative testing will enable many physicians to identify painful motions and loading. By integrating biomechanics with such testing, physicians may be aided in making functional diagnoses.

In a small study Delitto and colleagues (1995) suggested that appropriately classified back pain sufferers (those with functional classifications rather than tissue-specific diagnoses) do better with specific treatments. Further, as patients progress through the rehabilitation process, they seem to require different treatment approaches. For example, several studies suggest that manipulation can be beneficial for acute short-term troubles, while physical therapy and exercise approaches appear better for chronic conditions (Skargren, Carlsson, and Oberg, 1998). A major limitation of these studies is that none has assessed progressive treatments that change as the patient progresses through the rehabilitation process; rather, they all have assessed only single treatment approaches. Future studies must assess the efficacy of staged programs in which categorized patients follow progressive treatment involving several sequenced approaches.

In summary, the position suggesting that there is no detectable pathoanatomical basis for pain and activity intolerance in some patients and thus that they are functions of only psychosocial variables does not appear to be defendable. Improved tissue-based diagnosis, improved **provocative/functional diagnosis**, and better understanding of the interactions of psychological variables with pathoanatomical variables appear to have promise for helping to improve treatment outcomes.

Are Biomechanical Variables and Psychosocial Variables Distinct?

Are biomechanical variables and psychosocial variables distinct, or is there an interplay between them that, if understood, would underpin a more evidenced-based intervention program?

Most reports have made a clear separation between psychosocial and biomechanical factors. But is there any evidence that psychosocial factors could modulate musculoskeletal loading—or vice versa? First, consider that highly respected pain scientists present volumes of empirical data demonstrating that pain perception is modulated by sensory, neurophysiological, and psychological mechanisms, suggesting that the separation of the two for analysis is folly (e.g., Melzack and Wall, 1983). Teasell (1997) argued quite convincingly that, while psychological factors have been cited as being causative of pain and disability, in fact, psychological difficulties arise as the consequence of chronic pain (see also Gatchel, Polatin, and Mayer, 1995; Radanov et al., 1994) and disappear upon its resolution (see also Wallis, Lord, Bogduk, 1997). The evidence is compelling—pain and psychological variables appear to be linked.

Is there more direct evidence that psychosocial factors are inextricably linked with biomechanical factors? In a recent report, Marras and colleagues (2000) noted that certain personality factors, together with some psychosocial variables, appear to increase spinal loads by up to 27% in some personality types via muscular cocontraction. This appears to occur at moderate levels of loading, while biomechanical loading overrides any psychosocial effects under larger task demands. This conclusion was strengthened with a field study linking these general mechanistic observations with reported LBD in workers. In summary, LBDs appear to be associated with both loading and psychosocial factors, and these factors seem to be related and multifactorial. The same conclusion appears to be valid for many types of chronic pain conditions (Gamsa, 1990).

What Is the Significance of First-Time Injury Data for Cause and Prevention?

One of the best indicators of future back troubles is a previous history of back troubles (Bigos et al., 1991; Burton, Tillotson, Troup, 1989; Troup et al., 1987). This suggests that studies of first-time back trouble episodes may be quite revealing for causative factors. Burton, Tillotson, Symonds, et al. (1996) studied police officers in Northern Ireland wearing >8 kg (18 lb) of body armor in a jacket (this additional load was borne by the low back). This group demonstrated a shorter period of time to their first onset of pain when compared to officers in an English force that did not wear the body armor. The authors also found that spending more than two hours per day in a vehicle comprised a separate risk for first-time onset of LBP. Another survey (Troup, Martin, and Lloyd, 1981) noted that falls among employees across a variety of industries were a common cause of first-time onset and were associated with longer periods of sick leave and a greater propensity for recurrence than were injuries caused by other mechanisms. (Ligamentous damage resulting from this type of loading is discussed in chapter 4.) It is also interesting to note that personal factors appear to play some role in first-time occurrence. Biering-Sorensen (1984) tested 449 men and 479 women for a variety of physical characteristics and showed that those with larger amounts of spine mobility and less lumbar extensor muscle endurance (independent factors) had an increased occurrence of first-time back troubles. Luoto and colleagues (1995) reached similar conclusions. Muscular endurance, and not anthropometric variables, appears to be protective.

Some injuries just happen as the result of motor control errors. (This interesting mechanism is introduced in chapter 4, where we describe witnessing an injury using videoflouroscopy to view the spine.) These may be considered random events and may be more likely in people with poor motor control systems (Brereton and McGill, 1999).

How Do Biomechanical Factors Affect LBD?

Several approaches have provided evidence into the links between biomechanical factors and LBD. A few are summarized here.

Mechanical Loading and LBD: Field-Based Risk Factors

Of the epidemiological studies that have focused on kinematic and kinetic biomechanical factors, a few investigated loading of low back anatomical structures. These would be considered to be the strongest evidence. The majority of the studies, however, assessed indirect measures that are linked to spinal loading such as the presence of static work postures, frequent torso bending and twisting, lifting demands, pushing and/or pulling exertions, and exertion repetition. While tissue overload is the cause of tissue damage and related back troubles, these indirect measures of load merely act as surrogates. The attraction of using surrogate measures rather than direct tissue loads per se for epidemiological study is that they are simpler to quantify and survey in the field. However, trade-offs exist among methodological utility, biological reality, and robustness. Risk factors related to specific tissue-based injury mechanisms are found in chapter 4.

Several issues should be kept in mind when interpreting this literature. Virtually all reviews of the epidemiological literature (e.g., Pope, 1989; Andersson, 1991) have noted that specific job titles and types of work are associated with LBD (although LBD is defined differently in different studies). In particular, jobs characterized by manual handling of materials, sitting in vibrating vehicles, and remaining sedentary are all linked with LBD. However, this type of data does not reveal much about the links between specific characteristics of the work and the risk of suffering LBD; specifically, a dose–response relationship has not been elucidated. Furthermore, a review of 57 papers that surveyed low back disorders revealed no consistency between specific risk factors and the development of those disorders (Ferguson and Marras, 1997). This review demonstrated the large differences in the way surveillance was performed and in risk factor measurements. We are reminded once again that epidemiological approaches alone will not elucidate the biological pathway of the development of LBDs, a process that must be understood to develop optimal prevention and rehabilitation strategies.

As noted, the majority of specific risk factors that are addressed in the epidemiological literature (which is surprisingly sparse) are really surrogate factors, or indirect measures, of spine load. These surrogate factors are static work postures; seated work postures; frequent bending and twisting; lifting, pulling, and pushing; and vibration (especially seated).

- **Static work postures.** Research has suggested that work characterized by static postures is an LBD risk factor. While many studies have suggested a link with static work, the key paper on this topic was presented by Punnett and coworkers (1991), who reviewed 1995 back injury cases from an auto assembly plant. Analyzing jobs for postural and lifting requirements, they found that low back disorders were associated with postures that required maintaining mild trunk flexion (defined as the trunk flexed forward from 21 to 45°) (odds ratio = 4.9), postures involving maintaining severe trunk flexion (defined as the trunk being flexed forward greater than 45°) (odds ratio = 5.7), and postures involving trunk twisting or lateral bending greater than 20° (odds ratio = 5.9). Their results suggested that the risk of back injury increased with exposure to these deviated postures and with increased duration of exposure. Deviated postures greatly increase low back tissue loading, particularly when they must be held (Marras, Lavender, and Leurgens et al., 1993; McGill, 1997).

- **Seated work postures.** Kelsey (1975) linked the seated work posture to a greater risk of LBD. In a more recent study, Liira and colleagues (1996) suggested that although white collar (sedentary) workers who must sit for long periods have a greater risk of low back troubles (8% increase in odds risk), active blue collar workers gain some prophylactic effect from sitting down (14% reduction in odds risk). This suggests that variable work, and not too much of any single activity, may have merit in reducing mechanically induced low back troubles.

- **Frequent bending and twisting.** The U.S. Department of Labor report (1982) and many more studies (summarized by Andersson, 1981; Marras et al., 1995; Punnett et al., 1991; Snook, 1982) noted the increased risk of LBD from frequent bending and twisting. In fact, Marras documented in several studies the increased risk of LBD with higher torso velocities (e.g., Marras et al., 1993, 1995). (Note that this is isolated spine motion and not nonspecific torso motion.) While these studies did not examine a mechanism to explain a link with LBD, the associated motion within the spine will be shown in chapter 4 to form a pathomechanism for very specific disabling low back disorders.

- **Lifting, pulling, and pushing.** The National Institute for Occupational Safety and Health report (NIOSH, 1981) provides a good review linking activities requiring lifting, pushing, and pulling with increased risk of LBD.

- **Vibration.** Finally, vibration, particularly seated vibration, is linked to elevated rates of LBD (e.g., Kelsey, 1975; Pope, 1989).

While all of the risk factors noted here have been epidemiologically linked with an increased incidence of LBD, a subsequent section on tissue damage will provide insight into the mechanisms linking mechanical overload, the onset of pain and disability, and the natural history of these injuries as they pertain to job performance.

What Are the Lasting Physiological, Biomechanical, and Motor Changes to Which Injury Leads?

The following discussion of recent studies illustrates the substantial literature documenting specific performance deficits and subsequent anatomical changes in LBP populations.

- Several studies have documented a change in muscular function after injury (nicely summarized in Sterling, Jull, and Wright, 2001). These include, for example, studies of the delayed onset of specific torso muscles during sudden events (Hodges and Richardson, 1996, 1999) that impair the spine's ability to achieve protective stability during situations such as slips and falls, changes in torso agonist/antagonist activity during gait (Arendt-Nielson et al., 1995), and asymmetric muscle output during isokinetic torso extensor efforts (Grabiner, Koh, and Ghazawi, 1992) that alters spine tissue loading.
- Anatomical changes following low back injury include asymmetric atrophy in the multifidus (Hides et al., 1994) and fiber changes in the multifidus even five years after surgery (Rantanen et al., 1993). Further, in a very nice study of 108 patients with histories of chronic LBD ranging from 4 months to 20 years, Sihvonen and colleagues (1997) noted that 50% had disturbed joint motion and 75% of those with radiating pain had abnormal electromyograms to the medial spine extensor muscles.
- Finally, a recent study of those with a history of low back troubles has shown a wide variety of lingering deficits (McGill et al., in press). The back troubles were sufficient to cause work loss, but the subjects had been back to work for an average of 270 weeks. Generally, those with a history of troubles were heavier and had disturbances in the flexor/extensor strength ratio and the flexor/extensor endurance ratio together with the lateral bend endurance ratio, had diminished gross flexion range of lumbar motion, and had lingering motor difficulties compromising their ability to balance and bend down to pick up a light object. None of these changes could be considered to be good.

The broad implication of this work is that a history of low back trouble, even when a substantial amount of time has elapsed since the trouble, is associated with a variety of lingering deficits that would require a multidisciplinary intervention approach to diminish their presence. This collection of evidence is quite powerful in documenting pathoneuromechanical changes associated with chronic LBD. These changes are lasting years—not 6 to 12 weeks!

What Is the Optimal Amount of Loading for a Healthy Spine?

Lucid interpretation of the data in the epidemiological literature is limited by the fact that the levels at which tissue damage occurs remain obscure. Many are concerned with the known tissue damage that occurs with high magnitudes of load, repetition, and so on. For example, Herrin and colleagues (1986) found that musculoskeletal injuries were twice as likely if the worker's lumbar spine was exposed to compressive forces that exceeded 6800 newtons (predicted with a biomechanical model). On the other hand, several of the epidemiological studies have not been able to support a link between heavy work, when crudely measured, and the risk of LBD (e.g., Bigos et al., 1986; Porter, 1987). Mitigating factors appear to include repetition of similar movements and variety in work.

In the discussion of the U-shaped relationship between activity levels and LBD discussed in chapter 1, we saw that stress at optimal levels strengthens the system, while too little or too much is detrimental to health. Kelsey (1975) demonstrated a greater than expected increase in disc protrusions among sedentary workers. While many consider sitting a "low load" task, in fact it creates damaging conditions for the disc—the mechanism of which will be explained in chapter 4. This has obscured the emergence of a much clearer relationship among biomechanical loading, disc herniation, clinical impressions, and the ability to perform demanding work. Further, Videman and colleagues (1990) studied a cross-section of retired workers by comparing their LBD history with their MRI scans. A history of back pain and the visible parameters of spinal pathology were least prevalent in workers whose jobs had included moderate activity and most prevalent in workers with either sedentary or heavy work.

Many studies compare just two levels of work—for example, light versus heavy, light versus moderate, or moderate versus heavy. Because any relationship other than a straight line requires more than two points (or levels of activity or levels of loading), such studies cannot illustrate the U-shaped relationship. Consider the data compiled in a thorough epidemiological study reported by Liira and colleagues (1996) of LBD prevalence and physical work exposures. Even though the authors considered only the upper levels of excessive loading, they concluded that one quarter of excess back pain morbidity could be explained by physical work exposures. Real possibility remains that this is an underestimate.

Furthermore, the nature of LBD appears to be affected by the type of work. Videman and colleagues (1990) noted a tendency among those who had had sedentary careers to have marked disc degeneration in later years, while those who had performed heavy work (defined as not only lifting but also requiring large trunk motions) tended to have classic arthritic changes in the spine (stenosis, osteophytosis, etc.). In a similar study, Battie and colleagues (1995) reported an apparent contribution of genetic factors to various age-related changes in the spine using monozygotic twins, given significantly greater similarities in spinal changes than would be expected by chance.

Porter (1987, 1992) performed two studies furthering the notion of an optimal loading level for health. The first study tracked miners and nonminers treated at hospitals for back pain. While significantly more miners reported for back trouble compared with nonminers, significantly fewer were diagnosed with disc protrusions, while significantly more were reported to have stenosis and nerve root entrapment (conditions associated more with the arthritic spine according to the data of Videman, Nurminen, and Troup, 1990). The second study evaluated questionnaire results from 196 patients with symptomatic disc protrusion and 53 with root entrapment syndrome. They were asked about their history of heavy work between the ages of 15 and 20 years. Significantly more subjects with disc protrusion had done no heavy physical work in those early years. By contrast, more of those with nerve entrapment syndrome had done five full years of heavy work between 15 and 20 years of age. This collective data suggests that different work demands cause different spine conditions and perhaps that the optimum loading is different for different tissues. Nonetheless, the optimum activity appears to be varied work at a moderate level between sedentary and heavy work.

What constitutes an optimum load—a load that is not too much, not too little, not too repetitive, and not too prolonged? Currently, we are in need of assessment tools for determining optimum load to guide the intervention toward optimum health.

What Are the Links Between Personal Factors and LBD?

Some personal factors such as muscle endurance (not strength) and less spine range of motion (not more) are prophylactic for future back troubles. These will be addressed in part III, Low Back Rehabilitation. A few other personal factors are noted here.

A few specific personal factors appear to affect spine tissue tolerance according to the existing literature; age and gender are two examples. Jager and colleagues (1991) compiled the available literature on the tolerance of lumbar motion units to bear compressive load that passed their inclusion criteria. Their results revealed that when males and females were matched for age, females were able to sustain only approximately two thirds of the compressive loads of males. Furthermore, Jager and colleagues' data showed that within a given gender, the 60-year-old spine was able to tolerate only about two thirds of the load tolerated by the 20-year-old spine. Keep in mind that age and gender are very simple factors.

It appears that other personal factors, such as poor motor control system "fitness," can lead to a back injury during ordinarily benign tasks such as picking up a pencil from the floor. The modeling data of Cholewicki and McGill (1996) suggest that the spine can easily **buckle** during such a task. When the muscle forces are inherently low, a small motor error can cause rotation of a single spinal joint, placing all bending moment support responsibility on the passive tissues. Such scenarios do not constitute excessive tasks, but patients often report them to clinicians as the event that caused their injury. This phenomenon will not be found in the scientific literature, however. Many medical personnel would not record this event as the cause of injury since in many jurisdictions it would not be deemed a compensable injury. These types of injuries seem to be more influenced by the fitness of an individual's motor control system than by factors such as strength. McGill and colleagues (1995) noted that people differ in their ability to hold a load in their hands and breathe heavily. This is very significant since the muscles required to be continuously active to support the spine (and prevent buckling) are also used to breathe by rhythmically contracting. Those who must use their muscles to breathe in this way sacrifice spine stability. This is particularly noticeable in those with compromised lung elasticity (smokers, emphysemics, etc.). All of these deficient motor control mechanisms will heighten biomechanical susceptibility to injury or reinjury (Cholewicki and McGill, 1996) and are highly variable personal characteristics.

Additional factors other than simple load magnitude appear to modulate the risk of tissue damage. The mechanism of disc herniation provides an example. While disc herniations have been produced under controlled conditions (e.g., Gordon et al., 1991), they have not been produced consistently. Our lab has been able to consistently produce disc herniations by mimicking spine motion and load patterns seen in workers (Callaghan and McGill, 2001). Specifically, it appears that only a very modest amount of spine compression force is required (only 800-1000 N), but the spine specimen must be repeatedly flexed—mimicking repeated torso–spine flexion from continual bending to a fully flexed posture. The main relevance for this issue is that the way in which workers elect to move and bend will influence the risk of disc herniation. This highlights the need to examine how workers move in standardized tests at the beginning of their careers while they still have "virgin" backs, making it possible to determine cause and effect. Recent evidence suggests that those with a history of back troubles are more likely to lift flexing the spine and not the hips, increasing the risk of future back damage (McGill et al., in press).

What the Evidence Supports

In summary, it is interesting to consider why only some workers become patients. There is no question that damage to tissue can be caused by excessive loading, and damage causes pain. However, pain is a perception that is modulated by psychosocial variables in addition to physiological injury. Clearly both psychosocial and biomechanical variables are associated with LBD and are important in preventing low back injury and the ensuing chronicity; collectively the evidence from several scientific perspectives is overwhelming. The relative importance of either is often difficult to compare across studies as the metrics for each are different—biomechanical variables are reported in newtons, newton-meters, x cycles, and so forth, while psychosocial variables are reported in ordinal scales linked to perception (independent risk factors can be compared using odds ratios). Some influential reports have ignored biomechanical evidence and promoted psychosocial variables as being more important. However, no study of psychosocial variables has been able to conclusively establish causal links—only association. Some biomechanically based studies, together with the chronic pain literature, are strongly convincing in their establishment of both association and causality. Thank goodness! We can now proceed.

Back injury can begin with damage to one tissue, but this changes the biomechanical function of the joint. Tissue stresses change and other tissues become involved leading to progressive deterioration with time. Tissue damage does not always result from too high a load magnitude. In the case of disc herniation, repetitive motion, even in the absence of large loads, seems to be a significant causative mechanism. The notion that tissues heal within 6 to 12 weeks and that longer-lasting work intolerance has no pathoanatomical basis appears to be false; tissue injury data and the science of chronic pain mechanisms strongly suggest otherwise.

Understanding the role of biomechanical, psychosocial, and personal factors together with their interrelationships will build the foundation for better prevention in the future. The challenge is to develop variable tolerance guidelines, psychosocial guidelines, and higher-level medical practice codes.

On balance the evidence supports the following statements:

- Biomechanical factors are linked to both the incidence of first-time low back troubles and absenteeism, and subsequent episodes.
- Psychosocial factors appear to be important as well but may be more related to episodes after the initial back-related episode.
- Psychosocial and biomechanical factors appear to influence each other, both in terms of causation of work absence and in the course of recovery.
- The relationship between loading and LBDs appears to be a U-shaped function with the optimal loading being at a moderate level.
- Low back tissue damage can initiate a cascade of changes that may cause pain and intolerance to certain activities, and these changes may be disruptive for up to 10 years in an unfortunate few.
- Many types of tissue damage can escape detection in vivo. Even gross damage visible during dissection is often not visible on medical images. Thus, nonspecific diagnosis does not rule out the presence of mechanical damage and must not be used to imply that mechanical factors are not related to LBD.

References

Andersson, G.B. (1981) Epidemiologic aspects of low back pain in industry. *Spine,* 6: 53-60.

Andersson, G.B. (1991) The epidemiology of spinal disorders. In: J.W. Frymoyer (Ed.), *The adult spine: Principles and practice* (chapter 8). New York: Raven Press.

Arendt-Nielson, L., Graven-Neilson, T., Svarrer, H., and Svensson, P. (1995) The influence of low back pain on muscle activity and coordination during gait. *Pain,* 64: 231-240.

Battie, M.C., Haynor, D.R., Fisher, L.D., Gill, K., Gibbons, L.E., and Videman, T. (1995) Similarities in degenerative findings on magnetic resonance images of the lumbar spines of identical twins. *Journal of Bone and Joint Surgery,* 77-A: 1662-1670.

Biering-Sorensen, F. (1984) Physical measurements as risk indicators for low-back trouble over a one-year period. *Spine,* 9: 106-119.

Bigos, S.J., Battie, M.C., Spengler, D.M., Fisher, L.D., Fordyce, W.E., Hansson, T.H., Nachemson, A.L., and Wortley, M.D. (1991) A prospective study of work perceptions and psychosocial factors affecting the report of back injury. *Spine,* 16: 1-6.

Bigos, S.J., Spengler, D.M., Martin, N.A., Zeh, A., Fisher, L., Nachemson, A., and Wang, M.H. (1986) Back injuries in industry: A retrospective study. II. Injury factors. *Spine,* 11: 246-251.

Bogduk, N., Derby, R., April, C., Lous, S., and Schwarzer, R. (1996) Precision diagnosis in spinal pain. In: Campbell, J. (Ed.), *Pain 1996—An updated review* (pp. 313-323). Seattle: IASP Press.

Brereton, L., and McGill, S.M. (1999) Effects of physical fatigue and cognitive challenges on the potential for low back injury. *Human Movement Science* 18: 839-857.

Burton, A.K., Symonds, T.L., and Zinzen, E., et al. (1996) Is ergonomics intervention alone sufficient to limit musculoskeletal problems in nurses. *Occupational Medicine,* 47: 25-32.

Burton, A.K., Tillotson, K.M., Symonds, T.L., Burke, C., and Mathewson, T. (1996) Occupational risk factors for the first onset of low back trouble: A study serving police officers. *Spine,* 21: 2621.

Burton, A.K., Tillotson, K.M., and Troup, J.D.G. (1989) Prediction of low back trouble frequency in a working population. *Spine,* 14: 939-946.

Callaghan, J., and McGill, S.M. (2001) Intervertebral disc herniation: Studies on a porcine model exposed to highly repetitive flexion/extension motion with compressive force. *Clinical Biomechanics,* 16 (1): 28-37.

Cholewicki, J., and McGill, S.M. (1996) Mechanical stability of the in vivo lumbar spine: Implications for injury and chronic low back pain. *Clinical Biomechanics,* 11 (1): 1-15.

Delitto, A., Erhard, R.E., and Bowling, R.W. (1995) A treatment-based classification approach to acute low back syndrome: Identifying and staging patients for conservative treatment. *Physical Therapy,* 75: 470-489.

Ferguson, S.A., and Marras, W.S. (1997) A literature review of low back disorder surveillance measures and risk factors. *Clinical Biomechanics,* 12 (4): 211-226.

Fordyce, W.E. (Ed.) (1995) *Back pain in the workplace.* Seattle: IASP Press.

Fordyce, W.E. (1996) Response to Thompson/Merskey/Teasell letters. *Pain,* 65: 112-114.

Gamsa, A. (1990) Is emotional status a precipitator or a consequence of pain? *Pain,* 42: 183-195.

Gatchel, R.J., Polatin, P.B., and Mayer, T.G. (1995) The dominant role of psychosocial risk factors in the development of chronic low back pain disability. *Spine,* 20: 2702-2709.

Gordon, S.I., Yang, K.H., Mayer, P.J., Mace, A.H.J., Kish, V.I., and Radin, E.L. (1991) Mechanism of disc rupture—A preliminary report. *Spine,* 16: 450-456.

Grabiner, M.D., Koh, T.J., and Ghazawi, A.E. (1992) Decoupling of bilateral excitation in subjects with low back pain. *Spine,* 17: 1219-1223.

Hadler, N.M. (1991) Insuring against work incapacity from spinal disorders. In: J.W. Frymoyer (Ed.), *The adult spine.* New York: Raven Press.

Hadler, N. (2001) Editorial, The bane of the aging worker. *Spine,* 26 (12): 1309-1310.

Herrin, G.A., Jaraiedi, M., and Anderson, C.K. (1986) Prediction of overexertion injuries using biomechanical and psychophysical models. *American Industrial Hygiene Association Journal,* 47: 322-330.

Hides, J.A., Stokes, M.J., Saide, M., Jull, G.A., and Cooper, D.H. (1994) Evidence of lumbar multifidus muscle wasting ipsilateral to symptoms in patients with acute/subacute low back pain. *Spine,* 19: 165-177.

Hodges, P.W., and Richardson, C.A. (1996) Inefficient muscular stabilization of the lumbar spine associated with low back pain. *Spine,* 21: 2640-2650.

Hodges, P.W., and Richardson, C.A. (1999) Altered trunk muscle recruitment in people with low back pain with upper limb movement at different speeds. *Archives of Physical Medicine and Rehabilitation*, 80: 1005-1012.

Jager, M., Luttmann, A., and Laurig, W. (1991) Lumbar load during one-handed bricklaying. *International Journal of Industrial Ergonomics,* 8: 261-277.

Jonsson, H., Bring, G., Rauschning, W., Shalstedt, B. (1991) Hidden cervical spine injuries in traffic accident victims with skull fracures. *Journal of Spinal Disorders* 4(3): 251-263.

Kelsey, J.L. (1975) An epidemiological study of the relationship between occupations and acute herniated lumbar intervertebral discs. *International Journal of Epidemiology,* 4: 197-205.

Liira, J., Shannon, H.S., Chambers, L.W., and Haives, T.A. (1996) Long term back problems and physical work exposures in the 1990 Ontario health survey. *American Journal of Public Health,* 86 (3): 382-387.

Lord, S.M., Barnsley, L., Wallis, B.J., McDonald, G.J., and Bogduk, N. (1996) Percutaneous radiofrequency neurotomy for chronic cervical zygapophyseal joint pain. *New England Journal of Medicine,* 335: 1721-1726.

Luoto, S., Helioraara, M., Hurri, H., and Alavanta, M. (1995) Static back endurance and the risk of low back pain. *Clinical Biomechanics*, 10: 323-324.

Marras, W.S., Davis, K.G., Heaney, C.A., Maronitis, A.B., and Allread, W.G. (2000) The influence of psychosocial stress, gender, and personality on mechanical loading of the lumbar spine. *Spine* 25: 3045-3054.

Marras, W.S., Lavender, S.A., Leurgens, S.E., et al. (1993) The role of dynamic three-dimensional trunk motion in occupationally related low back disorders: The effects of workplace factors, trunk position and trunk motion characteristics on risk of injury. *Spine,* 18: 617-628.

Marras, W.S., Lavender, S.A., Leurgans, S.E., Fathallah, F.A., Ferguson, S.A., Allread, W.G., and Rajulu, S.L. (1995) Biomechanical risk factors for occupationally related low back disorders. *Ergonomics,* 38: 377-410.

McCall, I.W., Park, W.M., and O'Brien, J.P. (1979) Induced pain referral from posterior lumbar elements in normal subjects. *Spine,* 4: 441-446.

McGill, S.M. (1997) The biomechanics of low back injury: Implications on current practice in industry and the clinic. *Journal of Biomechanics,* 30: 465-475.

McGill, S.M., Grenier, S., Preuss, R.P., and Brown, S. (in press) Previous history of LBP with work loss is related to lingering effects in psychosocial, physiological, and biomechanical characteristics.

McGill, S.M., Sharratt, M.T., and Seguin, J.P. (1995) Loads on spinal tissues during simultaneous lifting and ventilatory challenge. *Ergonomics,* 38: 1772-1792.

Melzack, R., and Wall, P.D. (1983) *The challenge of pain.* New York: Basic Books.

Mendelson, G. (1982) Not cured by a verdict: Effect of a level settlement on compensation claimants. *Medical Journal of Australia*, 2: 219-230.

National Institute for Occupational Safety and Health (NIOSH). (1981) *Work practices guide for manual lifting.* Department of Health and Human Services (DHHS), NIOSH Publication No. 81-122.

Norman, R., Wells, R., Neumann, P., Frank, P., Shannon, H., and Kerr, M. (1998) A comparison of peak vs cumulative physical work exposure risk factors for the reporting of low back pain in the automotive industry. *Clinical Biomechanics,* 13: 561-573.

Pope, M.H. (1989) Risk indicators in low back pain. *Annals of Medicine,* 21: 387-392.

Porter, R.W. (1987) Does hard work prevent disc protrusion? *Clinical Biomechanics,* 2: 196-198.

Porter, R.W. (1992) Is hard work good for the back? The relationship between hard work and low back pain-related disorders. *International Journal of Industrial Ergonomics*, 9: 157-160.

Punnett, L., Fine, L.J., Keyerserling, W.M., Herrin, G.D., and Chaffin, D.A. (1991) Back disorders and non-neutral trunk postures of automobile assembly workers. *Scandinavian Journal of Work Environment and Health,* 17: 337-346.

Radanov, B.P., Sturzenegger, M., DeStefano, G., and Schinrig, A. (1994) Relationship between early somatic, radiological, cognitive and psychosocial findings and outcome during a one-year follow up in 117 patients suffering from common whiplash. *British Journal of Rheumatology,* 33: 442-448.

Rainville, J., Sobel, J.B., Hartigan, C., and Wright, A. (1997) The effect of compensation involvement on the reporting of pain and disability by patients referred for rehabilitation of chronic low back pain. *Spine,* 22: 2016-2024.

Rantanen, J., Hurme, M., Falck, B., et al. (1993) The lumbar multifidus muscle five years after surgery for a lumbar intervertebral disc herniation. *Spine,* 18: 568-574.

Schwarzer, A.C., Wang, S., O'Driscoll, D., Harrington, T., Bogduk, N., and Laurent, R., (1995) The ability of computed tomography to identify a painful zygapophysial joint in patients with low back pain. *Spine,* 20 (8): 907-912.

Sihvonen, T., Lindgren, K., Airaksinen, O., and Manninen, H. (1997) Movement disturbances of the lumbar spine and abnormal back muscle electromyographic findings in recurrent low back pain. *Spine,* 22: 289-295.

Skargren, E.I., Carlsson, P.G., and Oberg, B.E. (1998) One year follow-up comparison of the cost and effectiveness of chiropractic and physiotherapy as primary management for back pain. *Spine,* 23 (17): 1875-1884.

Snook, S.H. (1982) Low back pain in industry. In: White, A.A., and Gordon, S.L. (Eds.), *Symposium on idiopathic low back pain.* St. Louis: Mosby.

Sterling, M., Jull, G., and Wright, A., (2001) The effect of musculoskeletal pain on motor activity and control. *Journal of Pain,* 2 (3): 135-145.

Taylor, J.R., Twomey, L.T., and Corker, M. (1990) Bone and soft tissue injuries in post-mortem lumbar spines. *Paraplegia,* 28: 119-129.

Teasell, R.W. (1997) The denial of chronic pain. *Journal of Pain Research Management,* 2: 89-91.

Teasell, R.W., and Shapiro, A.P. (1998) Whiplash injuries: An update. *Journal of Pain Research Management,* 3: 81-90.

Thompson, Merskey, Teasell, Fordyce (1996) Letters published in *Pain,* 65: 111-114.

Troup, J.D.G., Foreman, T.K., Baxter, C.E., and Brown, D. (1987) The perception of back pain and the role of psychological tests of lifting capacity. *Spine,* 12: 645-657.

Troup, J.D.G., Martin, J.W., and Lloyd, D.C.E.F. (1981) Back pain in industry—A prospective study. *Spine,* 6: 61-69.

U.S. Department of Labor. (1982) Back injuries associated with lifting (Bulletin 2144). Washington, DC: Government Printing Office.

Vanharanta, H., Sacks, B.L., Spivey, M.A., et al. (1987) The relationship of pain provocation to lumbar disc deterioration as seen by CT/discography. *Spine,* 12: 295-298.

Videman, T., Nurminen, M., and Troup, J.D. (1990) Lumbar spinal pathology in cadaveric material in relation to history of back pain, occupation and physical loading. *Spine,* 15: 728-740.

Waddell, G. (1987) A new clinical model for the treatment of low back pain. *Spine,* 12: 632-644.

Waddell, G., Bircher, M., Finlayson, D., and Main, C.J. (1984) Symptoms and signs: Physical disease or illness behaviour? *British Medical Journal,* 289: 739-741.

Waddell, G., McCulloch, J.A., Kummell, E., and Venner, R.M. (1980) Non-organic signs in low back pain. *Spine,* 5: 117-125.

Wallis, B.J., Lord, S.M., and Bogduk, N. (1997) Resolution of psychological distress of whiplash patients following treatment by radiofrequency neurotomy: A randomized double-blind, placebo controlled study. *Pain,* 73: 15-22.

Werneke, M., and Hart, D. (2001) Centralization phenomenon as a prognostic factor for chronic low back pain and disability. *Spine,* 26 (7): 758-765.

White, A.A., and Gordon, S.L. (1982) Synopsis: Workshop on idiopathic low back pain. *Spine,* 7: 141-149.

Yingling, V.R., and McGill, S.M. (2000) Anterior shear of spinal motion segments: Kinematics, kinetics and resulting injuries observed in a porcine model. *Spine,* 24 (18): 1882-1889.

Functional Anatomy of the Lumbar Spine

CHAPTER 4

The average reader of this book will have already studied basic anatomy of the spine. This chapter begins by revisiting some anatomical features, possibly in a way not previously considered. Then these features will be related to normal function and injury mechanics to lay the foundation for the prevention and rehabilitation strategies that follow. I believe that clinicians and scientists alike who specialize in low back troubles do not devote sufficient effort to simply considering the anatomy. The answers to many questions relevant to the clinician can be found within an anatomical framework. The "mechanical" foundation for preventing and rehabilitating back troubles is contained here.

Upon completion of this chapter, you will understand the special anatomical features that have functional relevance. In addition you will have a better understanding of injury mechanisms, which will help ensure that you do not unknowingly include injury-exacerbating maneuvers in therapeutic exercise prescriptions.

The Vertebrae

As you undoubtedly know, the spine has twelve thoracic and five lumbar vertebrae. The construction of the vertebral bodies themselves may be likened to a barrel with round walls made of relatively stiff cortical bone (see figure 4.1). The top and bottom of the barrel are made of a more deformable cartilage plate (end plate) that is approximately .6 mm (.002 in.) thick but thinnest in the central region (Roberts, Menage, and Urban, 1989). The end plate is porous for the transport of nutrients such as oxygen and glucose, while the inside of the barrel is filled with cancellous bone. The trabecular arrangement within the cancellous bone is aligned with the stress trajectories that develop during activity. Three orientations dominate—one vertical and two oblique (Gallois and Japoit, 1925)(see figure 4.2, a-b).

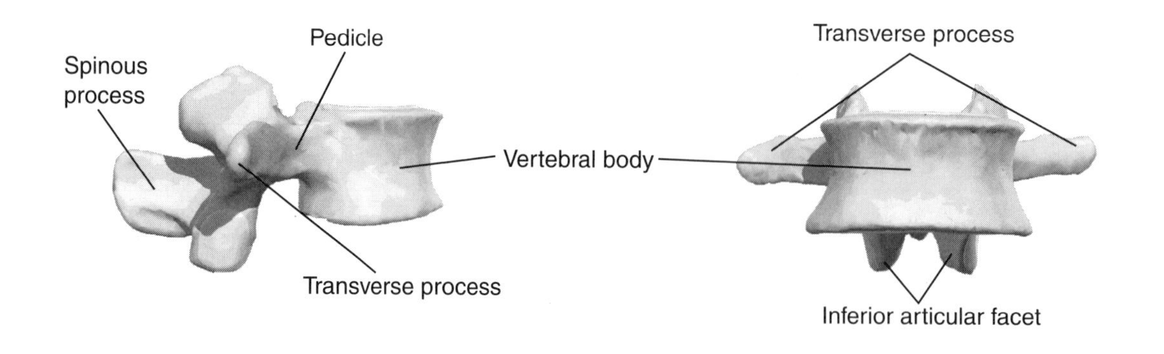

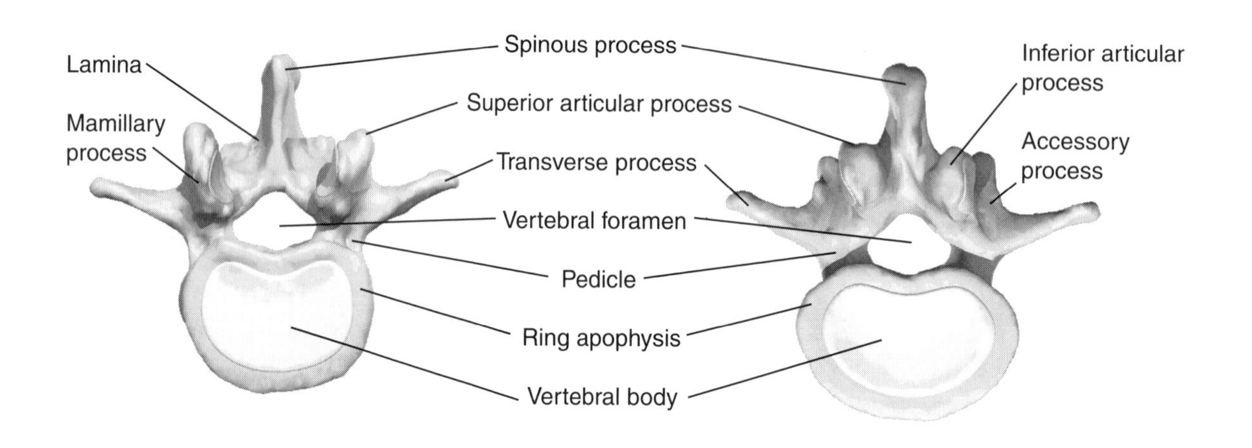

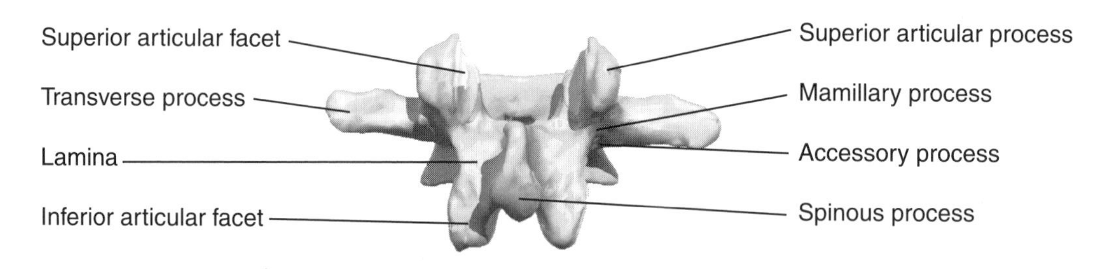

Figure 4.1 The parts of a typical lumbar vertebra.

Image courtesy of Primal Pictures.

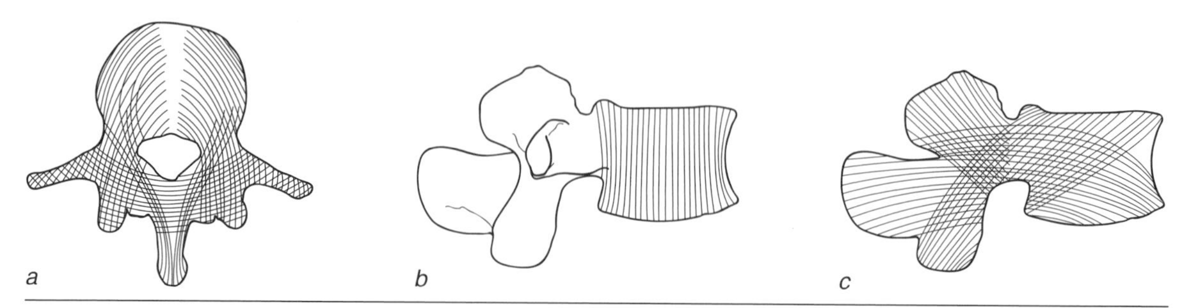

Figure 4.2 The arrangement of the trabeculae (first noted by Gallois and Japoit in 1925) is aligned with the dominant trajectories of stress. The three trabecular systems are shown in *a*, *b*, and *c*.

Vertebral Architecture and Load Bearing

The very special architecture of the vertebral bodies determines how they bear compressive load and fail under excessive loading. The walls of the vertebrae (or sides of the barrel) remain rigid upon compression, but the nucleus of the disc pressurizes (the classic work is by Nachemson, 1960, 1966) and causes the cartilaginous end plates of the vertebrae to bulge inward, seemingly to compress the cancellous bone (Brinckmann, Biggemann, and Hilweg, 1989). In fact, under compression the cancellous bone fails first (Gunning, Callaghan, and McGill, 2001), making it the determinant of failure tolerance of the spine (at least when the spine is not positioned at the end range of motion). It is difficult to injure the disc annulus this way (annular failure will be discussed later).

Althought this notion is contrary to the concept that the vertebral bodies are rigid, the functional interpretation of this anatomy suggests a very clever shock-absorbing and load-bearing system. Farfan (1973) proposed the notion that the vertebral bodies act as shock absorbers of the spine, although he based this more on vertebral body fluid flow than on end-plate bulging. He suggested that the discs were not the major shock absorbers of the spine, contrary to virtually any textbook on the subject! Since the nucleus is an incompressible fluid, bulging end plates suggest fluid expulsion from the vertebral bodies, specifically blood through the perivertebral sinuses (Roaf, 1960). This mechanism suggests a protective dissipation upon quasistatic and dynamic compressive loading of the spine. More vertebral body-based shock-absorbing mechanisms are documented subsequently. In summary, the common statement found in many textbooks that the discs are the shock absorbers of the spine now appears questionable; rather, the vertebral bodies appear to play a dominant role in performing this function.

Understanding Vertebral Mechanics

To truly appreciate vertebral behavior I encourage you to obtain a vertebra from a butcher (bovine or porcine is ideal). Hold the vertebra end plate to end plate between your thumb and finger and squeeze. If you have never done this before, you will be amazed by the deformation and elasticity. The vertebra experiences similar deformation as the incompressible nucleus of the disc presses over the central end plate during spine compression in vivo.

Deformable vertebrae is a new notion for many. How do the end plates bulge inward into seemingly rigid bone? The answer appears to be in the architecture of the cancellous bone. Vertebral cancellous bone structure is dominated by a system of columns of bone (shown in figure 4.2) that run vertically from end plate to end plate. The vertical columns are tied together with smaller transverse trabeculae. Upon axial compression, as the end plates bulge into the vertebral bodies, these columns experience compression and appear to bend. Under excessive compressive load, the bending columns will buckle as the smaller bony transverse trabeculae fracture, as documented by Fyhrie and Schaffler (1994) (see figure 4.3). In this way, the cancellous bone can rebound back to its original shape (at least 95% of the original unloaded shape) when the load is removed, even after suffering fracture and delamination of the transverse (trabeculae). This architecture appears to afford excellent elastic deformation, even after marked damage, and then to regain its original structure and function as it heals. Damaged cancellous fractures appear to heal quickly, given the small amount of osteogenic activity needed, at least when compared with the length of time needed to repair collagenous tissues.

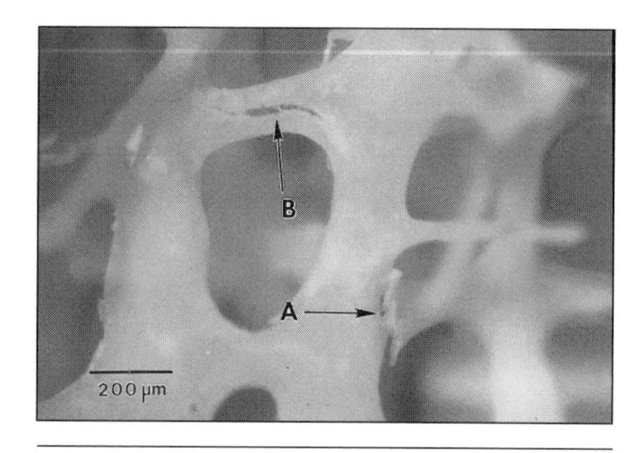

Figure 4.3 Under compressive loading, bulging of the end plate causes buckling stresses in the vertical trabeculae, which, when excessive, causes damage in the transverse trabeculae. Note the vertical (from compression) (A) and horizontal (from tension) (B) cracks in the transverse trabeculae.

Reprinted from *Bone*, Vol. 15(1), Fyhrie and Scheffler, "Failure mechanisms in human vertebral cancellous bone," 105-109, Copyright 1994, with permission of Elsevier Science.

Microfracturing of the trabeculae can occur with repetitive loading at levels well below the failure level from a single cycle of load. Lu and colleagues (2001) demonstrated that cyclic loading at 10% of ultimate failure load caused no damage or change in stiffness, but with 20,000 cycles of load at 20-30% of the ultimate failure load, both stiffness and energy absorbed at failure were decreased. Highly repetitive loads, even at quite low magnitudes, appear to cause microdamage.

The osteoporotic vertebra is characterized by mineral loss and declining bone density in the trabeculae. Because transverse trabeculae are far fewer in number than longitudinal trabeculae and because they are generally of smaller diameter, the transverse trabeculae specifically are the target for mechanical compromise with osteoporotic mineral loss (Silva and Gibson, 1997) (see figure 4.4). Interestingly, the same authors noted a higher tendency for the transverse trabeculae to disappear in females with greater incidence than in males. This loss in mechanical integrity of the transverse trabeculae has a great influence on the compressive strength of the vertebrae via the mechanism described earlier. Thus, the osteoporotic vertebra begins to slowly collapse when exposed to excessive load, with serial buckling of the columns of bone ultimately developing the classic wedge shape.

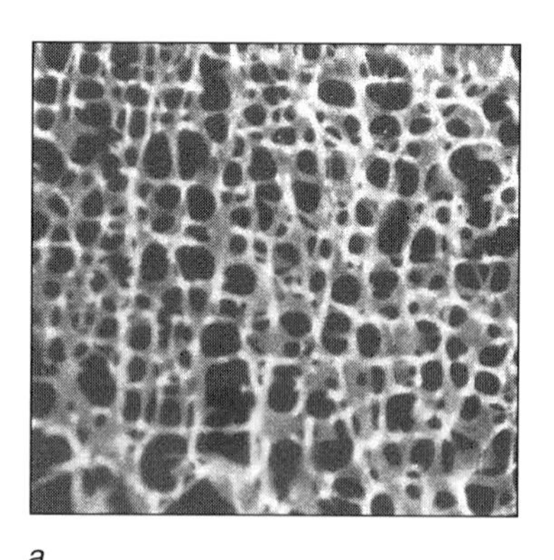
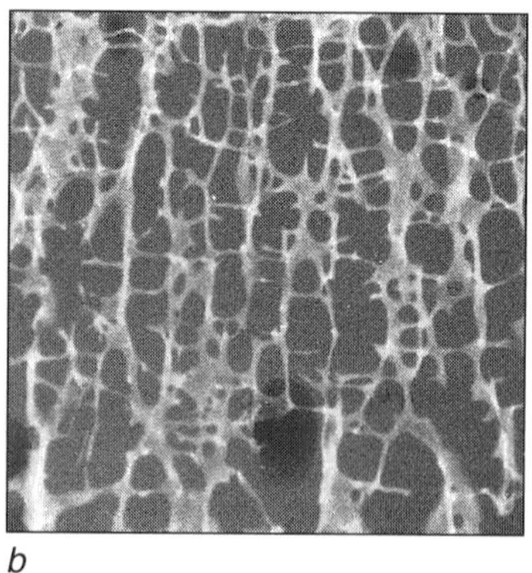

a *b*

Figure 4.4 With aging, the transverse or horizontal trabeculae thin and eventually lose their ability to support the vertical trabeculae, which can then buckle, causing vertebral collapse. *(a)* A healthy trabecular bone network from a 47-year-old woman. *(b)* Perforation in a horizontal trabecula in an elderly woman.

Reprinted from *Bone*, Vol. 21, Silva and Gibson, "Modeling the mechanical behaviour of vertebral trabecular bone: Effects of age-related changes in microstructure," 191-199, Copyright 1997, with permission from Elsevier Science.

It is interesting to contrast the other extreme of the bone density spectrum. The transverse trabeculae harvested from specimens who performed heavy work (in particular, weightlifters) were thick and dense. In addition, where the transverse trabeculae intersected with the vertical columns, the joints were characterized with heavy bony gusseting, similar to what a welder would weld to strengthen a right-angled joint. The transverse trabeculae appear to be crucial in determining compressive strength.

Yet another observed failure mechanism of the vertebral body is termed the "slow crush," in which extensive trabecular damage is observed without concomitant loss of stiffness or abrupt change in the load–deformation relationship (Gunning, Callaghan, and McGill, 2001) (see figure 4.5). Since slope change in the load–deformation relationship is often used to identify the yield point, or the initial tissue damage, this injury can go unnoticed. Interestingly enough, vertebrae failing under ever-increasing compressive load will gradually increase in stiffness, a testament to the wonderful architecture of the transverse and vertical trabeculae. Also interesting is the load-rate dependence of bone strength, as the end plate appears to fail first at low load rates while the vertebral bony elements fail at higher load rates (see figure 4.6).

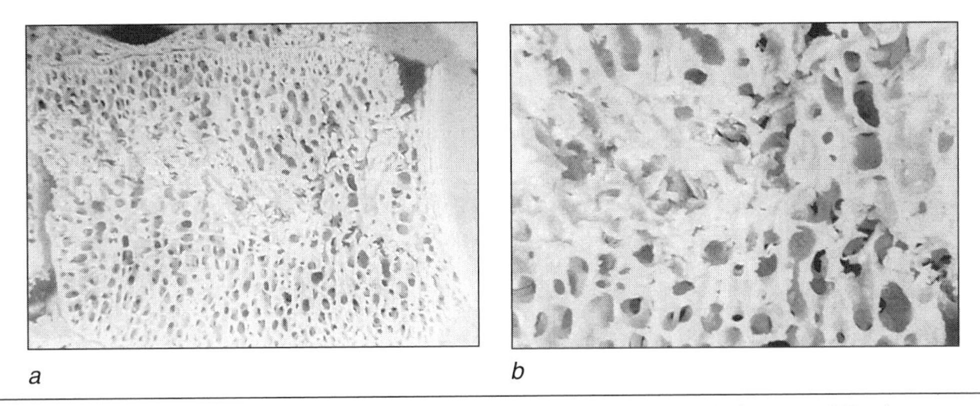

Figure 4.5 Massive trabecular damage found during the dissection and removal of marrow following an excessive "slow crush" compressive load *(a)*. Higher magnification of the crush fracture is shown in *(b)*. Even this massive fracture was not clear on X ray or any other examination methods.

Reprinted with permission of J. Gunning, J. Callaghan, and S. McGill, *Clinical Biomechanics,* 16(6), 2001. Copyright by Elsevier Science.

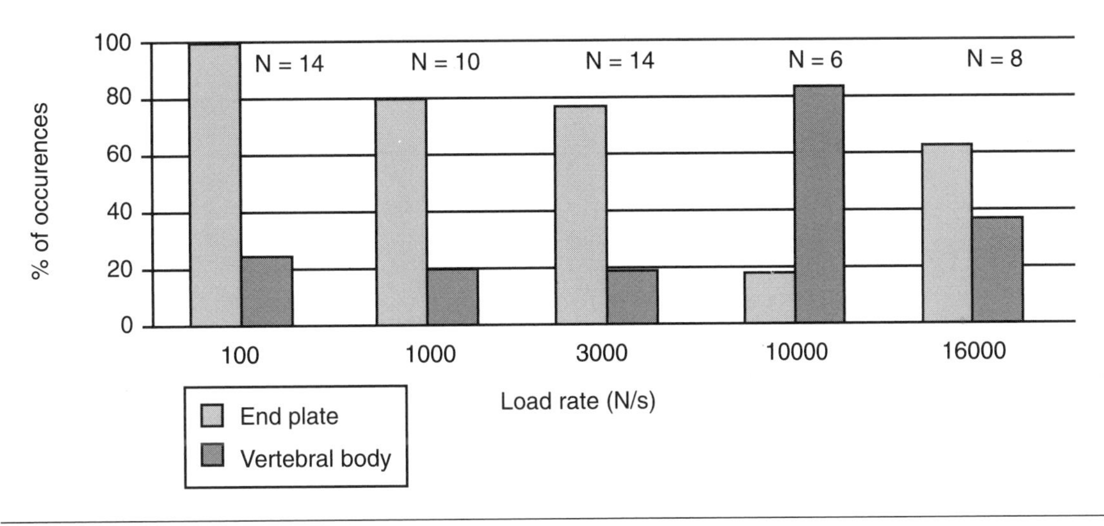

Figure 4.6 Compression injuries at different load rates. At low rates of compressive load the end plate appears to be the first structure to fail, but bone will fracture first under higher rates of load.

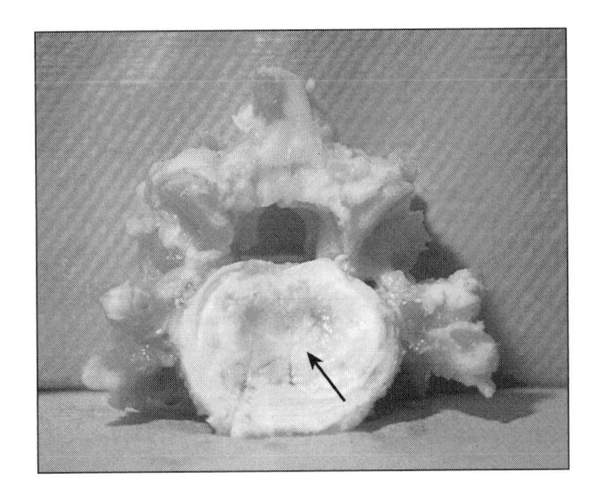

Both the disc and the vertebrae deform while supporting spinal loads. Under excessive compressive loading, the bulging of the end plates into the vertebral bodies also causes radial stresses in the end plate sufficient to cause fracture in a stellate pattern (see figure 4.7). These fractures, or cracks, in the end plate are sometimes sufficiently large to allow the liquid nucleus to squirt through the end plate into the vertebral body (McGill, 1997) (see figures 4.8 and 4.9). Sometimes a local area of bone collapses under the end plate to create a pit or crater that goes on to form the classic Schmorl's node (see figure 4.10, a-b). This type of injury is associated with compression of the spine

Figure 4.7 Stellate-patterned end-plate fracture (indicated with arrow) occurs as the nucleus is pressurized under compressive load, which causes it to bulge the end plate, imparting tensile stresses.

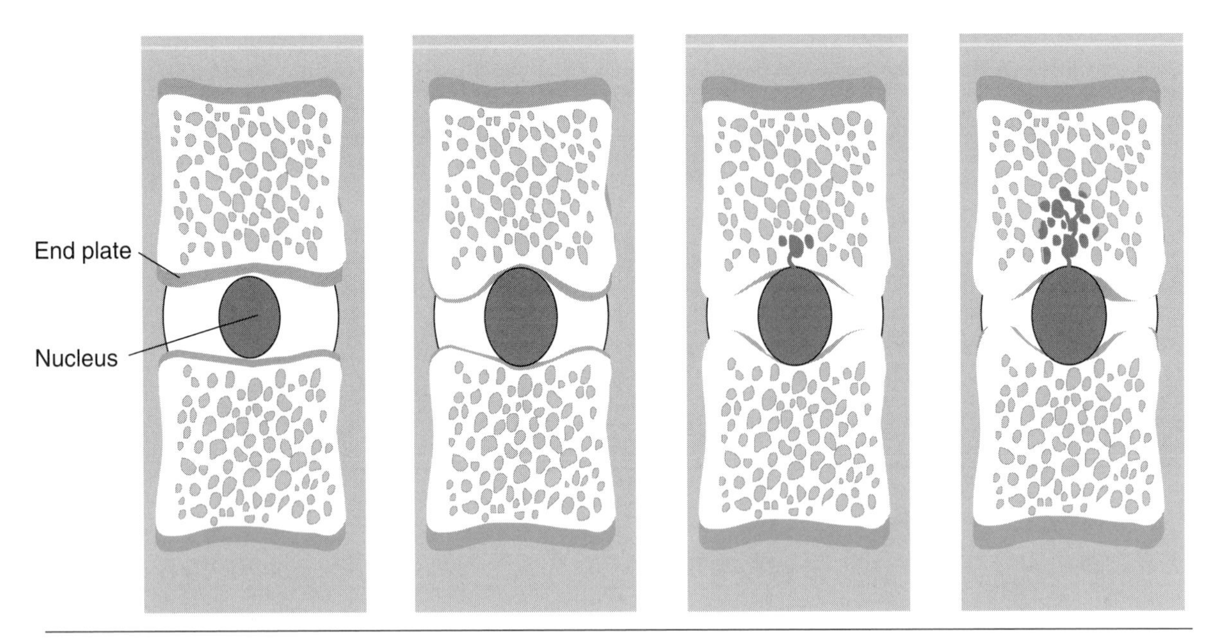

Figure 4.8 Under compressive loading the nucleus pressurizes, causing the end plate to bulge into the vertebral body. With excessive radial-tensile stress the end plate will fracture and the viscous nucleus will squirt through the crack into the vertebral body.

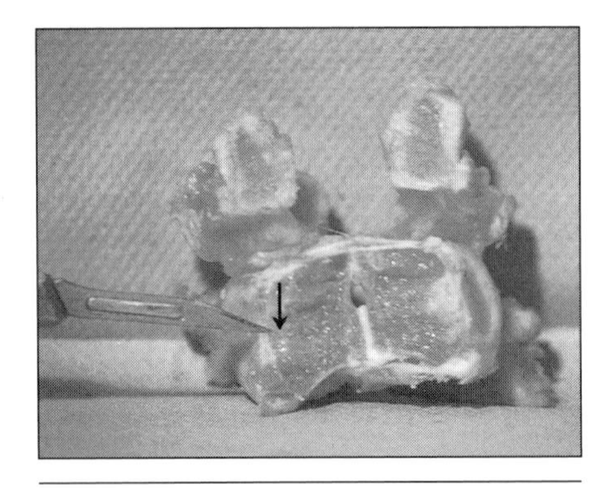

Figure 4.9 Following a more severe end-plate fracture, a portion of the nucleus has squirted through into the vertebral body (shown at the tip of the scalpel and the arrow). This is a porcine specimen.

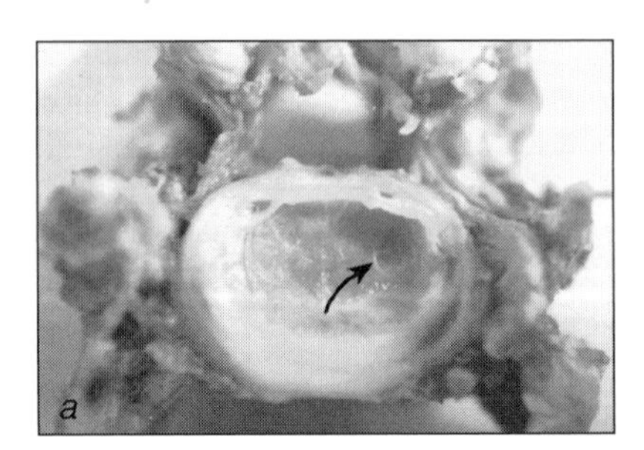

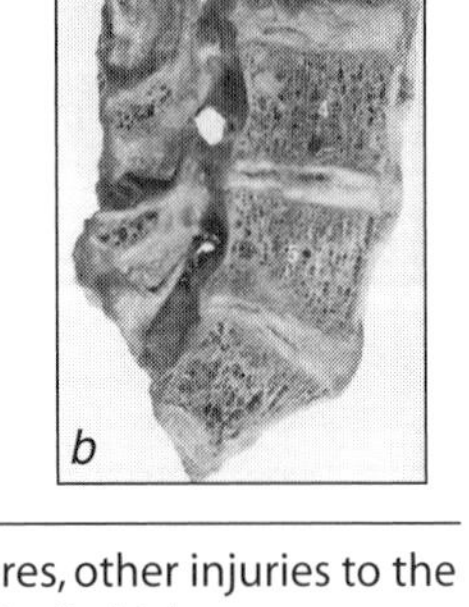

Figure 4.10 In addition to end-plate fractures, other injuries to the end plate under compressive load include "pits," which occur as trabucular bone fractures in small areas under the end plate *(a)*. These go on to form Schmorl's nodes. A Schmorl's node is shown at the tip of the arrow in *(b)*.

Adapted, by permission, from Kirkaldy-Willis and Burton, 1983, *Managing low back pain,* 3rd ed. (New York: Churchill Livingstone), 72.

when the spine is within the neutral range of motion (i.e., not flexed, bent, or twisted). In my experience, this type of compressive injury is very common and often misdiagnosed as a herniated disc due to the flattened interdiscal space seen on planar X rays. However, note that in end-plate fractures, the annulus of the disc remains intact. It is simply a case of the nucleus leaving the disc and progressing through the end plate into the cancellous core of the vertebra (sometimes referred to as a vertical herniation). Over the years we have compressed over 400 spinal units in a neutral posture, and all but two resulted in end-plate fractures as the primary tissue damage.

CLINICAL RELEVANCE

End-Plate Fractures

In my experience, end-plate fracture with the loss of nuclear fluids through the crack into the vertebral body (often forming Schmorl's nodes) is a very common compressive injury and perhaps the most misdiagnosed. Loss of the disc nucleus results in a flattened interdiscal space that when seen on planar X rays is usually diagnosed as a herniated disc or "degenerated disc." However, the annulus of the disc remains intact. It's simply a case of the nucleus squirting through the end-plate crack into the cancellous core of the vertebra. True disc herniation requires very special mechanical conditions that will be described shortly. When compressing spines in the lab, we hear an audible "pop" at the instant of end-plate fracture—exactly what patients report when they describe details of the event that resulted in their pain. I also strongly suspect that some fractures are somewhat benign in that no immediate severe pain occurs. Others, however, may be instantaneously acute; it depends on the biomechanical changes that accompany the fracture. If there is substantial loss of the nucleus from the disc (i.e., it is vertically herniated), then immediate loss of disc height and subsequent compromise of nerve root space will result. At this point the end-plate fracture will mimic the symptoms of true herniation—another reason for the common misdiagnosis.

Posterior Elements of the Vertebrae

The posterior elements of the vertebrae (pedicles, laminae, spinous processes, and facet joints) have a shell of cortical bone but contain a cancellous bony core in the thick sections. The transverse processes project laterally together with a superior and an inferior pair of facet joints (see figure 4.1). On the lateral surface of the bone that forms the superior facets are the accessory and mamillary processes that, together with the transverse process, are major attachment sites of the longissimus and iliocostalis extensor muscle groups (described later). The facet joints are typical synovial joints in that the articulating surfaces are covered with hyaline cartilage and are contained within a capsule. Fibroadipose enlargements, or meniscoids, are found around the rim of the facet, although mostly at the proximal and distal edges (Bogduk and Engel, 1984), which have been implicated as a possible structure that could "bind" and lock the facet joint (see figure 4.11, a-b).

The neural arch in general (pedicles and laminae) appears to be somewhat flexible. In fact, Bedzinski (1992) demonstrated flexibility of the **pars interarticularis** during flexion/extension of cadaveric spines, while Dickey and colleagues (1996) documented up to three-degree changes of the right pedicle with respect to the left pedicle during quite mild daily activities using pedicle screws in vivo. Failure of these elements together with facet damage, leading to **spondylolisthesis,** are sometimes blamed exclusively on anterior/posterior shear forces. However, a case could be made from epidemiological evidence in athletes such as gymnasts and Australian cricket bowlers (Hardcastle, Annear, and Foster, 1992) that the damage to these posterior elements may also be associated with full range of motion. The cyclic full spine flexion and extension in these sorts of activities fatigue the arch with repeated stress reversals.

On the other hand, there is no doubt that excessive shear forces cause injury to these posterior vertebral elements. Posterior shear of the superior vertebrae can lead to ligamentous damage but also failure in the vertebrae itself, as the end plate often

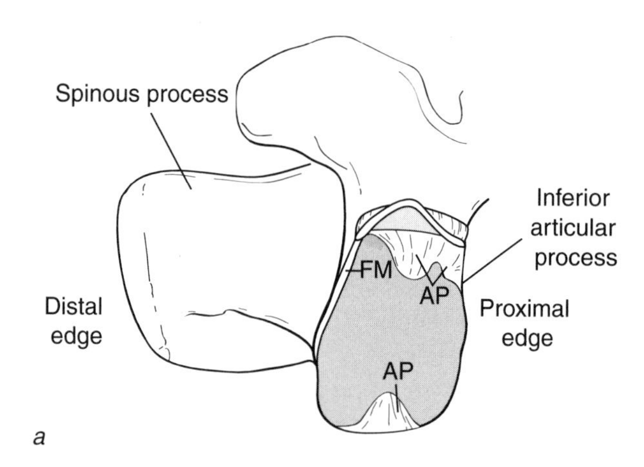

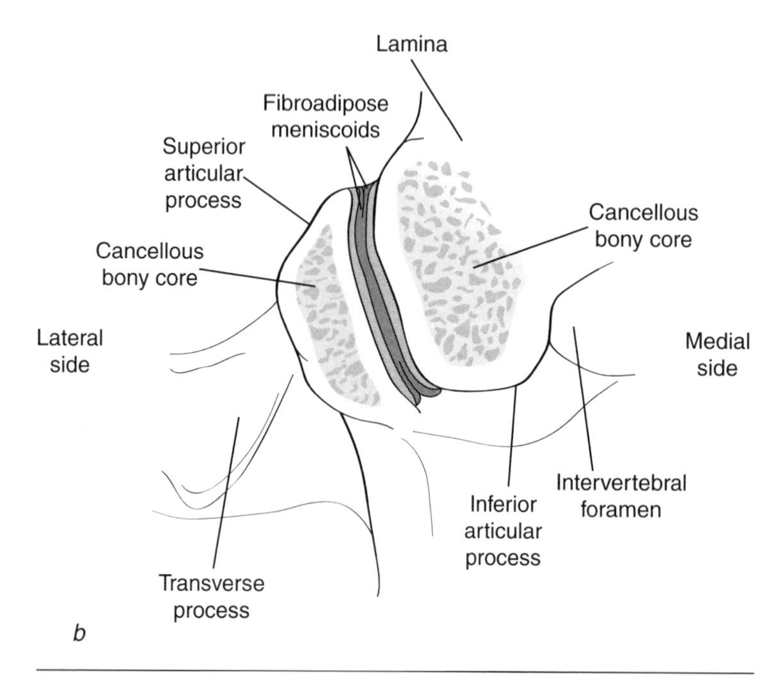

Figure 4.11 Lateral view of the inferior articular process *(a)*, revealing the facet, the fibroadipose meniscoids, and the adipose tissue pad, which have been implicated in joint binding. Cross section of the posterior view of the facet joint *(b)*, showing the positions of the fibro-adipose meniscoids and the adipose tissue pads in the joint.

avulses from the rest of the vertebral body (Yingling and McGill, 1999b) (see figure 4.12). Both our lab observations and discussions with international colleagues have reinforced our suspicion that this type of failure may be more common in the adolescent and geriatric spine than in the young and middle-aged adult spine. Further work is needed for confirmation.

Cripton and colleagues (1995) documented that anterior shear of the superior vertebra causes pars and facet fracture leading to spondylolisthesis with a typical tolerance of an adult lumbar spine of approximately 2000 N. Although similar injury mechanisms and tolerance values were observed in young porcine spine specimens (Yingling and McGill, 1999a, 1999b), the type of injury appeared to be modulated by loading rate. Specifically, anterior shear forces produced undefinable soft tissue injury at low load rates (100 N/s), but fractures of the pars, facet face, and vertebral body were observed at higher load rates (7000 N/s). Posterior shear forces applied at low load rates produced undefinable soft tissue failure and vertebral body fracture, while those forces at higher load rates produced wedge fractures and facet damage.

While shear tolerance of the vertebral motion unit apears to be in the range of 2000-2800 for one-time loading, Norman and colleagues (1998) noticed an increase in reported back pain in vivo in jobs that exposed workers to repetitive shear loads greater than 500 N. We consider these the best guidelines currently available.

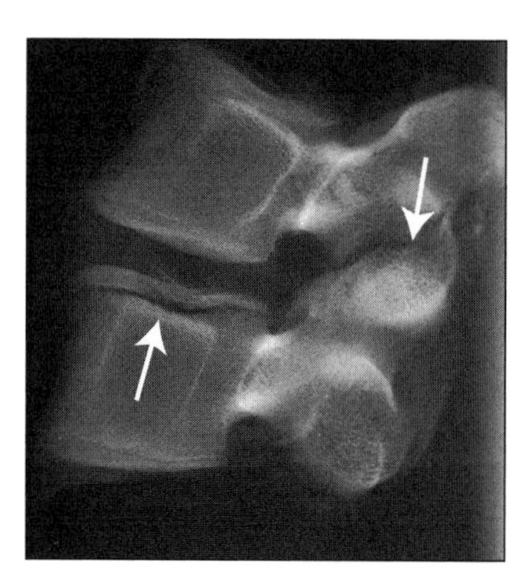

Figure 4.12 Shear injuries include fracture of the facet base and, on occasion, end-plate avulsion from the vertebrae.

The Posterior Elements of the Vertebrae Are Flexible

In "Understanding Vertebral Mechanics" on page 47, I recommended obtaining a vertebra from the butcher. Grasp the vertebral body in one hand and the whole neural arch in the other. Bend the arch up and down and note the flexibility. This flexural displacement occurs in each cycle of full spine flexion/extension common in events such as women's gymnastics (see figure 4.13). If this cycling continues, the stress reversals will eventually cause a fatigue crack in the pars. Further repetition will cause the crack to propagate through the full width of the pars, eventually resulting in fracture—and in vivo, the condition of spondylolisthesis.

Figure 4.13 Repetitive gymnastics moves such as this cause stress–strain reversals in the pars. In sufficient numbers they will result in fatigue fracture—leading to spondylolisthesis.

CLINICAL RELEVANCE

Neural Arch Fracture

Spondylolisthesis and neural arch defects are endemic among female gymnasts and cricket bowlers to name a few. Patients with spondylolisthesis generally do not do well with therapeutic exercise that takes the spine through the range of motion; rather, stability should be the rehabilitation objective. I have been involved in litigation cases in which clear spondylolisthesis existed but was alleged to be related to a specific event (for example, a recently occurring automobile accident). In surgery, however, there was evidence of substantial osteogenic activity, suggesting the injury was quite old. In fact, these patients (former gymnasts) must have had this damage while competing but were so fit and their spines so stable that they were able to remain in competition. Following retirement, having children, and losing fitness, a rather minor event became the instigator of their symptoms.

Intervertebral Disc

The intervertebral disc has three major components: the nucleus pulposus, annulus fibrosus, and end plates. The nucleus has a gel-like character with collagen fibrils suspended in a base of water and various mucopolysaccharides giving it both viscosity and some elastic response when perturbed in vitro. At the risk of sounding crude, the best way to describe a healthy nucleus is that it looks and feels like heavy phlegm. During in-vitro testing we have had it squirt out under pressure and literally stick to the wall. While there is no distinct border between the nucleus and the annulus, the lamellae of the annulus become more distinct, moving radially outward. The collagen fibers of each lamina are obliquely oriented (the obliquity runs in the opposite direction in each concentric lamella). The ends of the collagen fibers anchor into the vertebral body with Sharpey's fibers in the outermost lamellae, while the inner fibers attach to the end plate (the end plate was discussed in a previous section). The discs in cross section resemble a rounded triangle in the thoracic region and an ellipse in the lumbar region, suggesting **anisotropic** facilitation of twisting and bending.

Load-Bearing Abilities

The disc behaves as a hydrostatic structure that allows six degrees of motion between vertebrae. Its ability to bear load, however, depends on its shape and geometry. Due to the orientation of the collagen fibers within the concentric rings of the annulus, with one half of the fibers oblique to the other half, the annulus is able to resist loads when the disc is twisted. However, only half of the fibers are able to support this mode of loading, while the other half become disabled, resulting in a substantial loss of strength or ability to bear load. The annulus and the nucleus work together to support compressive load when the disc is subjected to bending and compression. Under spine compression the nucleus pressurizes, applying hydraulic forces to the end plates vertically and to the inner annulus laterally. This causes the annulus collagen fibers to bulge outward and become tensed. Years ago, Markolf and Morris (1974) elegantly demonstrated that a disc with the nucleus removed lost height but preserved much of its properties of axial stiffness, creep, and relaxation rates. The fact that the nucleus appears necessary to preserve disc height has implications for facet loading, shear stiffness, and ligament mechanics.

It is noteworthy that disc damage is most often accompanied by subdiscal bone damage (Gunning, Callaghan, and McGill, 2001). In fact, Keller and colleagues (1993) noted the interdependence of bone status and disc health. Recent evidence also suggests that excessive compression can lead to altered cell metabolism within the nucleus and increased rates of cell death (apoptosis) (Lotz and Chin, 2000). Thus, the evidence suggests that compressive loading involving lower compressive loads stimulates healthy bone (noted as a correlate of disc health) but that excessive loading leads to tissue breakdown.

Progressive Disc Injury

Consideration of progressive disc injury is in order here. A normal disc under compression deforms mainly the end plates vertically together with lesser outward bulging of the annulus (Brinckmann, Biggemann, and Hilweg, 1988). However, if little hydrostatic pressure is present, as in the case in which the nucleus has been lost through end-plate fracture or herniation, the outer annulus bulges outward and the inner annulus bulges inward during disc compression (see figure 4.14, a-b). This double convex bulging causes the laminae of the annulus to separate, or delaminate, and has

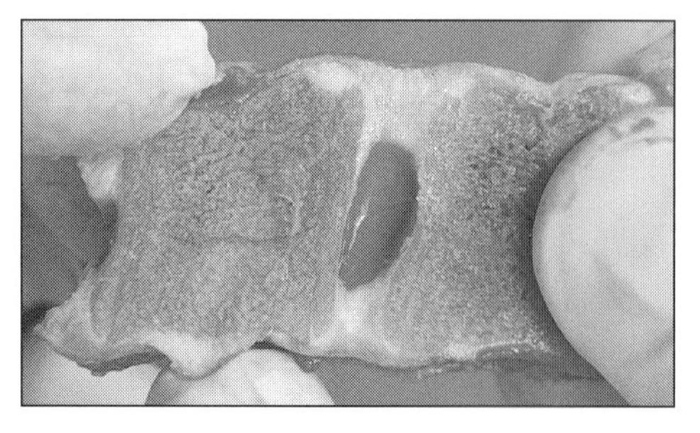

a

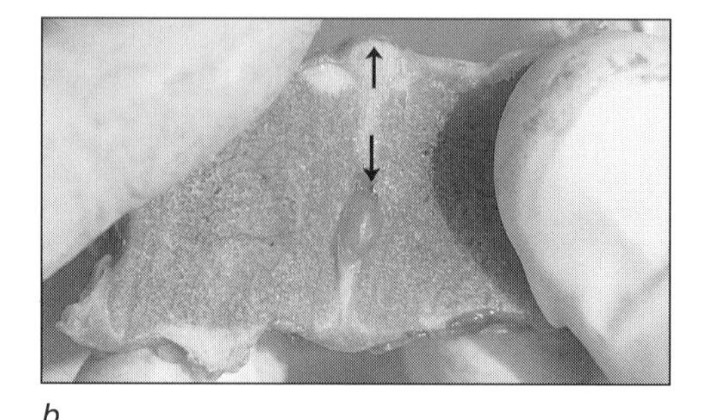

b

Figure 4.14 (*a*) In a healthy joint with a contained nucleus there is minimal annulus deformation under compressive load. (*b*) If the nucleus loses pressure (due to an end-plate fracture, for example), the annulus compresses, causing radial bulging both outward and inward (arrows) and producing delaminating stresses as the annulus layers are pulled apart.

been hypothesized as a pathway for nuclear material to leak through the lamellae layers and finally extrude, creating a frank herniated disc (Adams and Dolan, 1995).

Another interesting proposal has been recently put forth by Professor Adams (personal communication). He has suggested that a healthy disc builds interval pressures under compressive loads that are so high no nerve or vascular vessel could survive. Following initial end-plate damage, the disc can no longer build substantial pressure such that nerves and blood vessels are able to invade the disc. These are more possibilities to explain the increased vascularization of "degenerated" discs and their ability to generate pain.

From a review of the literature, one can make four general conclusions about annulus injury and the resulting bulging or herniation:

1. It would appear that the disc must be bent to the full end range of motion in order to herniate (Adams and Hutton, 1982).
2. Disc herniation is associated not only with extreme deviated posture, either fully flexed or bent, but also with repeated loading in the neighborhood of thousands of times, highlighting the role of fatigue as a mechanism of injury (Gordon et al., 1991; King, 1993).
3. Epidemiological data link herniation with sedentary occupations and the sitting posture (Videman, Nurminen, and Troup, 1990). In fact, Wilder and colleagues (1988) documented annular tears in young calf spines from prolonged simulated sitting postures and cyclic compressive loading (i.e., simulated truck driving).
4. Herniations tend to occur in younger spines (Adams and Hutton, 1985), meaning those with higher water content (Adams and Muir, 1976) and more hydraulic behavior. Older spines do not appear to exhibit classic extrusion of nuclear material but rather are characterized by delamination of the annulus layer and radial cracks that appear to progress with repeated loading (a nice review is provided by Goel, Monroe, et al., 1995).

A couple of years ago we sought the most potent mechanism leading to disc herniation. Given that it was critical to create a homogeneous cohort of specimens, we chose a pig spine model, controlling diet, physical activity, genetic makeup, disc degeneration, and so forth. We found that repeated flexion motion under simultaneous compressive loading was the easiest way to ensure herniation. In fact, it turned out that the numbers of cycles of flexion motion were more important than the actual magnitude of compressive load. While no herniations were produced with 260 N of compressive load and up to 85,000 flexion cycles, herniations were produced with 867 N of load and 22,000 to 28,000 cycles, and with 1472 N and only 5000 to 9500 cycles (Callaghan and McGill, 2001). Clearly, herniations are a function of repeated full flexion motion cycles with only a modest level of accompanying compressive load. In fact, we mimicked the lumbar motion and loading of a typical spine rehabilitation machine where the seated patient belts down the pelvis to isolate the lumbar spine and then extends the torso repetitively against a resistance over the midback. (Amazingly, some people are trained on this type of machine, even those with known disc herniations!) (See figure 4.15.) The time

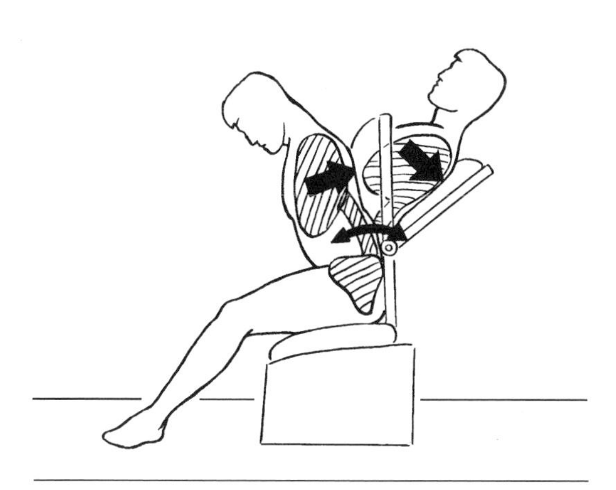

Figure 4.15 Rehabilitation devices such as this attempt to isolate lumbar motion by extending against a resistance pad but create simultaneous lumbar compression. We found that replicating the full range of motion from full flexion to neutral and using the compressive loads of these devices were powerful combinations to produce disc herniation.

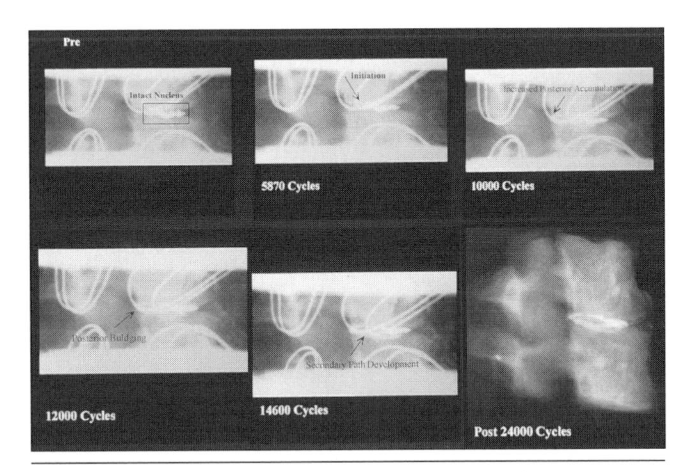

Figure 4.16 Serial radiographs showing the initial containment of the radiocontrast in the nucleus and, with repeated full flexion motion (with about 1400 N of compression), the progressive posterior tracking of the contrast until frank herniation.

Reprinted from *Clinical Biomechanics*, 16(1), J.P. Callaghan and S.M. McGill, "Intervertebral disc herniation: Studies on a porcine model exposed to highly repetitive flexion/extension motion with force," 28-37, 2001, with permission from Elsevier Science.

Figure 4.17 Repeated flexion can result in a frank herniation or a bulge, shown here in a specimen with the lamina removed to expose the posterior disc, which can impinge on nerves causing symptoms. Note that the nucleus was injected with a dye so that the bulge is more visible (arrow).

series radiographs (see figure 4.16) demonstrated the progression of the nuclear material tracking through the annulus with successive motion cycles. The herniated disc appears to result from cumulative trauma: even though we have crushed well over 400 vertebral motion segments, we have only once or twice observed a herniation without concomitant flexion cycles. Note that we are including both frank herniation and visible disc bulges (see figure 4.17) under this category of injury mechanism.

Also worth noting here is the intriguing hypothesis of Bogduk and Twomey who have suggested the possibility of an annular "sprain" similar to a sprain of the ankle ligaments (Bogduk and Twomey, 1991). They hypothesized that the outer layers of the annulus experience excessive strain under torsion. Given that these authors have presented evidence for the presence of nerve fibers particularly in this region, the annulus appears to be a good candidate for a source of pain.

CLINICAL RELEVANCE

Mechanisms of Annulus Failure (Herniation)

Damage to the annulus of the disc (herniation) appears to be associated with fully flexing the spine for a repeated or prolonged period of time. In fact, herniation of the disc seems almost impossible without full flexion. This has implications for exercise prescription particularly for flexion stretching and sit-ups or for activities such as prolonged sitting, all of which are characterized by a flexed spine. Some resistance exercise machines that take the spine to full flexion repeatedly must be reconsidered for those interested in sparing the posterior annulus portions of their discs.

Methodological Concerns for Testing Spine Motion Units

Conclusions drawn from experiments performed on spine motion units are limited by the choice of experimental techniques. A few of these issues are discussed here.

Position (Angular Rotation) Control Versus Load (or Torque) Control

The use of positional versus load control when loading vertebral motion units has been hotly debated for years (e.g., Goel, Wilder, et al.,1995). The issue at the heart of this methodological debate is, Which method more closely replicates in-vivo loading? The use of position control (e.g., a load rate of 500 mm/s or .16 in./s) in the in-vitro study increased the probability of producing an intervertebral disc herniation and resulted in more severe disc herniations (nuclear extrusion). This is due to the shape of the load–deformation relationship, which is highly nonlinear, progressing rapidly to a very steep final portion, and one could argue this is a rare occurrence in real life. Thus, constant velocity loading (position control) results in very rapid increases in load in the steep region. Torque control (e.g., a load rate of 500 N/s) proved to be less damaging and would be more representative of the in-vivo results found by Marras and Granata (1997). In addition, torque control would probably better represent possible protective mechanisms present in the body due to increased joint rotational stiffness.

Is an Animal Model Useful for Human Applications?

While human cadaveric material has provided a wealth of information in the area of spine injury, young, healthy human spines are extremely rare. Typical donors are elderly and sick; younger donors were either sick or had sustained substantial violent trauma. This is an important issue, as failure patterns in the spine are a function of biological age. For example, older degenerated discs are less susceptible to herniation (Adams and Hutton, 1982), rendering them unsuitable for investigations of herniation mechanics. Obtaining matched young, healthy human specimens for controlled study simply is not a reality.

For these reasons, we have chosen to use porcine cervical spines in lieu of human lumbar spines in many experiments. Porcine cervical spines present a healthy homogeneous sample of specimens, specifically providing scientific control over genetic makeup, age, weight, physical activity levels, and diet. While controlled experiments are only possible using animal models, the obvious liability is that of relevance for application to human mechanics and orthopedics; a comparison among the geometrical, anatomical, and functional aspects of the porcine cervical spine and human lumbar spine suggests they are reasonably biomechanically similar (Yingling, Callaghan, and McGill, 1999). Some have questioned the use of a quadruped cervical spine. While pigs are quadrupeds, supporting the weight of the head and additional loads imposed by muscular activity required to leverage the cantilevered head results in substantial compressive loads on the porcine cervical spine (see figure 4.18). This load bearing appears to be quite analogous to that of the human lumbar spine. The bipedal human cervical spine, however, does not have to endure the forces from continual extensor contraction, given minimal torques, as the head sits on top of the spine in the upright posture. Nonetheless, an animal preparation remains a model for the human spine and at some point must be calibrated and scaled for human application.

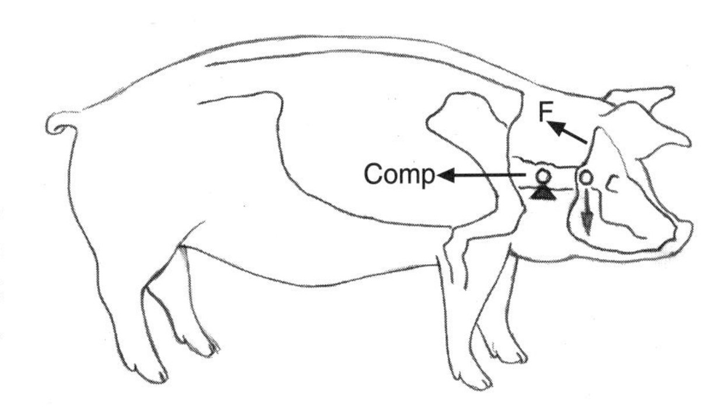

Figure 4.18 The pig spine shares many similar architectural features with the human lumbar spine. Consider the similar load-bearing requirements. The cantilevered quadruped neck is designed to bear large compressive forces (Comp) due to the extensor muscle forces (F) needed to support the weight of the head in this posture, together with limited range of motion. These are similar to the demands placed on the human lumbar spine when bending forward or even just standing.

Muscles

Traditional anatomical descriptions of the spine musculature have taken a posterior vantage point. This has hindered insight into the role of these muscles since many of the functionally relevant aspects are better viewed in the sagittal plane. (For a nice synopsis of the sagittal plane lines of action, see Bogduk, 1980; Macintosh and Bogduk, 1987.) Furthermore, many have developed their understanding of muscle function by simply interpreting the lines of action and region of attachment, assuming that the muscles act as straight-line cables. This may be misleading.

Understanding the function and purpose of each muscle requires two perspectives. First, knowledge of static muscle morphology is essential, though it may change over a range of motion. Second, knowledge of activation-time histories of the musculature must be obtained over a wide variety of movement and loading tasks. Muscles create force, but these forces play roles in moment production for movement and in stabilizing joints for safety and performance. Further understanding of the motor control system strategies chosen to support external loads and maintain stability requires the interpretation of anatomy, mechanics, and activation profiles. This section will enhance the discussion of anatomically based issues of the spine musculature and will blend the results of various electromyographic studies to help interpret function and the functional aspects of motor control.

Muscle Size

The physiologic cross-sectional area (PCSA) of muscle determines the force-producing potential, while the line of action and **moment arm** determine the effect of the force in moment production, stabilization, and so forth. It is erroneous to estimate muscle force based on muscle volume without accounting for fiber architecture or by taking transverse scans to measure anatomical cross-sectional areas (McGill, Patt, and Norman, 1988). Using areas directly obtained from CT or MRI slices has led to erroneous force estimates and interpretations of spine function. In such cases, since a large number of muscle fibers are not seen in a single transverse scan of a pennated muscle, muscle forces are underestimated. Thus, areas obtained from MRI or CT scans must be corrected for fiber architecture and scan plane obliquity (McGill, Santaguida, and Stevens, 1993).

In figure 4.19 transverse scans of one subject show the changing shape of the torso muscles over the thoracolumbar region, highlighting the need to combine transverse scan data with data documenting fiber architecture obtained from dissection. In this example, the thoracic extensors (longissimus thoracis and iliocostalis lumborum) seen at T9 provide an extensor moment at L4, even though

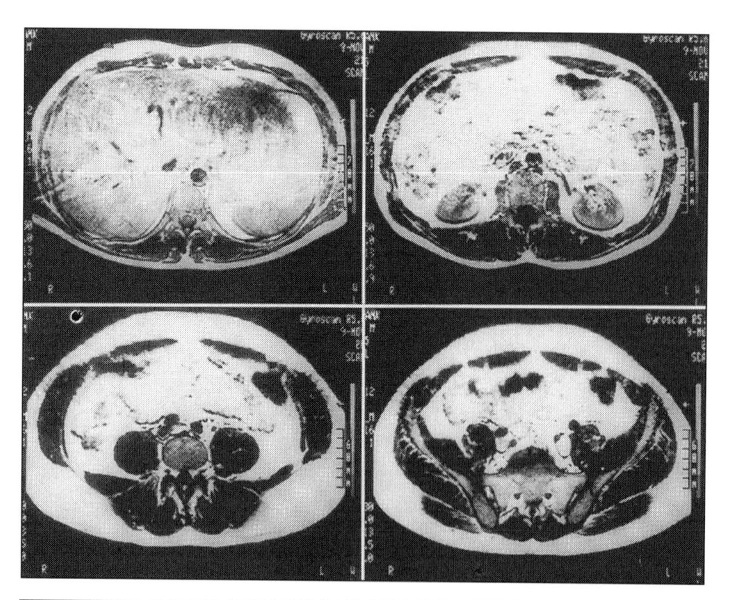

Figure 4.19 Transverse scans of one subject (supine) at the levels of T9, L1, L4, and S1, showing the musculature in cross-section. Note that many muscles seen at more superior levels pass tendons over the lower levels transmitting force. This illustrates the error in simply using a single scan to estimate muscle force and moment potential.

Reprinted from *Clinical Biomechanics,* 8, S.M. McGill, L. Santaguida, and J. Stevens, "Measure of the trunk musculature from T6 to L5 using MRI scans of 15 young males corrected for muscle fibre orientation," 171, 1993, with permission from Elsevier Science.

they are not seen in the L4 scan. Only their tendons overlie the L4 extensors. A sagittal plane schematic shows the errors in measuring muscle cross-sectional area from a single transverse slice—which has caused some to underestimate the potential of the muscle.

Raw muscle physiological cross-sectional areas and moment arms (McGill, Santaguida, and Stevens, 1993) are provided in appendix A.1. Areas corrected for oblique lines of action are shown in table 4.1 for some selected muscles at several levels of the thoraco-lumbar spine. Guidelines for estimating true physiological areas are provided in McGill and colleagues (1988). Other recent sources of raw muscle geometry obtained from MRI for both males and females are found in Marras and colleagues (2001).

Moment arms of the abdominal musculature are generally obtained from CT- or MRI-based studies. Generally, subjects lie supine or prone within the MRI or CT scanner, and the distance from their spine to the muscle centroid is measured. Lying on their backs in this posture causes the abdominal contents to collapse posteriorly under gravity (McGill, Juker, and Axler, 1996). In real life and when standing, the abdominals are pushed away from the spine by the visceral contents. Recently, research has shown that CT/MRI studies of the abdominal muscle moment arms obtained from subjects in the supine posture underestimated the true values by 30%.

In summary, understanding the force and mechanical potential of muscles requires an appreciation for the curving line of action, which is best obtained in the anatomy lab. But, unfortunately, these specimens are usually atrophied, eliminating them as a source for muscle size estimates. Muscle areas obtained from various medical imaging techniques need to be corrected to account for fiber architecture and contractile components that do not appear in the particular scan level (for example, only the tendon passes the level). Further, moment arms for muscle lines of action from subjects who are lying down need to be adjusted for application to upright postures maintained in real life.

Table 4.1 Corrected Muscle Cross-Sectional Areas

A few examples of corrected cross-sectional areas, anterior/posterior moment arms, and lateral moment arms perpendicular to the muscle fiber line of action using the cosines listed in McGill, Patt, and Norman (1988). These are the values that should be used in biomechanical models rather than the uncorrected values obtained directly from scan slices.

Muscle		Cross-sectional area (mm)^2	Moment arms (ant/post) (mm)	Moment arms (lateral) (mm)
Longissimus pars lumborum*	L3-L4	644	51	17
Quadratus lumborum	L1-L2	358	31	43
	L2-L3	507	32	55
	L3-L4	582	29	59
	L4-L5	328	16	39
External oblique	L3-L4	1121	17	110
Internal oblique	L3-L4	1154	20	89

*The laminae of longissimus pars lumborum at the L4-L5 level would have been listed here by virtue of their cosines but were not, as they could not be distinguished on all scan slices.

Muscle Groups

This section describes specific muscle groups from a functional perspective and introduces some issues fundamental for understanding injury avoidance and the choice of appropriate rehabilitation approach.

Rotatores and Intertransversarii

Many anatomical textbooks describe the function of the small rotator muscles of the spine, which attach to adjacent vertebrae, as creating axial twisting torque. This is consistent with their nomenclature (rotatores). Similarly, the intertransversarii are often assigned the role of lateral flexion. These proposals have several problems. First, these small muscles (see figure 4.20) have such small physiological cross-sectional areas that

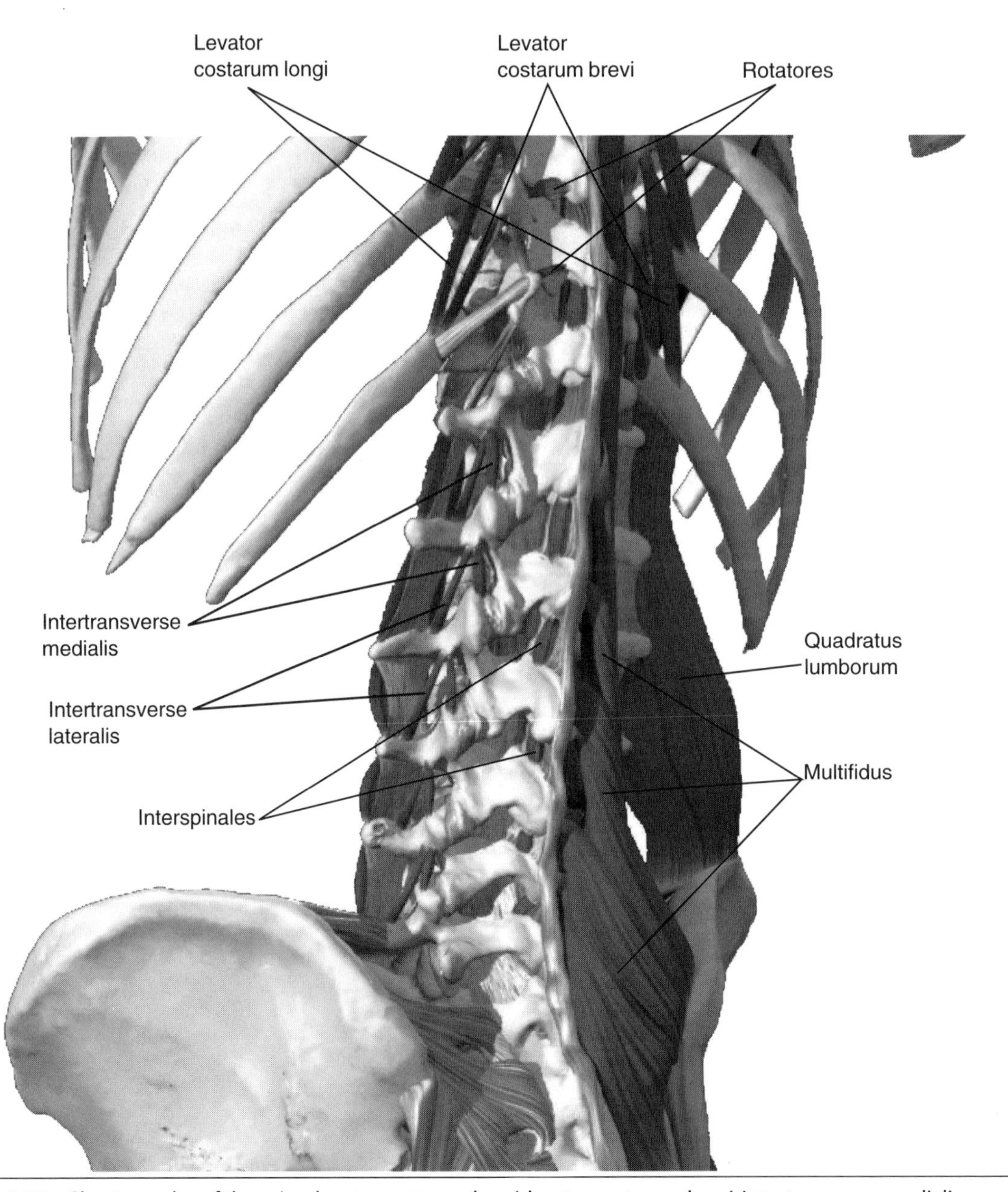

Figure 4.20 Short muscles of the spine: levator costarum longi, levator costarum brevi, intertransverse medialis, intertransverse lateralis, rotatores, and interspinalis. They have been described erroneously as creating axial twisting torque.

Image courtesy of Primal Pictures.

they can generate only a few newtons of force, and second, they work through such a small moment arm that their total contribution to rotational axial twisting and bending torque is minimal. For these reasons, I believe they serve another function.

The rotatores and intertransversarii muscles are highly rich in muscle spindles, approximately 4.5 to 7.3 times more rich than the multifidus (Nitz and Peck, 1986). This evidence suggests that they may function as length transducers or vertebral position sensors at every thoracic and lumbar joint. In some EMG experiments we performed a number of years ago, we placed indwelling electrodes very close to the vertebrae. In one case we had a strong suspicion that the electrode was in a rotator muscle. The subject attempted to perform isometric twisting efforts with the spine untwisted (or constrained in a neutral posture) in both directions but produced no EMG activity from the rotator—only the usual activity in the abdominal obliques, and so on. However, when the subject tried to twist in one direction (with minimal muscular effort), there was no response, while in the other direction, there was major activity. This particular rotator seemed not to be activated to create axial twisting torque but rather in response to twisted position change. Thus, its activity was elicited as a function of twisted position—which was not consistent with the role of creating torque to twist the spine. This is stronger evidence that these muscles are not rotators at all but function as position transducers in the spine proprioception system.

CLINICAL RELEVANCE

Manual Therapy and the Function of the Rotatores and Intertransversarii

We now suspect that the rotatores and intertransversarii are actually length transducers and thereby position sensors, sensing the positioning of each spinal motion unit. These structures are likely affected during various types of manual therapy with the joint at end range of motion (a posture used in chiropractic technique, for example).

Extensors: Longissimus, Iliocostalis, and Multifidus Groups

The major extensors of the thoraco-lumbar spine are the longissimus, iliocostalis, and multifidus groups. Although the longissimus and iliocostalis groups are often separated in anatomy books, it may be more enlightening in a functional context to recognize the thoracic portions of both of these muscles as one group and the lumbar portions as another group. The lumbar and thoracic portions are architecturally (Bogduk, 1980) and functionally different (McGill and Norman, 1987). Bogduk (1980) partitions the lumbar and thoracic portions of these muscles into longissimus thoracis pars lumborum and pars thoracis, and iliocostalis lumborum pars lumborum and pars thoracis.

These two functional groups (pars lumborum, which attach to lumbar vertebrae, and pars thoracis, which attach to thoracic vertebrae) form quite a marvelous architecture for several reasons and are discussed in a functional context with this distinction (i.e., pars lumborum vs. pars thoracic). Fiber typing studies note differences between the lumbar and thoracic sections: the thoracic sections contain approximately 75% slow-twitch fibers, while lumbar sections are generally evenly mixed (Sirca and Kostevc, 1985). The pars thoracis components of these two muscles attach to the ribs and vertebral components and have relatively short contractile fibers with long tendons that run parallel to the spine to their origins on the posterior surface of the sacrum and medial border of the iliac crests (see figure 4.21). Furthermore, their line of action over the lower thoracic and lumbar region is just underneath the fascia, such that forces in these muscles have the greatest possible moment arm and therefore produce the greatest amount of extensor moment with a minimum of compressive penalty to the spine (see figure 4.22). When seen on a transverse MRI or CT scan at a lumbar level, pars thoracis tendons have the greatest extensor moment arm, overlying the lumbar bulk—often over 10 cm (4 in.) (McGill, Patt, and Norman, 1988, 1993) (see figure 4.23).

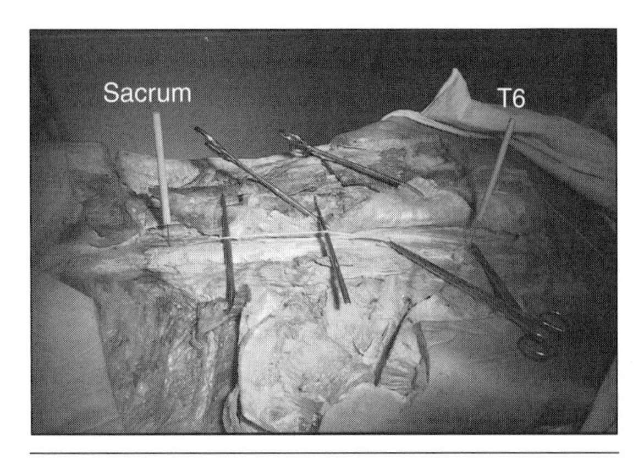

Figure 4.21 An isolated bundle of longissimus thoracis pars thoracis (inserting on the ribs at T6), with tendons lifted by probes, course over the full lumbar spine to their sacral origin. They have a very large extensor moment arm (just underneath the skin).

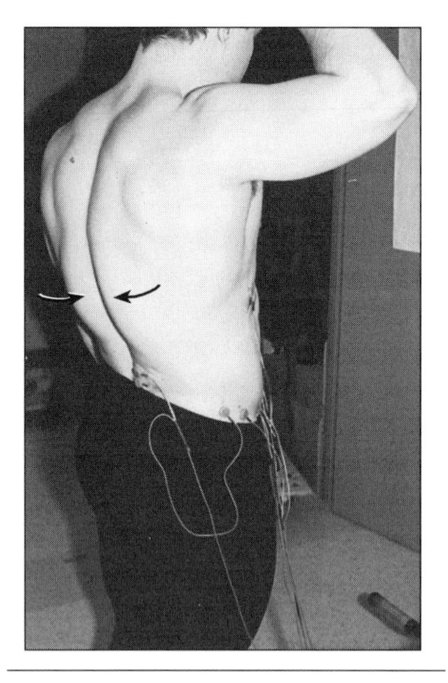

Figure 4.22 This world-class power lifter exemplifies the hypertrophied bulk of the iliocostalis and longissimus muscles seen in trained lifters. This muscle bulk is in the thoracic region, but the tendons span the entire lumbar spine.

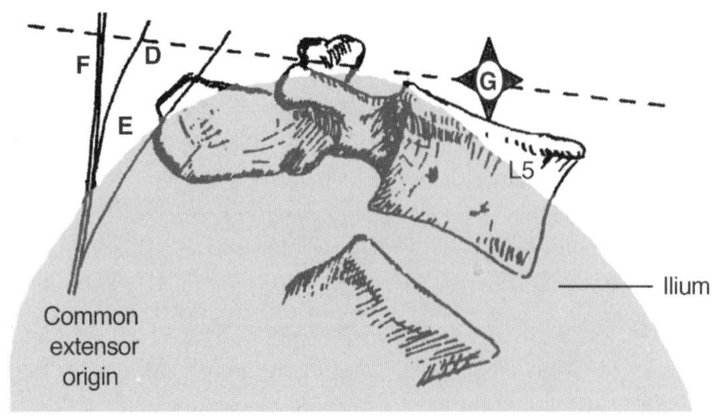

Figure 4.23 The moments arms of the pars lumborum portions of iliocostalis and longissimus (A and B). The large moment arm of the pars thoracis iliocostalis and longissimus muscles (C) gives them their ability to create major lumbar extensor moments. The lines of action of the pars lumborum portion of longissimus (D), the pars lumborum portion of iliocostalis (E), and the pars thoracis portion of longissimus and iliocostalis (F) join at their common point of origin at the sacral spine. The mechanical fulcrum or the axis about which muscles create moments is indicated by the diamond shape (G).

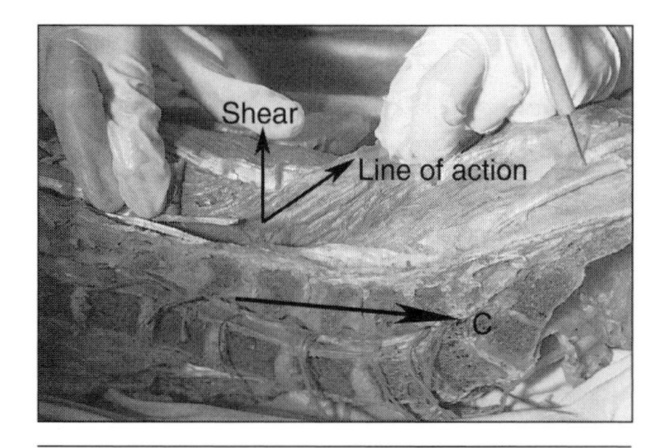

Figure 4.24 Iliocostalis lumborum pars lumborum and longissimus thoracis pars lumborum originate over the posterior surface of the sacrum, follow a very superficial pathway, and then dive obliquely to their vertebral attachments. This oblique orientation creates posterior shear (S) forces and extensor moments on each successive superior vertebrae. The compressive axis (C) is indicated.

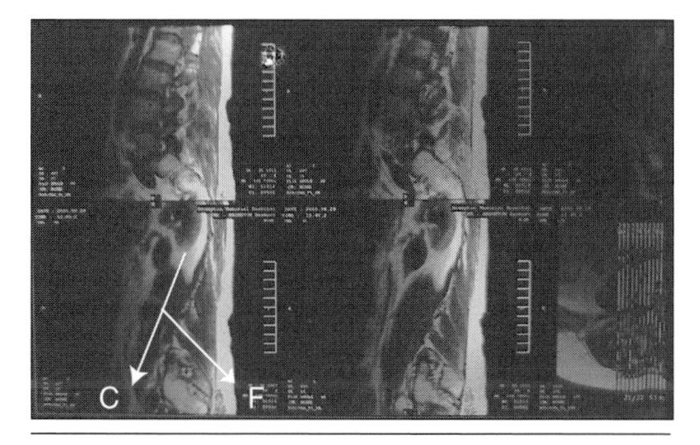

Figure 4.25 The oblique angle of the lumbar portions of longissimus and iliocostalis are seen in vivo in this MRI picture. Their line of force (F) is shown relative to the compressive axis (C).

The lumbar components of these muscles (iliocostalis lumborum pars lumborum and longissimus thoracis pars lumborum) are very different anatomically and functionally from their thoracic namesakes. They connect to the mamillary, accessory, and transverse processes of the lumbar vertebrae and originate, once again, over the posterior sacrum and medial aspect of the iliac crest. Each vertebra is connected bilaterally with separate laminae of these muscles (see figure 4.24). Their line of action is not parallel to the compressive axis of the spine but rather has a posterior and caudal direction that causes them to generate posterior shear forces together with extensor moment on the superior vertebrae (see figure 4.25). These posterior shear forces support any anterior reaction shear forces of the upper vertebrae that are produced as the upper body is flexed forward in a typical lifting posture. It is important to clarify that this flexion of the torso is accomplished through hip rotation, not lumbar flexion. These muscles lose their oblique line of action and reorient to the compressive axis of the spine with lumbar flexion (McGill, Hughson, and Parks, 2000) so that a flexed spine is unable to resist damaging shear forces (see figure 4.26, a-d). This possible injury mechanism, together with activation profiles during clinically relevant activities, is addressed in a later section.

The multifidus muscles perform quite a different function from those of the longissimus and iliocostalis groups, particularly in the lumbar region where they attach posterior spines of adjacent vertebrae or span two or three segments (see figure 4.27). Their line of action tends to be parallel to the compressive axis or in some cases runs anteriorly and caudal in an oblique direction. The major mechanically relevant feature of the multifidii, however, is that since they span only a few joints, their forces affect only local areas of the spine. Therefore, the multifidus muscles are involved in producing extensor torque (together with very small amounts of twisting and side-bending torque) but only provide the ability for corrections, or moment support, at specific joints that may be the foci of stresses. Interestingly, the multifidus muscles appear to have quite low muscle spindle density—certainly less than the iliocostalis or longissimus muscles (Amonoo-Kuofi, 1983). This may be due to their more medial location and subsequent smaller length excursions (see table 5.3 for muscle length changes, which were assessed using a number of extreme postures that are depicted in figure 5.1 on page 91). An injury mechanism involving inappropriate neural activation signals to the multifidus is proposed in chapter 5, using an example of injury observed in the laboratory. It is also worth noting here, given the recent emphasis on multifidus, that some people have considered more lateral portions of the extensors to be multifidus. This has presented some problems in both functional interpretation and rehabilitation.

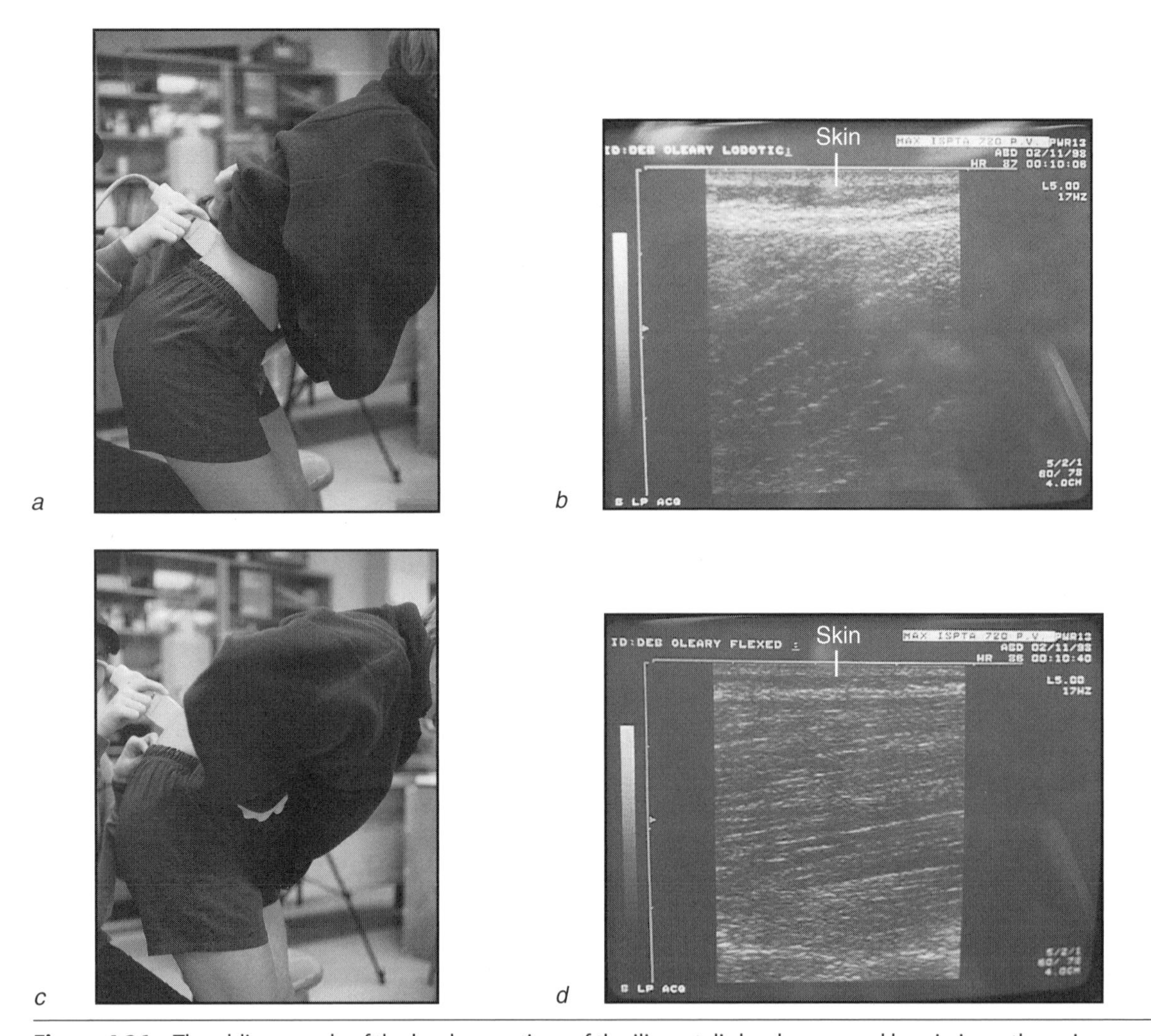

Figure 4.26 The oblique angle of the lumbar portions of the iliocostalis lumborum and longissimus thoracis protects the spine against large anterior shear forces. However, this ability is a function of spine curvature. A neutral spine *(a)* and the oblique angle of these muscles as viewed with an ultrasound imager *(b)*. The loss of this angle with spine flexion *(c)* so that anterior shear forces cannot be counteracted *(d)*. This ensures shear stability and is another reason to consider adopting a neutral spine during flexed weight-holding tasks.

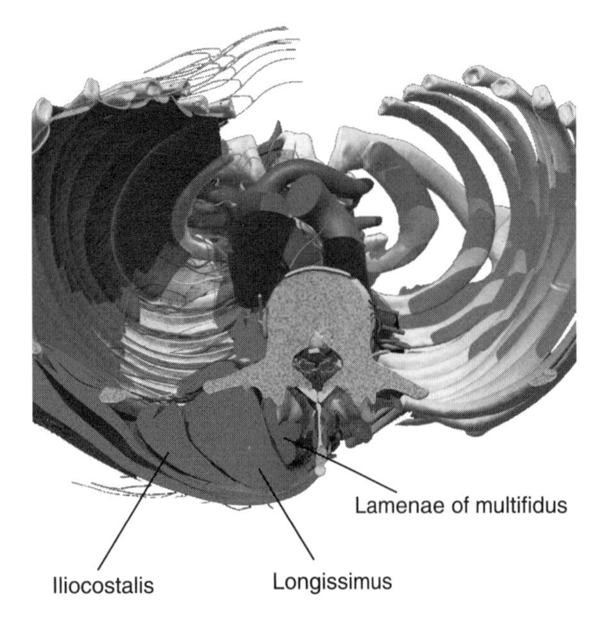

Figure 4.27 Multifidus is actually a series of laminae that can span one to three vertebral segments. Their lines of action do not support anterior shear of the superior vertebrae but actually contribute to it. Examining their cross-sectional area reveals that the multifidus is a relatively small lumbar extensor.

Image courtesy of Primal Pictures.

Table 4.2 Muscle Lengths in Centimeters (Including Tendon Length) Obtained From the Upright Standing Position and From Various Extreme Postures*

	Upright standing	60° flexion	25° lateral bending	10° twist	Combined[a]
R rectus abdominis	30.1	19.8[b]	27.4	30.0	17.7
L rectus abdominis	30.1	19.8[b]	32.6	30.5	22.9[b]
R external oblique 1	16.7	14.4	12.6[b]	15.5	13.3
L external oblique 1	16.7	14.4	21.3[b]	18.0	19.2
R external oblique 2	15.8	12.9	10.7[b]	15.6	9.5[b]
L external oblique 2	15.8	12.9	21.0[b]	16.2	18.2
R internal oblique 1	11.1	9.9	9.1	12.2	7.6[b]
L internal oblique 1	11.1	9.9	14.1[b]	10.0	12.2
R internal oblique 2	10.3	8.8	7.6[b]	11.2	6.9[b]
L internal oblique 2	10.3	8.8	13.2[b]	9.5	10.7
R psoas lumborum (L1)	15.1	20.0[b]	14.2	15.4	18.7[b]
L psoas lumborum (L1)	15.1	20.0	16.8	15.7	20.7[b]
R psoas lumborum (L2)	12.5	16.4[b]	11.8	12.9	15.6[b]
L psoas lumborum (L2)	12.5	16.4[b]	13.8	13.0	17.0[b]
R psoas lumborum (L3)	9.9	12.9[b]	9.5	10.0	12.5[b]
L psoas lumborum (L3)	9.9	12.9[b]	10.9	9.8	13.3[b]
R psoas lumborum (L4)	7.5	9.4[b]	7.4	7.6	9.3[b]
L psoas lumborum (L4)	7.5	9.4[b]	8.0	7.4	9.5[b]
R iliocostalis lumborum	23.0	29.7[b]	18.9	23.5	26.4
L iliocostalis lumborum	23.0	29.7[b]	25.5	22.8	31.8[b]

(continued)

Table 4.2 *(continued)*

	Upright standing	60° flexion	25° lateral bending	10° twist	Combined[a]
R longissimus thoracis	27.5	33.7[b]	25.4	27.6	31.1
L longissimus thoracis	27.5	33.7[b]	28.8	27.4	34.8[b]
R quadratus lumborum	14.6	18.2[b]	11.9	14.4	14.9
L quadratus lumborum	14.6	18.2	17.4	15.1	20.9[b]
R latissimus dorsi (L5)	29.4	32.1	26.8	29.8	29.0
L latissimus dorsi (L5)	29.4	32.1	31.5	29.1	34.6
R multifidus 1	5.3	7.3[b]	5.2	5.1	7.1[b]
L multifidus 1	5.3	7.3[b]	5.4	5.5	7.5[b]
R multifidus 2	5.1	7.2[b]	5.0	5.1	7.01[b]
L multifidus 2	5.1	7.2[b]	5.2	5.1	7.2
R psoas (L1)	29.2	28.6	28.1	29.0	27.2
L psoas (L1)	29.2	28.6	30.2	29.5	29.6
R psoas (L2)	25.8	25.3	25.1	25.7	24.4
L psoas (L2)	25.8	25.3	26.5	26.0	26.1
R psoas (L3)	22.1	21.8	21.7	22.0	21.2
L psoas (L3)	22.1	21.8	22.5	22.2	22.3
R psoas (L4)	18.7	18.6	18.6	18.7	18.4
L psoas (L4)	18.7	18.6	18.9	18.8	18.9

*The range of extreme postures used to assess muscle length changes is illustrated in figure 5.1 on page 91.
[a]Combinations of 60° flexion, 25° right lateral bend, 10° counterclockwise twist.
[b]Muscle lengths that differ by more than 20% from those obtained during upright standing.

Reprinted, by permission, from S.M. McGill, 1991, "Kinetic potential of the lumbar trunk musculature about three orthogonal orthopaedic axes in extreme positions," *Spine* 19(8): 809-815. ©Lippincott, Williams, and Wilkins.

A Note on Latissimus Dorsi

Latissimus dorsi is involved in lumbar extensor moment generation and stabilization due to its origin at each lumbar spinous process via the lumbodorsal fascia and insertion on the humerus (see figure 4.28). During pulling and lifting motion, the latissimus is active (see chapter 5), which has implications for its role and how it is trained for functional motion patterns.

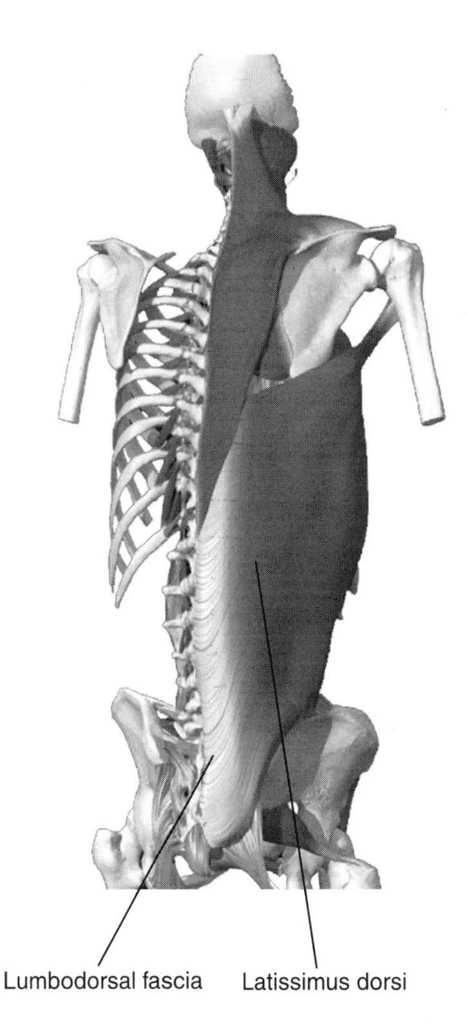

Figure 4.28 Latissimus dorsi originates from each lumbar spinous process via the lumbodorsal fascia and inserts on the humerus to perform both lumbar extension and stabilization roles.

Image courtesy of Primal Pictures.

CLINICAL RELEVANCE

Exercise for the Extensor Muscles of the Low Back

Research has shown that the thoracic extensors (longissimus thoracis pars thoracis and iliocostalis lumborum pars thoracis) that attach in the thoracic region are actually the most efficient lumbar extensors since they have the largest moment arms as they course over the lumbar region. For this reason, it is time to revisit the clinical practice of "isolating muscle groups," in this case the lumbar extensors for the lumbar spine. Specifically, what are referred to as the lumbar extensors (located in the lumbar region) contribute only a portion of the total lumbar extensor moment. Training of the lumbar extensor mechanism must involve the extensors that attach to the thoracic vertebrae, whose bulk of contractile fibers lies in the thoracic region but whose tendons pass over the lumbar region and have the greatest mechanical advantage of all lumbar muscles. Thus, exercises to isolate the lumbar muscles cannot be justified from an anatomical basis nor from a motor control perspective in which all "players in the orchestra" must be challenged during training.

(continued)

Another important clinical issue involves the anatomical features of the extensors. While the lumbar sections of the longissimus and iliocostalis muscles that attach to the lumbar vertebrae create extensor torque, they also produce large posterior shear forces to support the shearing loads that develop during torso flexion postures. Some therapists unknowingly disable these shear force protectors by having patients fully flex their spines during exercises, creating **myoelectric** quiescence in these muscles, or by recommending the pelvic tilt during flexing activities such as lifting. Discussion of this functional anatomy is critical for developing the strategies for injury prevention and rehabilitation described later in this book.

Abdominal Muscles

In this section we will consider several important aspects of lumbar mechanics in which the abdominal muscles are involved.

Abdominal Fascia

The abdominal fascia contains the rectus abdominis and connects laterally to the aponeurosis of the three layers of the abdominal wall. Its functional significance is made more important by connections of the aponeurosis with pectoralis major, together with fascial elements that cross the midline to transmit force to the fascia (and abdominal muscles) on the opposite side of the abdomen (Porterfield and DeRosa, 1998) (see figure 4.29, a-b). Such anatomical features underpin and justify exercises (detailed later) that integrate movement patterns that simultaneously challenge the abdominals, the spine, and the shoulder musculature.

Rectus Abdominis

While many classic anatomy texts consider the abdominal wall to be an important flexor of the trunk, the rectus abdominis appears to be the major trunk flexor—and the most active during sit-ups and curl-ups (Juker et al., 1998). Muscle activation amplitudes obtained from both intramuscular and surface electrodes over a variety of tasks are shown in table 5.4 (page 93). It is interesting to consider why the rectus abdominis is partitioned into sections rather than being a single long muscle, given that the sections share a common nerve supply and that a single long muscle would have the advantage of broadening the force–length relationship over a greater range of length change. Perhaps a single muscle would bulk upon shortening, compressing the viscera, or be stiff and resistant to bending. Not only does the sectioned rectus abdominis limit bulking upon shortening, but the sections also have a bead effect, which allows bending at each tendon to facilitate torso flexion/extension or abdominal distension or contraction as the visceral contents change volume (M. Belanger, University of Quebec at Montreal, personal communication, 1996).

The "beaded" rectus also performs another role—the lateral transmission of forces from the oblique muscles forming a continuous hoop around the abdomen (Porterfield and DeRosa, 1998). The intermuscular tendons and fascia prevent the fibers of rectus from being ripped apart laterally from these hoop stresses. This aspect of abdominal mechanics is elucidated further in the next section, which discusses the forces developed in the oblique muscles.

Another clinical issue is the controversy regarding upper and lower abdominals. While the obliques are regionally activated (and have functional separation between upper and lower regions), all sections of the rectus are activated together at similar levels during flexor torque generation. A significant functional separation does not

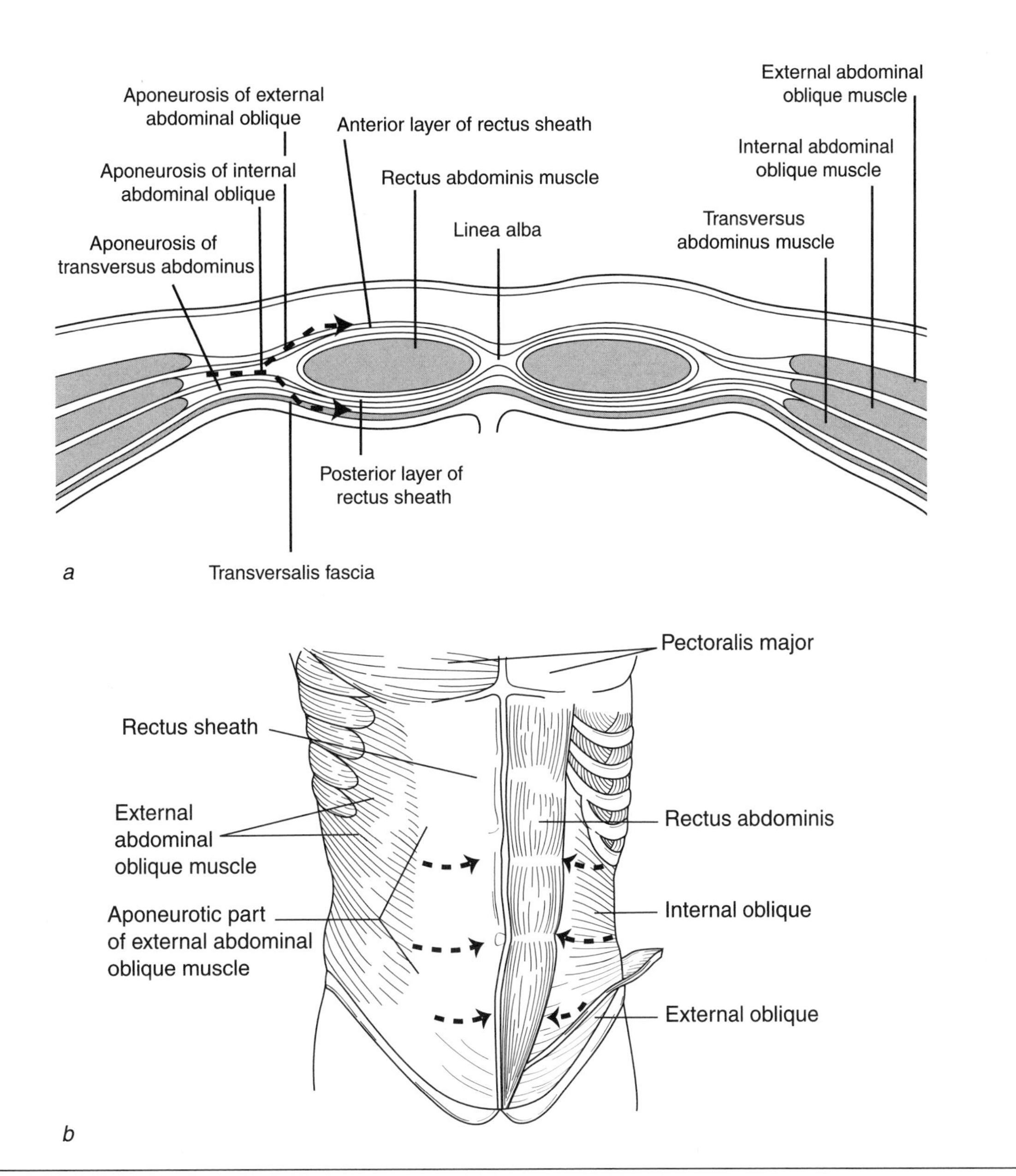

Figure 4.29 The abdominal fascia connects the obliques of the abdominal wall with rectus abdominis *(a)* and, to a lesser extent, pectoralis major and helps to transmit hoop stresses around the abdomen *(b)*.

appear to exist between upper and lower rectus (Lehman and McGill, 2001) in most people. Research reporting differences in the upper and lower rectus activation sometimes suffers from the absence of normalization of the EMG signal during processing. Briefly, researchers have used raw amplitudes of myoelectric activity (in millivolts) to conclude that there is more, or less, activity relative to other sections of the muscle, but the magnitudes are affected by local conductivity characteristics. Thus, amplitudes must be normalized to a standardized contraction and expressed as a percentage of this activity (rather than in millivolts) (see figure 4.30). Researchers may also have inadvertently monitored pyramidalis (an optional muscle at the base of the rectus), which would cloud interpretation.

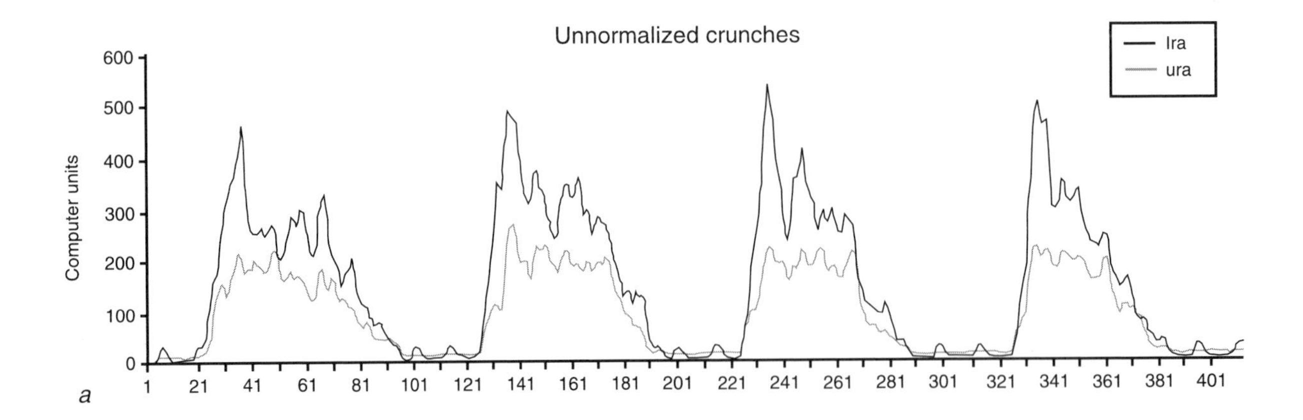

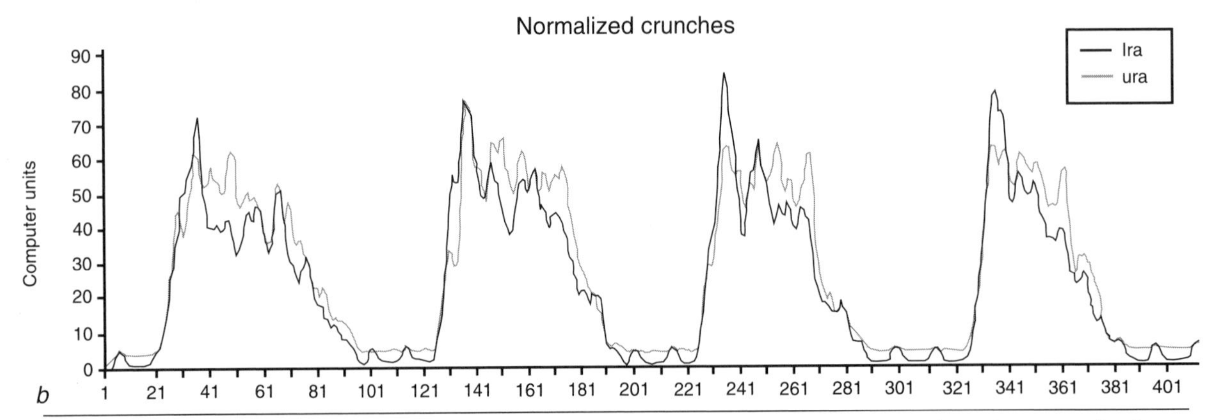

Figure 4.30 Studies that have reported an upper and lower rectus abdominis generally evaluated unnormalized EMG signals—for example, in the study of men performing crunches (*a*). However, when the signals are normalized properly, the apparent difference disappears (*b*). While there are regional activation zones in the abdominal obliques, there is little evidence to suggest that a functional "upper and lower" rectus abdominis exists in most people.

Reprinted, by permission, from G. Lehman and S.M. McGill, 1999, *Journal of Manipulative Physiological, Therapeutics.*

CLINICAL RELEVANCE

Upper and Lower Rectus

A distinct upper and lower rectus does not exist in most people (although some individuals may have the ability to preferentially activate one section slightly differently from the other in select activities). Thus, training the rectus can be accomplished with a single exercise. This is not true for the obliques, as they have several neural compartments—lateral, medial, upper, and lower.

Abdominal Wall

The three layers of the abdominal wall (external oblique, internal oblique, and transverse abdominis) perform several functions. All three are involved in flexion and appear to have their flexor potential enhanced because of their attachment to the linea semilunaris (see figure 4.31) (McGill, 1996), which redirects the oblique muscle forces down the rectus sheath to effectively increase the flexor moment arm. The obliques are involved in torso twisting (McGill, 1991a, 1991b) and lateral bend (McGill, 1992) and appear to play a role in lumbar stabilization since they increase their activity, to a small degree, when the spine is placed under pure axial compression (Juker, McGill,

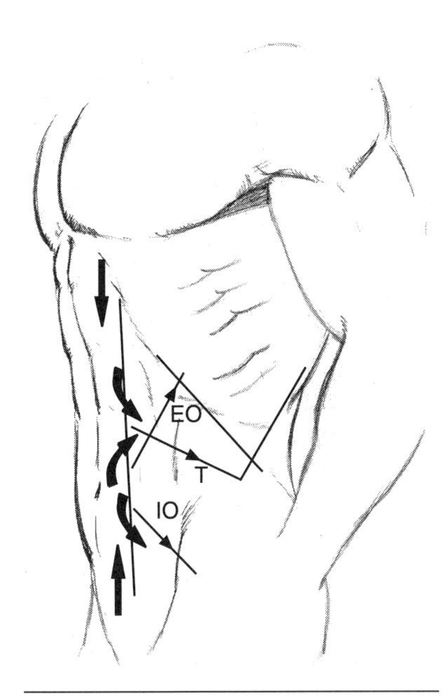

Figure 4.31 The oblique muscles (EO: anterior portion of external oblique; IO: anterior portion of internal oblique; T: transverse abdominis) transmit force along their fiber lengths and then redirect force along rectus abdominis via their attachment to the linea semilunaris to enhance their effective flexor moment arm.

Reprinted from *Journal of Biomechanics,* 29(7), S.M. McGill, "A revised anatomical model of the abdominal musculature for torso flexion efforts," 973-977, 1996, with permission from Elsevier Science.

and Kropf, 1998). (This functional notion will be developed later in this chapter.) The obliques are also involved in challenged lung ventilation, assisting with "active expiration" (Henke et al., 1988). The important functional implication of active expiration will be discussed in a subsequent section.

Researchers have focused a lot of attention on the transverse abdominis for two reasons. First, some believe it is involved in spine stability through its beltlike containment of the abdomen and also in the generation of intra-abdominal pressure (IAP). Richardson and colleagues (1999) nicely summarized the second reason for the focused attention on this muscle: its aberrant recruitment during ballistic arm movements in those with chronic low back conditions. Specifically, studies revealed that transversus has a delayed onset of activation time in those with back troubles, prior to a rapid arm movement—the hypothesis being that the trunk must first be made stiff and stable. No doubt this is evidence that an aberrant motor pattern exists in rapid activity; however, the question remains as to whether this is important in more normally paced activities. Delayed onset confirms motor deficits, but a 10- to 30-ms onset delay would be irrelevant during a normal movement in which these muscles are continually cocontracted to ensure stability. It is also interesting that, contrary to popular thought, the transversus and internal oblique are very similar in their fiber orientation. In our very limited intramuscular EMG electrode work (Juker et al., 1998) we saw a high degree of coupling between these two muscles in a wide variety of nonballistic exertions (no doubt some myoelectric cross-talk existed). Nonetheless, several groups (Cresswell and colleagues, 1994, and Hodges and Richardson, 1996, are the most experienced) have noted the extra activity of transversus when IAP is elevated. The combination of transverse activation (and almost always concomitant oblique activity) with elevated IAP enhances stability, without question. In fact, Cholewicki and colleagues (1999) recently concluded that building IAP on its own adds spine stability.

In the broader functional perspective, the components of the abdominals (rectus, obliques, and transversus) work together but also independently. While a variety of sources have provided myoelectric evidence from a variety of tasks, an anatomical/functional interpretation is needed. As noted earlier, the obliques differentially activate to create twisting torque and can enhance flexion torque. Rectus is primarily a flexor—in fact, those people who have a great deal of motor control in the abdominals can preferentially activate each muscle (see figure 4.32, a through d). Finally, some have suggested an upper and lower partitioning of the abdominals. As previously mentioned, this impression is probably an artifact resulting from poor electromyographic technique; there does not appear to be a functional upper and lower rectus (Lehman and McGill, 2001; Vera-Garcia, Grenier, and McGill, 2000). On the other hand, regional differences do exist in the obliques; some sections can be preferentially recruited both medially and laterally together with upper and lower portions. Finally, the obliques, together with transverse abdominis, form a containing hoop around the entire abdomen with the anterior of the hoop composed of the abdominal fascia and the posterior composed of the lumbodorsal fascia. The resulting "hoop stresses" and stiffness assist with spine stability.

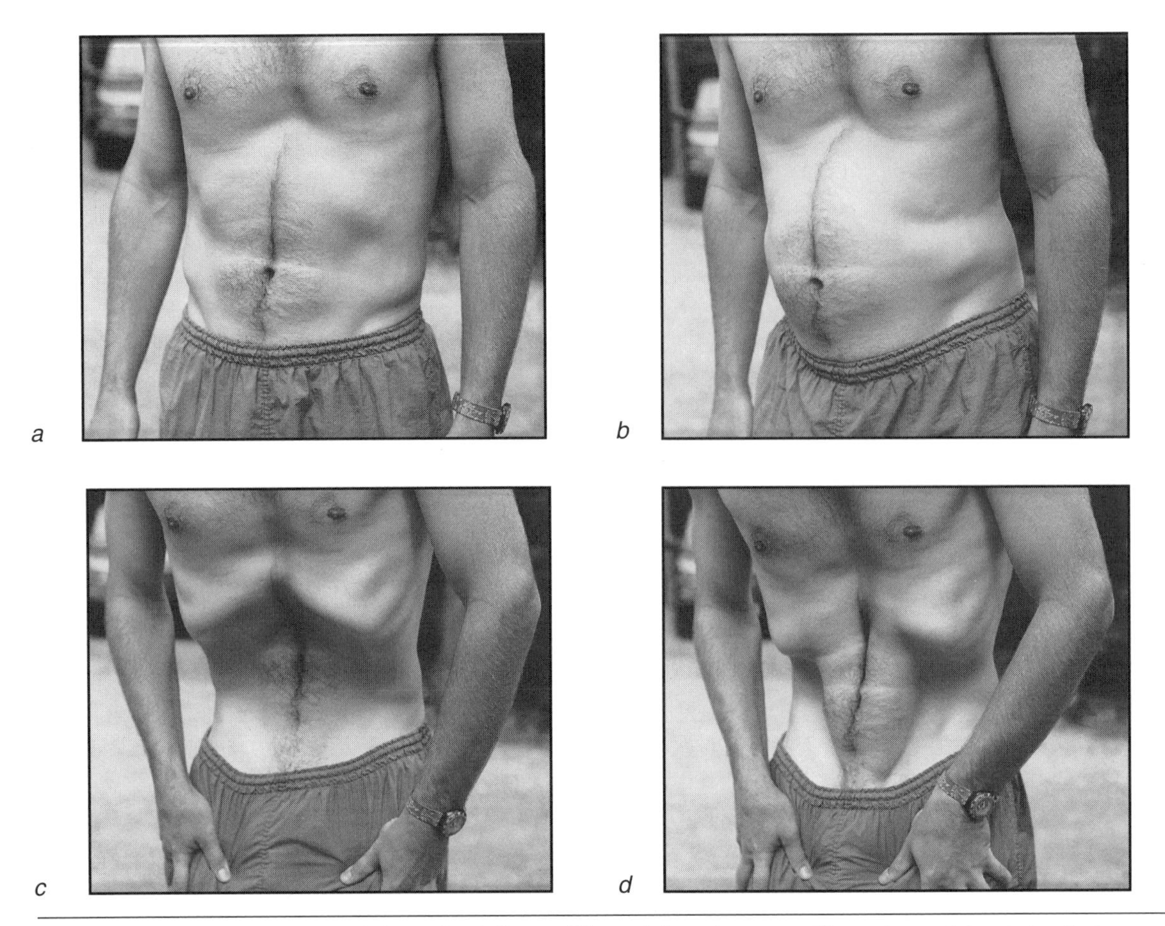

Figure 4.32 Some trained people have the ability to differentially activate specific portions of the abdominal musculature. This sequence shows *(a)* an inactive abdominal wall, *(b)* the abdominal wall "ballooned," and *(c)* contraction of transverse abdominis, which draws and "hollows" the wall. Placing the hands on the thighs and pushing *(d)* allows a flexor moment to develop where good muscular control is able to activate just the rectus abdominis with the previously activated transverse and little oblique activity.

CLINICAL RELEVANCE

Abdominal Muscle Exercises

The functional divisions of the abdominal muscles justify the need for several exercise techniques to challenge them in all of their roles: moment generation, spine stability, and heavy breathing. While the obliques are regionally activated, with several neural compartments, there appears to be no functional separation of upper and lower rectus abdominis. Thus, a curl-up exercise activates all portions of the rectus abdominis. However, upper and lower portions of the oblique abdominal muscles are activated separately depending on the demands placed on the torso. Finally, detailed examination of the fascial connections reveals force transmission among the shoulder musculature, the spine, and the abdominal muscles, justifying exercises incorporating larger movement patterns for the advanced patients who can bear the higher loads.

Psoas

The psoas, a muscle that crosses the spine and hip, is unique for many reasons. While it has been claimed to be a major stabilizer of the lumbar spine, I believe that this claim needs interpretation. Although the psoas complex attaches to T12 and to every lumbar vertebra on its course over the pelvic ring (see figure 4.33), its activation profile (see Juker, McGill, and Kropf, 1998; Juker, McGill, Kropf, et al., 1998; and Andersson, et al., 1995, for indwelling EMG data) is not consistent with that of a spine stabilizer (in the purest sense) but rather indicates that the role of the psoas is purely as a hip flexor. During our work to implant intramuscular electrodes in this muscle, we performed the insertion technique several times on ourselves. The first time the electrodes were in my own psoas, I tried several low back exertions trying to activate it (including flexion, extension, side bending, etc.). None of these really caused the psoas to fire. Simply raising the leg while standing, with hip flexion, caused massive activation, clearly indicating that the psoas is a hip flexor.

After this pilot work, in a larger study we found that any task requiring hip flexion involved the psoas. Although we obtained this impression from studying the activation profile of the psoas, other considerations stem from its architecture. Why does the psoas traverse the entire lumbar spine and, in fact, course all the way to the lower thoracic spine? Why not let just the iliacus perform hip flexion? If only the iliacus were to flex the hip, the pelvis would be torqued into an anterior pelvic tilt, forcing the lumbar spine into extension. These forces are buttressed by the psoas, which adds stiffness between the pelvis and the lumbar spine. In effect, it can be thought of as a spine stabilizer but only in the presence of significant hip flexor torque. Also, an activated and stiffened psoas will contribute some shear stiffness to the lumbar motion segment—but once again, only when hip flexor torque is required. The fact is that the psoas and iliacus are two separate muscles (see Santaguida and McGill, 1995), functionally, architecturally, and neurally. There is no such thing as an iliopsoas muscle!

In addition, some clinical discussion has been centered around the issue of whether the psoas is an internal or external rotator of the femur/hip. Although it has some

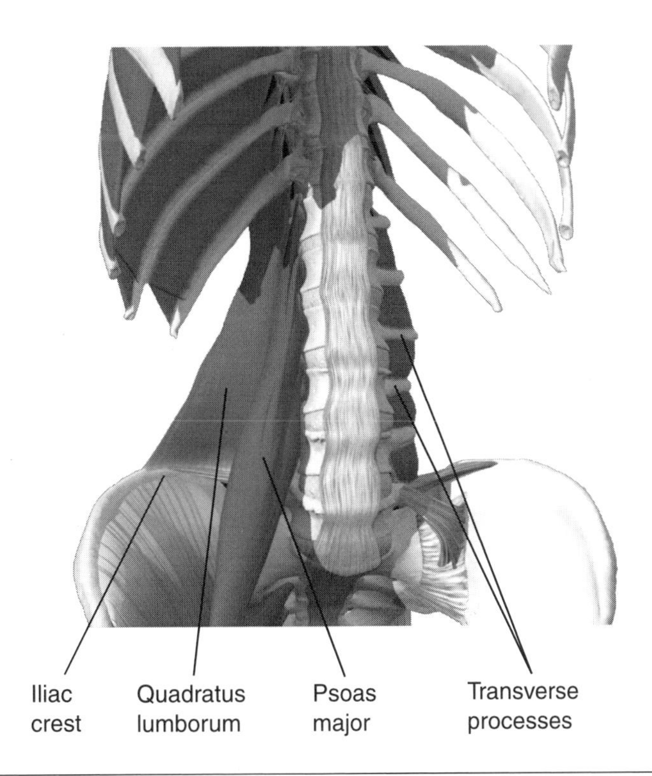

Figure 4.33 The psoas attaches to each lumbar vertebra (lateral vertebral body and transverse process) where each of these laminae of psoas fuse and form a common tendon that courses through the iliopectineal notch to the femur. The quadratus lumborum attaches each transverse process with the ribs and iliac crest, forming the guy wire support system.

Image courtesy of Primal Pictures.

architectural advantage to externally rotate the hip, Juker, McGill, Kropf, et al. (1998) observed only small activation bias during hip rotation tasks. However, this may have been due to the need for significant hip stabilization, resulting in substantial hip cocontraction. In a recent study, we examined a matched cohort of workers—half of whom had a history of disabling low back troubles, while the other half had never missed work. Those who had a history, but were asymptomatic at the time of the test, had a significant loss of hip extension and hip internal rotation (but more external rotation). This is an interesting observation, given much clinical discussion regarding the "tight" psoas. Even though we do not fully understand the neuromechanics, this muscle as a clinical concern is worth studying further.

CLINICAL RELEVANCE

Psoas Function

Myoelectric evidence and anatomical analysis suggest that the psoas major acts primarily to flex the hip, that its activation is minimally linked to spine demands, and that it imposes substantial lumbar spine compression when activated. Caution is advised when training this muscle due to the substantial spine compression penalty that is imposed on the spine when the psoas is activated.

Quadratus Lumborum

The quadratus lumborum (QL) is another special muscle for several reasons. First, the architecture of this muscle suits a stabilizing role by attaching to each lumbar vertebra, effectively buttressing adjacent vertebrae bilaterally, together with attachments to the pelvis and rib cage (see figure 4.33). Specifically, the fibers of the QL cross-link the vertebrae and have a large lateral moment arm via the transverse process attachments. Thus, by its design the QL could buttress shear instability and be effective in stabilizing all loading modes. Typically, under compressive load the first mode of buckling instability is lateral (Lucas and Bresler, 1961); the QL can play a significant role in local lateral buttressing. Also, the QL hardly changes length during any spine motion (see table 4.2 on page 65 and McGill, 1991b), suggesting that it contracts virtually isometrically. Further insight into its special function comes from an earlier observation that the motor control system involves this muscle together with the abdominal wall when stability is required in the absence of major moment demands.

The muscle appears to be active during a variety of flexion-dominant, extensor-dominant, and lateral bending tasks. (Note that myoelectric access to the QL is quite tricky, and it is difficult to confirm where the intramuscular electrodes are within the muscle. Certainly our techniques on this muscle were not very precise. In addition, they tend to migrate upon contraction, further clouding interpretation of the signal.) Andersson and colleagues (1996) found that the QL did not relax with the lumbar extensors during the flexion/relaxation phenomenon. The flexion/relaxation phenomenon is an interesting task since there are no substantial lateral or twisting torques and the extensor torque appears to be supported passively, further suggesting some stabilizing role for the QL.

In another experiment (note again that our laboratory techniques to obtain QL activation were rather imprecise at the time), subjects stood upright with a bucket in each hand. We incrementally increased the load in each bucket (resulting in progressively more spine compression). Our data suggest that the QL increased its activation level (together with the obliques) as more stability was required (McGill, Juker, and Kropf, 1996b). This task forms a special situation since only compressive loading is applied to the spine in the absence of any bending moments. In summary, the strength of the evidence from several perspectives leaves one to conclude that the QL performs a very special stabilizing role for the lumbar spine in a wide variety of tasks.

CLINICAL RELEVANCE

Quadratus Lumborum

The quadratus lumborum appears to be highly involved with stabilization of the lumbar spine, together with other muscles, suggesting that a clinical focus of this muscle is warranted. Exercises emphasizing activation of the quadratus lumborum while sparing the spine are described in the rehabilitation section.

Muscle Summary

This section has provided an overview of the roles of the muscles of the trunk in supporting postures and moving and stabilizing the lumbar spine. The posterior muscles were presented in four large functional groups. The deepest group (the small rotators) appear to act as position sensors rather than as torque generators. The more superficial extensors (multifidus and iliocostalis lumborum and longissimus thoracis) fall into three categories to

1. generate large extension moments over the entire lumbar region,
2. generate posterior shear, or
3. affect and control only one or two lumbar segments.

The roles of the abdominal muscles in trunk flexion and in trunk stabilization were highlighted together with the roles of the psoas and quadratus lumborum. Clearly, many muscles play a large role in protecting the low back from injury. Chapter 13 applies these findings to exercise regimens for individuals with low back pain.

Ligaments

When the lumbar spine is neither flexed nor extended **(neutral lordosis),** only muscle contributions need be considered in the mechanics to support the spine. However, as the spine flexes, bends, and twists, passive tissues are stressed; the resultant forces of those tissues change the interpretation of injury exacerbation and/or the discussion of clinical issues. For this reason, I introduce the mechanics of passive tissues in this section, followed by some examples illustrating their effects on clinical mechanics. Also fascinating is the distribution of mechanoreceptors documented in every lumbar ligament and fascia. Solomonow and colleagues' (2000) recent evidence suggests a significant proprioceptive role for spinal ligaments.

Longitudinal Ligaments

The vertebrae are joined to form the spinal column by two ribbonlike ligaments, the anterior longitudinal and the posterior longitudinal ligaments, which assist in restricting excessive flexion and extension (see figure 4.34). Both ligaments have bony attachments to the vertebral bodies and collagenous attachments to the annulus. Very little evidence exists for the presence of mechanoreceptors in these ligaments. Posterior to the spinal cord is the ligamentum flavum, which is characterized by a composition of approximately 80% elastin and 20% collagen, signifying a very special function for this ligament. It has been proposed that this highly elastic structure, which is under pretension throughout all levels of flexion, acts as a barrier to material that could buckle and encroach on the cord in some regions of the range of motion.

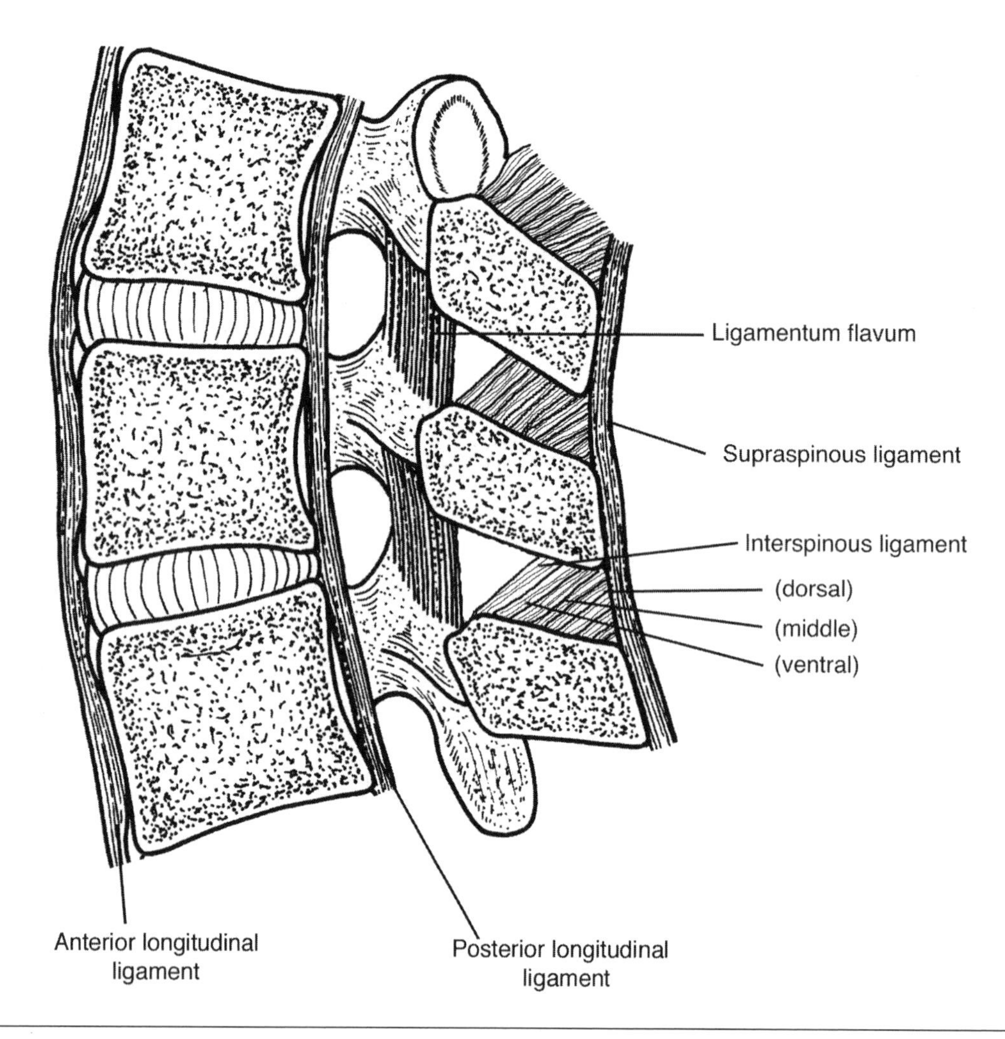

Figure 4.34 Major lumbar ligaments. Note the controversy surrounding the interspinous ligament in figures 4.35 and 4.36.

Adapted, by permission, from J. Watkins, 1999, *Structure and Function of the Musculoskeletal System.* (Champaign, IL: Human Kinetics), 149.

Interspinous and Superspinous Ligaments

The interspinous and superspinous ligaments are often classed as a single structure in anatomy texts, although functionally they appear to have quite different roles. The interspinous ligaments connect adjacent posterior spines but are not oriented parallel to the compressive axis of the spine. Rather, they have a large angle of obliquity, which has been a point of contention. For years most anatomy books have shown these ligaments with an oblique angle that would cause posterior shear of the superior vertebrae (see figure 4.35). This error is believed to have originated around the turn of the century with the hypothesis being that an artist held the vertebral section upside down when drawing; other artists simply copied the previous art rather than look at a spine. This was corrected by Heylings (see figure 4.36), noting that indeed these ligaments have the obliquity to resist posterior shear of the superior vertebrae—but also impose anterior shear forces during full flexion (Heylings, 1978). While many anatomy textbooks suggest that this ligament serves to protect against excessive flexion

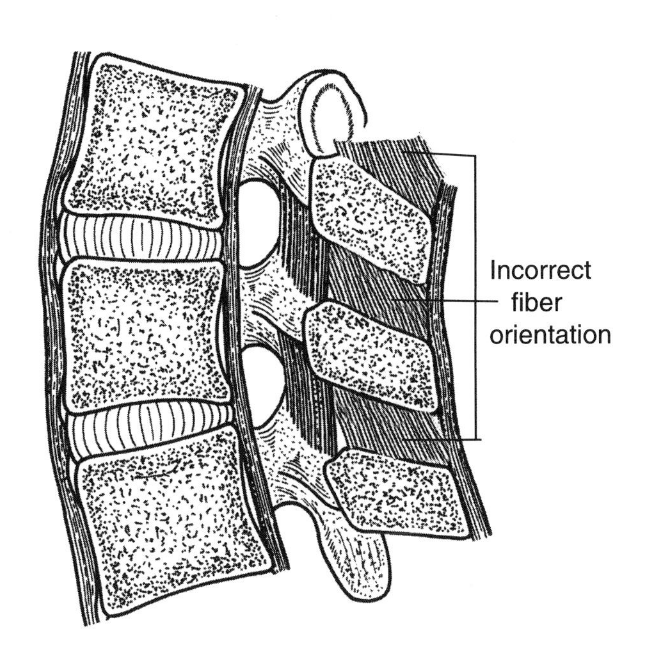

Figure 4.35 For most of the past 100 years many anatomical artists have drawn the interspinous ligament upside down, as shown in this example. Such drawings have caused the ligament function to be misinterpreted as that of a supporter of anterior shear, which is incorrect. The original figure of this example, from J. Watkins, 1999, *Structure and Fu ction of the Musculoskeletal System,* Champaign, IL: Human Kinetics, was drawn correctly but it has been altered to illustrate the incorrect rendering many artists have drawn in the past.

Adapted, by permission, from J. Watkins, 1999, *Structure and Function of the Musculoskeletal System.* (Champaign, IL: Human Kinetics), 149.

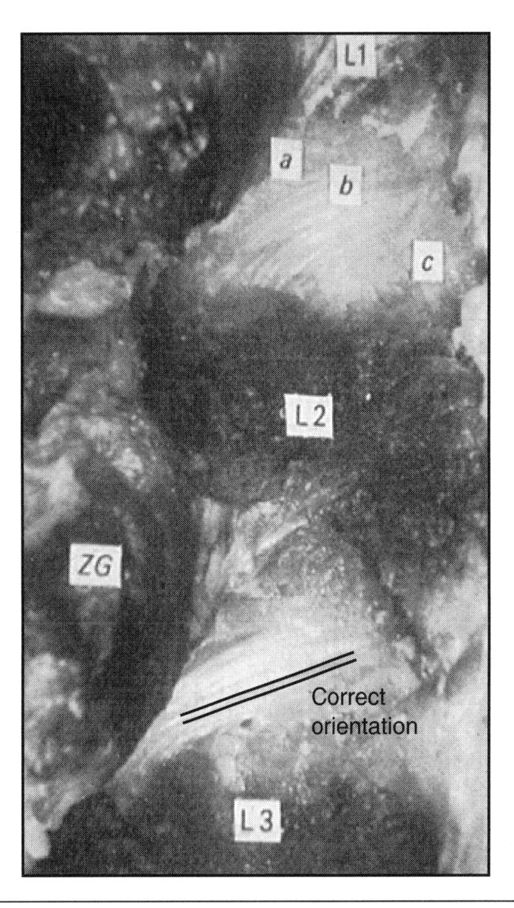

Figure 4.36 The interspinous ligament runs obliquely to the compressive axis and thus has limited capacity to check flexion rotation of the superior vertebrae. Rather, the interspinous ligament may act as a collateral ligament, controlling vertebral rotation and imposing anterior shear forces on the superior vertebrae. (ZG) Zagapophyseal joint or facet joint, (L2), (L3) lumbar spinous processes.

Adapted, by permission, from D. Heylings, 1978, "Supraspinous and interspinous ligaments of the human lumbar spine," *Journal of Anatomy* 125(1): 129. Copyright Blackwell Publishing.

(based on erroneous anatomy), I do not believe this notion is correct. Heylings (1978) suggested that the ligament acts like a collateral ligament similar to those in the knee, whereby the ligament controls the vertebral rotation as it follows an arc throughout the flexion range. This in turn helps the facet joints remain in contact, gliding with rotation. Furthermore, with its oblique line of action, the interspinous ligament protects against posterior shearing of the superior vertebrae and is implicated in an injury scenario discussed later in this chapter.

In contrast to the interspinous ligament, the superspinous ligament is aligned more or less parallel to the compressive axis of the spine, connecting the tips of the posterior spines. It appears to provide resistance against excessive forward flexion. Finally, both supraspinous and interspinous ligaments have an extensive network of free nerve endings (type IV receptors) (Yahia, Newman, and Rivard, 1988) together with Ruffini corpuscles and Pacinian corpuscles (Yahia et al., 1988; Jiang et al., 1995). Both Yahia and colleagues (1988) and Solomonow and colleagues (2000) suggest a proprioceptive role for the ligaments to prevent excessive strain in fully flexed postures and—given their architecture—quite possibly when under excessive shear load.

CLINICAL RELEVANCE

Interspinous Ligament Architecture

The interspinous ligament appears to provide a collateral action as it guides the sliding motion of the facet joints and checks posterior shear of the superior vertebra. The flip side of this functional role is that during full flexion the superior vertebra is sheared anteriorly, often adding to the reaction shear forces produced in a forward bending posture. Therapeutic exercise recommending full spine flexion stretches must consider the resultant shearing forces imposed on the joint by interspinous ligament strain. Too often, even those patients with shear pathology—for example, those with spondylolisthesis—are prescribed flexion stretches. This appears to be ill-advised.

Other Ligaments in the Thoraco-Lumbar Spine

Other ligaments in the thoraco-lumbar spine include the intertransverse ligaments and the facet capsule.

- **Intertransverse ligaments.** These ligaments span the transverse processes and have been argued to be sheets of connective tissue rather than true ligaments (Bogduk and Twomey, 1991). In fact, Bogduk and Twomey suggest that the intertransverse ligament membrane forms a septum between the anterior and posterior musculature that is an embryological holdover from the development of these two sections of muscle.
- **Facet capsule.** The facet capsule consists of connective tissue with bands that restrict both the joint flexion and the distraction of the facet surfaces that result from axial twisting. The ligaments that form the capsule have been documented to be rich in proprioceptive organs—Pacinian and Ruffini corpuscles (Cavanaugh et al., 1996; McLain and Pickar, 1998)—and have been observed to respond to multidirectional stress (Jiang et al., 1995), at least in cats.

Normal Ligament Mechanics and Injury Mechanics

Determining the roles of ligaments has involved qualitative interpretation using their attachments and lines of action together with functional tests in which successive ligaments were cut and the joint motion reassessed. Early studies attempting to determine the amount of relative contribution of each ligament to restricting flexion were performed on cadaveric preparations that were not preconditioned prior to testing. This usually entails a loading regimen so that the cadaveric specimens better reflect in vivo behavior. Specifically, they did not take into account the fact that, upon death, discs, being hydrophilic, increase their water content and consequently their height. The swollen discs in cadaveric specimens produced an artificial preload on the ligaments closest to the disc, inappropriately suggesting that the capsular and longitudinal ligaments are more important in resisting flexion than they actually are in vivo. For this reason, these early data describing the functional roles of various ligaments were incorrect. The work of Sharma and colleagues (1995) showed that the major ligaments for resisting flexion are the supraspinous ligaments. Those early studies showing the posterior longitudinal ligament and the capsular ligaments are important for resisting flexion did not employ the necessary preconditioning discussed above.

Mechanical failure of the ligaments is worth considering. King (1993) noted that soft tissue injuries are common during high-energy, traumatic events such as automobile collisions. Our own observations on pig and human specimens loaded at slow load rates in bending and shear modes suggest that excessive tension in the longitudinal ligaments causes avulsion or bony failure near the ligament attachment site. Noyes and colleagues (1994) noted that slow strain rates (0.66%/s) produced more ligament avulsion injuries, while fast strain rates (66%/s) resulted in more mid-ligamentous failure, at least in monkey knee ligaments. The clinical results by Rissanen (1960), however, showed that approximately 20% of randomly selected cadaveric spines possessed visibly ruptured lumbar interspinous ligaments, while the dorsal and ventral portions of the interspinous ligaments and the supraspinous ligaments were intact. These could be considered to represent the living population.

Given the oblique fiber direction of the interspinous complex (see figure 4.36, page 77), a very likely scenario of interspinous ligament damage is falling and landing on one's behind, driving the pelvis forward on impact and creating a posterior shearing of the lumbar joints when the spine is fully flexed. The interspinous ligament is a major load-bearing tissue in this example of high-energy loading characterized by anterior shear displacement combined with full flexion. Considering the available data, I believe that damage to the ligaments of the spine, particularly the interspinous complex, is not likely during lifting or other normal occupational activities. Rather, ligament damage seems to occur primarily during traumatic events, as described earlier. The subsequent joint laxity is well known to accelerate arthritic changes (Kirkaldy-Willis and Burton, 1992). What has been said in reference to the knee joint, "ligament damage marks the beginning of the end," is also applicable to the spine in terms of being the initiator of the cascade of degenerative change.

CLINICAL RELEVANCE

Patterns of Ligament Injury

Torn spinal ligaments appear to be the result of ballistic loading—particularly slips and falls or traumatic sporting activity with the spine at its end range of motion. Those with recently developing spine symptoms and accompanying spine instability can often recount prior incidents in which the ligaments could have been damaged. If the traumatic event occurs at work but the delayed sequellae develop later during an event at home, is this compensable? Arguing these types of questions requires a solid understanding of injury mechanisms.

Lumbodorsal Fascia (LDF)

While a functional interpretation of the lumbodorsal fascia (LDF) (also called the thoraco-lumbar fascia by some) is provided later, a short anatomical description is given here. First, the fascia has bony attachments on the tips of the spinous process (except the shorter L5 in many individuals) and to the posterior-superior iliac spines (PSIS). Some fascial connections cross the midline, suggesting some force transmission, thus completing the hoop around the abdomen with the previously described abdominal fascia anteriorly. The transverse abdominis and internal oblique muscles obtain their posterior attachment to the fascia, as does the latissimus dorsi over the upper regions of the fascia. The fascia, in wrapping around the back, forms a compartment around the lumbar extensors (multifidus and pars lumborum groups of iliocostalis and longissimus) and has been implicated in compartment syndrome (Carr et al., 1985; Styf, 1987) (see figure 4.37). Recent studies attribute various mechanical roles to the lumbodorsal fascia (LDF). In fact, some have recommended lifting techniques based on these hypotheses. However, are they consistent with experimental evidence? Gracovetsky and colleagues (1981) originally suggested that lateral forces generated by the internal oblique and transverse abdominis muscles are transmitted to the LDF

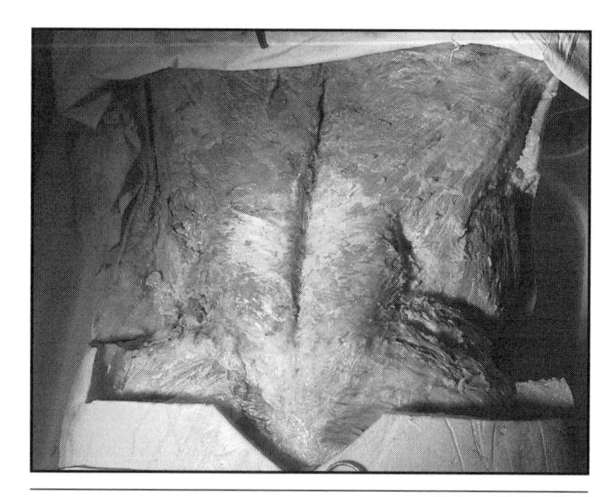

Figure 4.37 Collagen fiber arrangement in the LDF binds the lumbar extensor muscles and tendons from the thoracic muscles together as they course to the sacral attachments. Thus, one of the functions of the LDF appears to be acting as an extensor muscle retinaculum—and a natural abdominal-back belt.

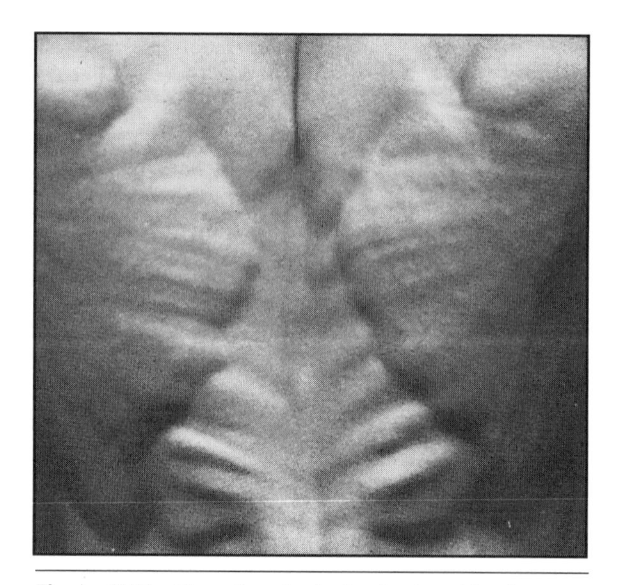

Figure 4.38 Stress lines in the lumbodorsal fascia indicate that the latissimus dorsi is the dominant force activator—at least in this example.

via their attachments to the lateral border, and that the fascia could support substantial extensor moments. They hypothesized that this lateral tension on the LDF increased longitudinal tension by virtue of the collagen fiber obliquity in the LDF, causing the posterior spinous processes to move together, resulting in lumbar extension. This proposed sequence of events formed an attractive proposition because the LDF has the largest moment arm of all the extensor tissues. As a result, any extensor forces within the LDF would impose the smallest compressive penalty to vertebral components of the spine.

However, three studies, all published about the same time, collectively challenge the viability of this hypothesis: Tesh and colleagues (1987), who performed mechanical tests on cadaveric material; Macintosh and colleagues (1987), who recognized the anatomical inconsistencies with the abdominal activation; and McGill and Norman (1988), who tested the viability of LDF involvement with the latissimus dorsi as well as with the abdominals (see figure 4.38). These collective works show that the LDF is not a significant active extensor of the spine. Nonetheless, the LDF is a strong tissue with a well-developed lattice of collagen fibers, suggesting that its function may be that of an extensor muscle retinaculum (Bogduk and Macintosh, 1984), or nature's back belt. In addition, the fascia does contain both Ruffini and Pacinian corpuscles together with diffuse innervation (Yahia et al., 1998). The tendons of longissimus thoracis and iliocostalis lumborum pass under the LDF to their sacral and iliac attachments. It appears that the LDF may provide a form of retinacular "strapping" for the low back musculature. Finally, the abdominal wall and the latissimus dorsi forces add tension to the fascia and stiffness to the spine to prevent specific types of unstable behavior and tissue damage (explained in chapter 6 on spine stability).

CLINICAL RELEVANCE

Lumbodorsal Fascia Anatomy

No evidence justifies specific lifting techniques to involve the LDF for extension of the spine. However, activation of the latissimus dorsi and the deep abdominal obliques contributes stiffening (and stabilizing) forces to the lumbar spine via the fascia (guidelines for activating these muscles are provided in chapter 13). Furthermore, the LDF appears to act as a retinaculum and probably fulfills a proprioceptive function. It is part of a "hoop" around the abdomen, which consists of the LDF posteriorly, the abdominal fascia anteriorly, and the active abdominal muscles laterally; the three together complete the stabilizing corset (see figure 4.39).

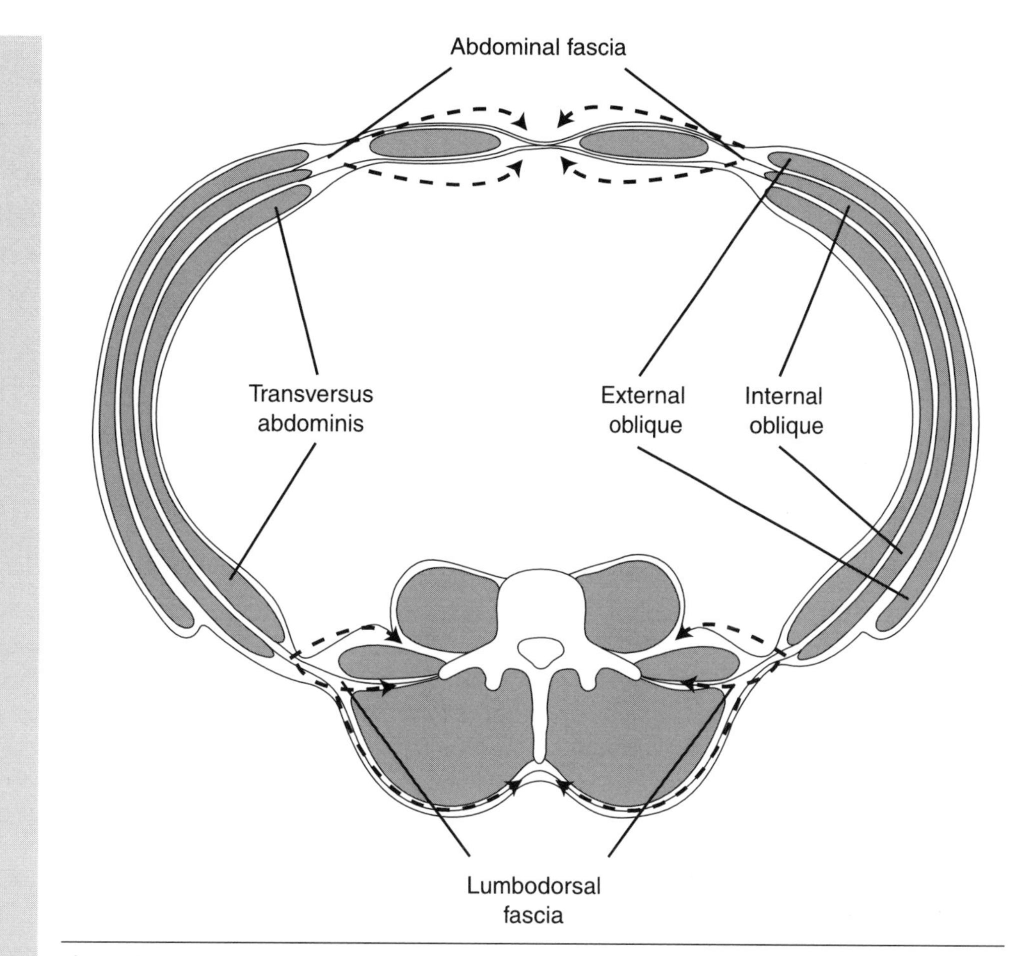

Figure 4.39 The abdominal fascia, anteriorly, and the LDF, posteriorly, are passive parts of the abdominal hoop. The lateral active musculature (transverse abdominis and internal oblique) serves to tension the hoop (dashed arrows).

Clinically Relevant Aspects of Pain and Anatomic Structure

Recall from the introduction of this book that tissue damage can alter the biomechanics of a spinal joint and that once the biomechanics have changed any innervated tissue can be the candidate for symptoms. Pain originates with the free nerve endings of the various pain receptors that typically form small nerve fibers. As noted in Guyton (1981), not all of the small fibers originate in pain receptors: some originate in organs sensitive to temperature, pressure, or other "touching" sensations. Pain may also be initiated at higher levels in the pain pathway: Howe and colleagues (1977) demonstrated that mechanical pressures on the dorsal root ganglion produce discharges for up to 25 minutes following the removal of the mechanical pressure. In addition, Cavanaugh (1995) showed that nerve endings are sensitive to chemical mediators released during tissue damage and inflammation. Some studies have attempted to examine inflammatory processes by the injection of various chemicals. For example, Ozaktay and colleagues (1994) injected carrageenan into the region of nerve receptors around the facet joints of rabbits and reported that the discharge from the pressure-sensitive neurons lasted for three hours. This finding suggests that tissue damage producing inflammatory processes may contribute to a long-lasting muscle spasm. In a recent summary Cavanaugh (1995) presented evidence to document the possible role of various chemical mediators that sensitize various components along the pain pathway so that pain is produced during events that are normally nonnoxious. Much remains to be understood.

Bogduk (1983) provided an excellent review of the innervation of lumbar tissues. For example, the facet is well innervated with a variety of low- and high-threshold nerves, suggesting both pain and nociceptive functions. Free nerve endings have also been observed around the superficial layers of intervertebral discs.

Tissue-Specific Types of Pain

I have had some personal experience with direct mechanical irritation of specific low back tissues and the resultant pain. Admittedly, these results are limited, given the single subject (myself) and the subjective nature of the observations, but I believe they are worth reporting nonetheless. I obtained these insights from indwelling EMG experiments in which needles were used to implant fine wire EMG electrodes in the psoas, quadratus lumborum, multifidus, and three layers of the abdominal wall. Many people have experienced the burning sensation as the needle penetrates the skin. This is cutaneous pain, as the application of both ice and a hot material feel similar. As the needle applies pressure to, and punctures through, the lumbodorsal fascia, the pain is felt as a scraping sensation and sometimes as an electric current. The same sort of pain is felt as the needle progresses through the different sheaths between the layers of muscle of the abdominal wall. It is interesting to note that fibromyalgic patients sometimes report this scratchy type of muscularly located pain—it is very consistent with epimysium/fascia irritation. Once the needle was inside the muscle, no pain was perceived, just an occasional feeling of mechanical pressure. As the needle touched the peritoneum of the abdominal cavity in any location, a general intestinally sick feeling was produced in the abdominal region, focused anteriorly to a small area just below the naval. As the needle touched the bone of the vertebra, even with very light pressure, a very pointed and "boring" pain was produced, similar to the pain experienced on being kicked in the shins. Once again, the reader must realize that these are the experiences of a single person. Nonetheless, they do provide crude qualitative insight into the types of pain produced in specific tissues.

Can Pain Descriptors Provide a Reliable Diagnosis?

Pain is clearly produced from tissue irritation, particularly mechanical overload. Some have argued that some tissues may or may not be candidates for sources of pain based on the presence or absence of nociceptive nerve endings. This may be a diversionary argument. If the overload is sufficient as to damage tissue and produce biomechanical change in the joint, then the loading patterns of other tissues are disturbed. Thus, even though one tissue may not be capable of producing pain, if it is damaged sufficiently to shift load to another suitably innervated tissue, pain may result. For example, innervated annulus and disc end plates may be sources of pain as a consequence of end-plate fracture or annular herniation. But end-plate fracture can cause significant disc height loss, which can lead to nerve entrapment, complex joint instability, and subsequent facet overload, and so on. Once the biomechanics of the joint have been altered, it is no longer fruitful to attempt to diagnose specific tissue damage; the picture is complex. Functional diagnosis is the only feasible option.

A Final Note

In this chapter, a rudimentary anatomical/biomechanical knowledge is assumed on the part of the reader. Using this foundation, some anatomical features were reviewed that are not often considered or discussed in classic anatomical texts. Serious discussion of anatomy must involve function and, by extension, must consider biomechanics and motor control. Hopefully, the functional discussions throughout this chapter have stimulated you to give more consideration to the architecture of the lumbar spine. The challenge for the scientist and clinician alike is to become conversant with the functional implications of the anatomy, which will guide decisions to develop the most appropriate prevention programs for the uninjured and the best treatments for patients.

References

Adams, M., and Dolan, P. (1995) Recent advances in lumbar spinal mechanics and their clinical significance. *Clinical Biomechanics,* 10 (1): 3.

Adams, M.A., and Hutton, W.C. (1985) Gradual disc prolapse. *Spine,* 10: 524.

Adams, M.A., and Hutton, W.C. (1982) Prolapsed intervertebral disc: A hyperflexion injury. *Spine,* 7: 184.

Adams, P., and Muir, H. (1976) Qualitative changes with age of proteoglycans of human lumbar discs. *Annals of the Rheumatic Diseases*, 35: 289.

Amonoo-Kuofi, H.S. (1983) The density of muscle spindles in the medial, intermediate, and lateral columns of human intrinsic postvertebral muscles. *Journal of Anatomy,* 136: 509-519.

Andersson, E.A., Oddsson, L., Grundstrom, O.M., Nilsson, J., and Thorstensson, A. (1996) EMG activities of the quadratus lumborum and erector spinae muscles during flexion-relaxation and other motor tasks. *Clinical Biomechanics,* 11: 392-400.

Andersson, E., Oddsson, L., Grundstrom, H., and Thorstensson, A. (1995) The role of the psoas and iliacus muscles for stability and movement of the lumbar spine, pelvis and hip. *Scandinavian Journal of Medicine and Science in Sports,* 5: 10-16.

Bedzinski, R. (1992). Application of speckle photography methods to the investigations of deformation of the vertebral arch. In: Little, E.G. (Ed.), *Experimental mechanics*. New York: Elsevier.

Bogduk, N. (1980) A reappraisal of the anatomy of the human lumbar erector spinae. *Journal of Anatomy,* 131 (3): 525.

Bogduk, N. (1983) The innervation of the lumbar spine. *Spine,* 8: 286.

Bogduk, N., and Engel, R. (1984) The menisci of the lumbar zygapophyseal joints: A review of their anatomy and clinical significance. *Spine,* 9: 454.

Bogduk, N., and Macintosh, J.E. (1984) The applied anatomy of the thoracolumbar fascia. *Spine,* 9: 164.

Bogduk, N., and Twomey, L. (1991) Clinical anatomy of the lumbar spine (2nd ed.). New York: Churchill Livingstone.

Brinckmann, P., Biggemann, M., and Hilweg, D. (1988) Fatigue fracture of human lumbar vertebrae. *Clinical Biomechanics,* 3 (Suppl. 1): S1-S23.

Brinckmann, P., Biggemann, M., and Hilweg, D. (1989) Prediction of the compressive strength of human lumbar vertebrae. *Clinical Biomechanics,* 4 (Suppl. 2).

Carr, D., Gilbertson, L., Frymeyer, J., Krag, M., and Pope, M. (1985) Lumbar paraspinal compartment syndrome: A case report with physiologic and anatomic studies. *Spine,* 10: 816.

Callaghan, J.P., and McGill, S.M. (2001) Intervertebral disc herniation: Studies on a porcine model exposed to highly repetitive flexion/extension motion with compressive force. *Clinical Biomechanics,* 16 (1): 28-37.

Cavanaugh, J.M. (1995) Neural mechanisms of lumbar pain. *Spine,* 20 (16): 1804.

Cavanaugh, J.M., Ozaktay, C.A., Yamashita, T., and Hing, A.I. (1996) Lumbar facet pain: Biomechanics, neuroanatomy and neurophysiology. *Journal of Biomechanics*, 29: 1117-1129.

Cholewicki, J., Juluru, K., and McGill, S.M. (1999) The intra-abdominal pressure mechanism for stabilizing the lumbar spine. *Journal of Biomechanics,* 32 (1): 13-17.

Cresswell, A.G., Oddsson, L., and Thorstensson, A. (1994) The influence of sudden perturbations on trunk muscle activity in intraabdominal pressure while standing. *Experimental Brain Research,* 98: 336-341.

Cripton, P., Berlemen, U., Visarino, H., Begeman, P.C., Nolte, L.P., and Prasad, P. (1995) Response of the lumbar spine due to shear loading. In: *Injury prevention through biomechanics* (p. 111). Detroit: Wayne State University.

Dickey, J.P., Pierrynowski, M.R., and Bednar, D.A. (1996) Deformation of vertebrae in vivo—Implications for facet joint loads and spinous process pin instrumentation for measuring sequential spinal kinematics. Presented at the Canadian Orthopaedic Research Society, Quebec City, May 25.

Farfan, H.F. (1973) *Mechanical disorders of the low back.* Philadelphia: Lea and Febiger.

Fyhrie, D.P., and Schaffler, M.B. (1994) Failure mechanisms in human vertebral cancellous bone. *Bone,* 15 (1): 105-109.

Gallois, J., and Japoit, T. (1925) Architecture intérieure des vertèbres du point de vue statique et physiologique. *Rev Chir* (Paris), 63: 688.

Goel, V.K., Monroe, B.T., Gilbertson, L.G., and Brinckmann, P. (1995) Interlaminar shear stresses and laminae-separation in a disc: Finite element analysis of the L3-L4 motion segment subjected to axial compressive loads. *Spine,* 20 (6): 689.

Goel, V.K., Wilder, D.G., Pope, M.H., and Edwards, W.T. (1995) Controversy: Biomechanical testing of the spine: Load controlled versus displacement controlled analysis. *Spine,* 20: 2354-2357.

Gordon, S.J., et al. (1991) Mechanism of disc rupture—A preliminary report. *Spine,* 16: 450.

Gracovetsky, S., Farfan, H.F., and Lamy, C. (1981) Mechanism of the lumbar spine. *Spine,* 6 (1): 249.

Gunning, J.L., Callaghan, J.P., and McGill, S.M. (2001) The role of prior loading history and spinal posture on the compressive tolerance and type of failure in the spine using a porcine trauma model. *Clinical Biomechanics,* 16 (6): 471-480.

Guyton, A.C. (1981) Sensory receptors and their basic mechanisms of action. In: *Textbook of medical physiology* (6th ed., p. 588). Philadelphia: W.B. Saunders.

Hardcastle, P., Annear, P., and Foster, D. (1992) Spinal abnormalities in young fast bowlers. *Journal of Bone and Joint Surgery,* 74B (3): 421.

Henke, K.G., Sharratt, M.T., Pegelow, D., and Dempsey, J.A. (1988) Regulation of end-expiratory lung volume during exercise. *Journal of Applied Physiology,* 64: 135-146.

Heylings, D.J.A. (1978) Supraspinous and interspinous ligaments of the human lumbar spine. *Journal of Anatomy,* 123: 127.

Hodges, P.W., and Richardson, C.A. (1996) Inefficient muscular stabilisation of the lumbar spine associated with low back pain: A motor control evaluation of transversus abdominis. *Spine,* 21: 2640-2650.

Howe, J.F., Loeser, J.D., and Calvin, W.H. (1977) Mechanosensitivity of dorsal root ganglia and chronically ignored axons: A physiological basis for the radicular pain of nerve root compression. *Pain,* 3: 25.

Jiang, H.J., Russell, G., Raso, J., Moreau, M.J., Hill, D.J., and Bagnall, K.M. (1995) The nature and distribution of the innervation of human supraspinal and interspinal ligaments. *Spine,* 20: 869-876.

Juker, D., McGill, S.M., and Kropf, P. (1998) Quantitative intramuscular myoelectric activity of lumbar portions of psoas and the abdominal wall during cycling. *Journal of Applied Biomechanics,* 14 (4): 428-438.

Juker, D., McGill, S.M., Kropf, P., and Steffen, T. (1998). Quantitative intramuscular myoelectric activity of lumbar portions of psoas and the abdominal wall during a wide variety of tasks. *Medicine and Science in Sports and Exercise,* 30 (2): 301-310.

Keller, T.S., Ziv, I., Moeljanto, E., and Spengler, D.M. (1993) Interdependence of lumbar disc and subdiscal bone properties: A report of the normal and degenerated spine. *Journal of Spinal Disorders,* 6 (2): 106-113.

King, A.I. (1993) Injury to the thoraco-lumbar spine and pelvis. In: Nahum and Melvin (Eds.), *Accidental injury, biomechanics and prevention.* New York: Springer.

Kirkaldy-Willis, W.H., and Burton, C.V. (1992) *Managing low back pain* (3rd ed.). New York: Churchill Livingstone.

Lehman, G., and McGill, S.M. (2001) Quantification of the differences in EMG magnitude between upper and lower rectus abdominis during selected trunk exercises. *Physical Therapy,* 81: 1096-1101.

Lotz, J.C., and Chin, J.R. (2000) Intervertebral disc cell death is dependent on the magnitude and duration of spinal loading. *Spine,* 25 (12): 1477-1483.

Lu, W.W., Luk, K.D.K., Cheung, K.M.C., Fang, D., Holmes, A.D., and Leong, J.C.Y. (2001) Energy absorption of human vertebral body under fatigue loading. In: Abstracts, International Society for Study of the Lumbar Spine, Edinburgh, Scotland, June 19-23.

Lucas, D., and Bresler, B. (1961) Stability of the ligamentous spine. Tech. Report No. 40, Biomechanics Laboratory, University of California, San Francisco.

Macintosh, J.E., and Bogduk, N. (1987) The morphology of the lumbar erector spinae. *Spine,* 12 (7): 658.

Macintosh, J.E., Bogduk, N., and Gracovetsky, S. (1987) The biomechanics of the thoracolumbar fascia. *Clinical Biomechanics,* 2: 78.

Markolf, K.L., and Morris, J.M. (1974) The structural components of the intervertebral disc. *Journal of Bone and Joint Surgery,* 56A (4): 675.

Marras, W.S., and Granata, K.P. (1997) Changes in trunk dynamics and spine loading during repeated exertions. *Spine,* 22: 2564-2570.

Marras, W.S., Jorgensen, M.J., Granata, K.P., and Waind, B. (2001) Female and male trunk geometry: Size and prediction of the spine loading trunk muscles derived from MRI. *Clinical Biomechanics,* 16: 38-46.

McGill, S.M. (1991a) Electromyographic activity of the abdominal and low back musculature during the generation of isometric and dynamic axial trunk torque: Implications for lumbar mechanics. *Journal of Orthopaedic Research,* 9: 91.

McGill, S.M. (1991b) The kinetic potential of the lumbar trunk musculature about three orthogonal orthopaedic axes in extreme postures. *Spine,* 16 (7): 809-815.

McGill, S.M. (1992) A myoelectrically based dynamic 3-D model to predict loads on lumbar spine tissues during lateral bending. *Journal of Biomechanics,* 25 (4): 395.

McGill, S.M. (1996) A revised anatomical model of the abdominal musculature for torso flexion efforts. *Journal of Biomechanics,* 29 (7): 973-977.

McGill, S.M. (1997). Invited paper: Biomechanics of low back injury: Implications on current practice and the clinic. *Journal of Biomechanics,* 30 (5): 465-475.

McGill, S.M., Juker, D., and Axler, C. (1996). Correcting trunk muscle geometry obtained from MRI and CT scans of supine postures for use in standing postures. *Journal of Biomechanics,* 29 (5): 643-646.

McGill, S.M., Juker, D., and Kropf, P. (1996a). Appropriately placed surface EMG electrodes reflect deep muscle activity (psoas, quadratus lumborum, abdominal wall) in the lumbar spine. *Journal of Biomechanics,* 29 (11): 1503-1507.

McGill, S.M., Juker, D., and Kropf, P. (1996b). Quantitative intramuscular myoelectric activity of quadratus lumborum during a wide variety of tasks. *Clinical Biomechanics,* 11 (3): 170-172.

McGill, S.M., Hughson, R.L., and Parks, K. (2000) Changes in lumbar lordosis modify the role of the extensor muscles. *Clinical Biomechanics,* 15 (1): 777-780.

McGill, S.M., and Norman, R.W. (1987) Effects of an anatomically detailed erector spinae model on L4/L5 disc compression and shear. *Journal of Biomechanics,* 20 (6): 591.

McGill, S.M., and Norman, R.W. (1988) The potential of lumbodorsal fascia forces to generate back extension moments during squat lifts. *Journal of Biomedical Engineering,* 10: 312.

McGill, S.M., Patt, N., and Norman, R.W. (1988) Measurement of the trunk musculature of active males using CT scan radiography: Duplications for force and moment generating capacity about the L4/L5 joint. *Journal of Biomechanics,* 21 (4): 329.

McGill, S.M., Santaguida, L., and Stevens, J. (1993) Measurement of the trunk musculature from T6 to L5 using MRI scans of 15 young males corrected for muscle fibre orientation. *Clinical Biomechanics,* 8: 171.

McLain, R.F., and Pickar, J.G. (1998) Mechanoreceptor endings in human thoracic and lumbar facet joints. *Spine,* 23: 168-173.

Nachemson, A.L. (1960) Lumbar interdiscal pressure. *Acta Orthopaedica Scandinavica,* (Suppl. 43).

Nachemson, A. (1966) The load on lumbar discs in different positions of the body. *Clin Rel Res,* 45: 107.

Nitz, A.J., and Peck, D. (1986) Comparison of muscle spindle concentrations in large and small human epaxial muscles acting in parallel combinations. *American Surgeon,* 52: 273-277.

Norman, R.W., Wells, R., Neumann, P., Frank, J., Shannon, H., and Kerr, M. (1998) A comparison of peak vs cumulative physical work exposure factors for the reporting of low back pain in the automotive industry. *Clinical Biomechanics,* 13: 561-573.

Noyes, F.R., De Lucas, J.L., and Torvik, P.J. (1994) Biomechanics of ligament failure: An analysis of strain-rate sensitivity and mechanisms of failure in primates. *Journal of Bone and Joint Surgery*, 56A: 236.

Ozaktay, A.C., et al. (1994) Effects of carrageenan induced inflammation in rabbit lumbar facet joint capsule and adjacent tissue. *Journal of Neuroscience Research*, 20: 355.

Porterfield, J.A., and DeRosa, C. (1998) *Mechanical low back pain: Perspectives in functional anatomy*. Philadelphia: W.B. Saunders.

Richardson, C., Jull, G., Hodges, P., and Hides, J. (1999) Therapeutic exercise for spinal segmental stabilization in low back pain. Edinburgh, Scotland: Churchill Livingstone.

Rissanen, P.M. (1960) The surgical anatomy and pathology of the supraspinous and interspinous ligaments of the lumbar spine with special reference to ligament ruptures. *Acta Orthopaedic Scandinavica,* (Suppl. 46).

Roaf, R. (1960) A study of the mechanics of spinal injuries. *Journal of Bone and Joint Surgery*, 42B: 810.

Roberts, S., Menage, J., and Urban, J.P.G. (1989) Biochemical and structural properties of the cartilage end-plate and its relationship to the intervertebral disc. *Spine*, 14: 166.

Santaguida, P., and McGill, S.M. (1995) The psoas major muscle: A three dimensional mechanical modelling study with respect to the spine based on MRI measurement. *Journal of Biomechanics,* 28: 339-345.

Sharma, M., Langrama, N.A., and Rodriguez, J. (1995) Role of ligaments and facets in lumbar spine stability. *Spine*, 20 (8): 887.

Silva, M.J., and Gibson, L.J. (1997) Modeling the mechanical behaviour of vertebral trabecular bone: Effects of age-related changes in microstructure. *Bone,* 21: 191-199.

Sirca, A., and Kostevc, V. (1985) The fibre type composition of thoracic and lumbar paravertebral muscles in man. *Journal of Anatomy*, 141: 131.

Solomonow, M., Zhou, B., Harris, M., Lu, Y., and Baratta, R.V. (2000) The ligamento-muscular stabilizing system of the spine. *Spine*, 23: 2552-2562.

Styf, J. (1987) Pressure in the erector spinae muscle during exercise. *Spine*, 12: 675.

Tesh, K.M., Dunn, J., and Evans, J.H. (1987) The abdominal muscles and vertebral stability. *Spine*, 12 (5): 501.

Vera-Garcia, F.J., Grenier, S.G., and McGill, S.M. (2000) Abdominal response during curl-ups on both stable and labile surfaces. *Physical Therapy*, 80 (6): 564-569.

Videman, T., Nurminen, M., and Troup, J.D.G. (1990) Lumbar spinal pathology in cadaveric material in relation to history of back pain, occupation and physical loading. *Spine*, 15 (8): 728.

Wilder, D.G., Pope, M.H., and Frymoyer, J.W. (1988) The biomechanics of lumbar disc herniation and the effect of overload and instability. *Journal of Spinal Disorders*, 1 (1): 16.

Yahia, L.H., Newman, N., and Rivard, C.H. (1988) Neurohistology of the lumbar spine. *Acta Orthopaedic Scandinavica*, 59: 508-512.

Yingling, V.R., Callaghan, J.P., and McGill, S.M. (1999) The porcine cervical spine as a reasonable model of the human lumbar spine: An anatomical, geometrical and functional comparison. *Journal of Spinal Disorders*, 12 (5): 415-423.

Yingling, V.R., and McGill, S.M. (1999a) Mechanical properties and failure mechanics of the spine under posterior shear load: Observations from a porcine model. *Journal of Spinal Disorders*, 12 (6): 501-508.

Yingling, V.R., and McGill, S.M. (1999b) Anterior shear of spinal motion segments: Kinematics, kinetics and resulting injuries observed in a porcine model. *Spine*, 24 (18): 1882-1889.

Normal and Injury Mechanics of the Lumbar Spine

Chapter 4 described the tissues, or the anatomical parts, and their role in function; this chapter will describe the normal mechanics of the whole lumbar spine. Since most biomechanics texts provide descriptions of spine motion, I will address it only briefly here. Instead, I will focus on the functional implications of that motion, which are far more important. I will also explain injury mechanisms and the changes that follow injury. Controversy remains as to whether these changes are a consequence of injury or in fact play a causative role.

Upon completion of this chapter, you will be able to explain the role of tissues in various tasks and consequently identify back-sparing techniques. In addition, you will understand the changes that follow injury, which have an impact on functional ability and rehabilitation decisions.

Kinematic Properties of the Thoraco-Lumbar Spine

The ranges of thoracic and lumbar segmental motion about the three principal axes (shown in table 5.1) demonstrate the greater flexion, extension, and lateral bending capability of the lumbar region and the relatively greater twisting capability of the thoracic region. While the segmental ranges shown in the table are population averages, keep in mind that a large variability exists among people, among age groups (McGill, Yingling, and Peach, 1999), and among segments within an individual.

Joint stiffness values convey the amount of translational and rotational deformation of a spine section under the application of force or moment. The average stiffness values (shown in table 5.2 compiled by Ashton-Miller and Schultz, 1988) document the translational stiffness of the spine in a neutral posture; they indicate greater stiffness under compression than under shear loads. In rotational modes greater stiffness occurs during axial torsion than during rotation about the flexion/extension and lateral bending axes. While generally the range of motion decreases with age, certain injuries, particularly to the disc, can increase the range of motion in bending and shear

Table 5.1 Range of Motion of Each Spine Level (in Degrees)

Level	Flexion	Flexion and extension combined	Extension	Lateral bending	Axial twist
T1-2		4		6	9
T2-3		4		6	8
T3-4		4		6	8
T4-5		4		6	8
T5-6		4		6	8
T6-7		5		6	8
T7-8		6		6	8
T8-9		6		6	7
T9-10		6		6	4
T10-11		9		7	2
T11-12		12		9	2
T12-L1		12		8	2
L1-2	8		5	6	2
L2-3	10		3	6	2
L3-4	12		1	8	2
L4-5	13		2	6	2
L5-S1	9		5	3	5

All data are from White and Panjabi (1978), except flexion and extension lumbar data, which are from Pearcy et al. (1984) and Pearcy and Tibrewal (1984).

translation (Spencer, Miller, and Shultz, 1985). Kirkaldy-Willis and Burton (1992) implicated these large unstable movements in facet joint derangement. Recent data have quantified the increase in the range of motion about all three spine axes as disc degeneration proceeds from grade I to grades III and IV. Radial tears of the annulus are most prevalent in these stages. But this extra motion is replaced by extreme loss of motion in grade V discs, which are characterized by collapse and osteophyte formation (Tanaka et al., 2001).

Table 5.2 Average Stiffness Values for the Adult Human Spine

	Comp.	Shear		Bending		Axial torsion
Spine level		**Ant./Post.**	**Lat.**	**Flex./Ext.**	**Lat.**	
T1-T12	1250	86/87	101	155/189	172	149
L1-L5	667	145/143	132	80/166	92	395
L5-S1	1000	78/72	97	120/172	206	264

Compression and shear values are given in newtons per millimeter and bending and axial torsion in newton-meters per radian.
T1-T2 data from White and Panjabi (1978).
L1-L5 data from Schultz et al. (1979) and Berkson, Nachemson, and Shultz (1979).
L5-S1 data from McGlashen et al. (1987).

As the spine moves in three dimensions (flexion/extension, lateral bend, and twist) the alignment of muscle vectors change with respect to the vertebral orthopedic axes. This causes the role of the various muscles to change. Sometimes their relative contribution to producing a specific moment changes along with the resultant joint compression and shear. Muscle lengths and their moment potential as a function of spine posture are shown in table 4.2 (page 65) and table 5.3. A range of extreme postures is shown in figure 5.1. Some muscles close to the spine obviously do not undergo great length excursions.

CLINICAL RELEVANCE

Motion Palpation—Pathology or Normal Asymmetry?

Stiffness asymmetries when bending to the right compared to the left and when twisting clockwise compared to counterclockwise at specific vertebral levels are not uncommon. This finding is of great importance to the clinician who may sometimes suspect pathology at a specific location but is simply experiencing normal anatomical asymmetry.

Recent work by Ross and colleagues (1999) exemplifies the peril in assuming that a joint with an asymmetric feel upon **palpation** is pathological. Clinicians who hold to a typical motion palpation philosophy will often identify an abnormal feeling at a specific spinal level as the target for therapy. Sometimes the asymmetrical stiffness may simply be asymmetric skeletal anatomy—perhaps a single facet with a unique angular orientation. Obviously, such a joint would be resistive to any "mobilizing" therapy.

Table 5.3 Moment Potential (N · m) of Some Representative Muscles in Various Postures Shown in Figure 5.1

Muscle	Force	Upright standing			60° flexion			25° R lateral bending			10° twist			Combined[a]		
	(N)	F[b]	B[c]	T[d]	F	B	T	F	B	T	F	B	T	F	B	T
R rectus abdominis	350	-28	17	6	-35	17	12	-28	21	8	-29	15	4	-33	20	15
R external oblique (1)	315	-7	27	9	-7	24	15	-7	28	11	-8	28	6	-1	20	22
R internal oblique (1)	280	-15	19	-16	-9	8	-31	-13	18	-20	-14	17	-18	-2	2	-33
R pars lumborum (L1)	455	36	23	9	35	23	8	36	24	9	3	24	5	38	23	7
R pars lumborum (L4)	595	23	23	-7	23	24	-4	23	23	-8	23	22	-8	23	25	-6
R iliocostalis lumborum	210	18	15	-1	18	14	2	18	15	-1	18	15	-2	18	15	1
R longissimus thoracis	280	24	10	-3	22	10	0	24	13	-3	24	12	-5	24	11	-1
R quadratus lumborum	175	5	11	5	7	14	1	5	12	6	6	13	4	7	14	0
R latissimus dorsi (L5)	140	9	5	-3	8	7	-4	8	6	-5	9	5	-5	8	7	-6
R multifidus	98	5	1	3	5	1	2	5	1	3	5	1	2	5	1	2
R psoas (L1)	154	1	9	3	-5	9	2	1	9	4	1	6	3	-4	8	4

(continued)

Table 5.3 *(continued)*

Only a portion of some muscles are represented; for example, the laminae of psoas to L1 is shown rather than the whole psoas.

[a]Combination of 60° flexion, 25° right lateral bending, and 10° counterclockwise twist in one posture; [b]F = flexion; [c]B = lateral bend; [d]T = axial twist

Reprinted, by permission, from S.M. McGill, 1991, "Kinetic potential of the lumbar trunk musculature about three orthogonal orthopaedic axes in extreme positions," *Spine* 19(8): 809-815. ©Lippincott, Williams, and Wilkins.

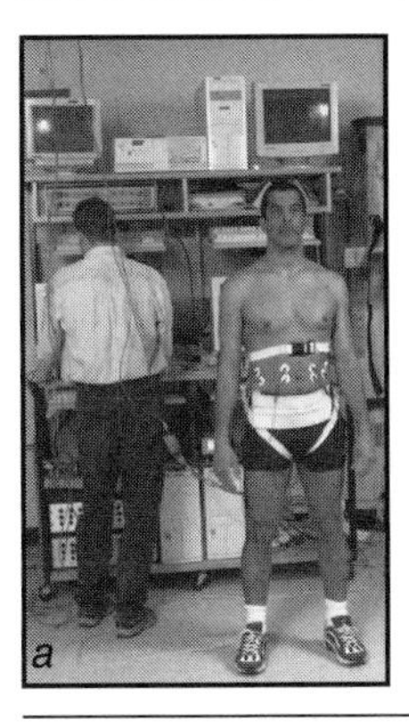

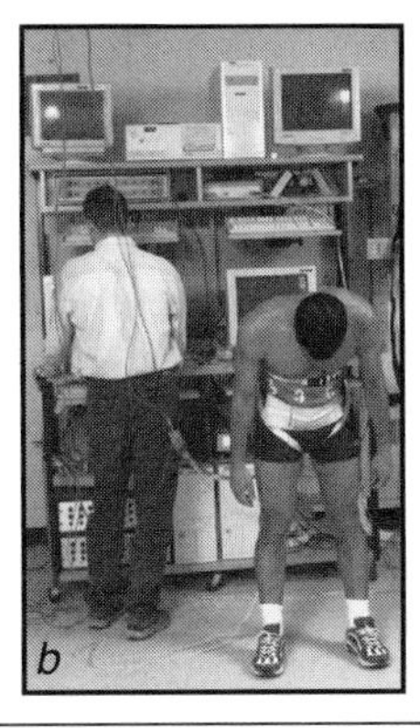

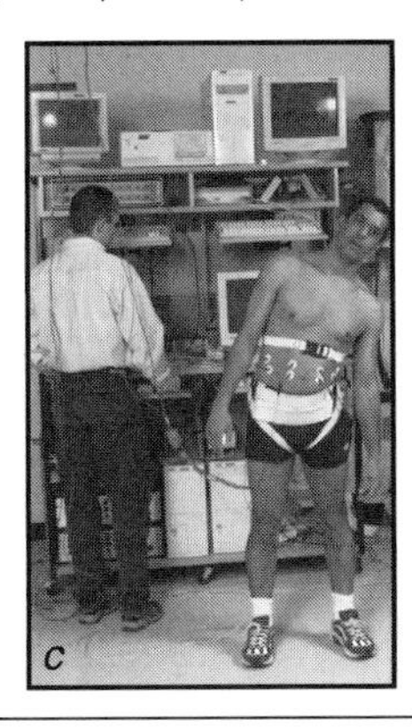

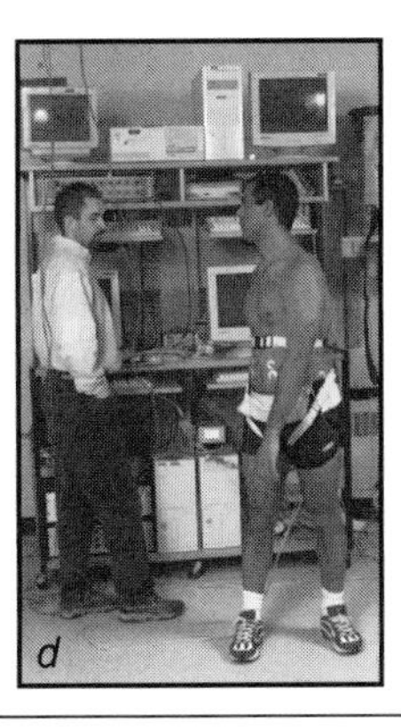

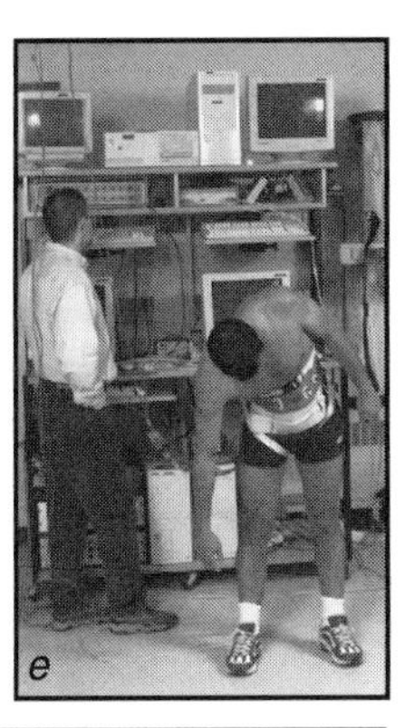

Figure 5.1 A range of extreme postures were chosen to assess muscle length changes (table 4.2) and their potential to produce three dimensional moments (table 5.3). Postures depicted are *(a)* upright standing, *(b)* 60° flexion, *(c)* 25° R lateral bending, *(d)* 10° twist, and *(e)* combined.

CLINICAL RELEVANCE

Lumbo-Pelvic Rhythm

The typical description of torso flexion suggests that the first 60° of torso flexion takes place in the lumbar spine, while any further flexion is accomplished by flexion about the hips (see figure 5.2). Although this notion is very popular in

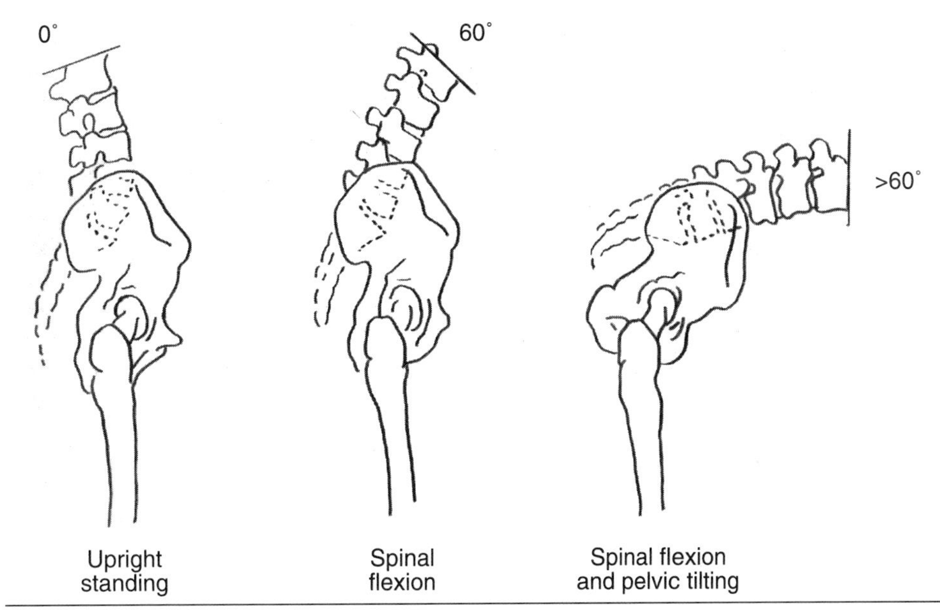

Figure 5.2 The lumbo-pelvic rhythm is the textbook description of how people bend over. The first 60° takes place in the lumbar spine (flexing), while further rotation of the torso is accomplished with rotation about the hip. We have never measured this sequence in anyone—from professional athlete to back patient.

(continued)

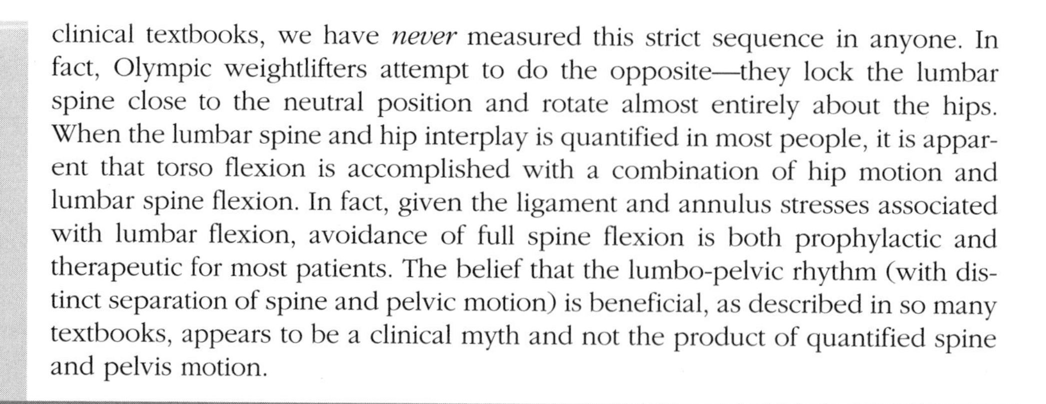
clinical textbooks, we have *never* measured this strict sequence in anyone. In fact, Olympic weightlifters attempt to do the opposite—they lock the lumbar spine close to the neutral position and rotate almost entirely about the hips. When the lumbar spine and hip interplay is quantified in most people, it is apparent that torso flexion is accomplished with a combination of hip motion and lumbar spine flexion. In fact, given the ligament and annulus stresses associated with lumbar flexion, avoidance of full spine flexion is both prophylactic and therapeutic for most patients. The belief that the lumbo-pelvic rhythm (with distinct separation of spine and pelvic motion) is beneficial, as described in so many textbooks, appears to be a clinical myth and not the product of quantified spine and pelvis motion.

Kinetics and Normal Lumbar Spine Mechanics

Interpretation of the function of the anatomical components of the lumbar spine requires

- analysis of their architecture and neural activation of muscles, and
- knowledge of forces in the individual tissues (both active and passive) during a wide variety of tasks.

This information is crucial for understanding how tissue overloading and injuries occur and also for optimizing treatment strategies for specific spine injury. Table 5.4 and figure 5.3 provide activation levels quantified with surface and intramuscular EMG for a variety of torso muscles and over a variety of activities. These will be referred to throughout this book. This section will address several issues and controversies about the functional interpretation of the thoraco-lumbar anatomy. Given the inability of the clinician and scientist to measure individual tissue forces in vivo, the only tenable option is to use sophisticated measurement techniques to collect various biological signals from living subjects and integrate them with sophisticated modeling approaches to estimate tissue loads. In chapter 2, I described briefly the technique we used to assess the various issues in this chapter (the virtual spine).

Standing and Bending Forward

Several studies over the years have shown the flexion/relaxation phenomenon, or the apparent myoelectric silence of the low back extensor muscles during a standing-to-full-flexion maneuver. The hypothesis has been that, as full flexion occurs, either the extensors shut down their neural drive by reflex or the passive tissues simply take the load as they strain under full flexion. A study by McGill and Kippers (1994) using the virtual spine approach described in chapter 2 quantified individual tissue forces, thus adding more insight into the understanding of this task. As one bends forward, the spine flexes and the extensors undergo eccentric contraction. As full flexion is approached, the passive tissues rapidly take over moment production, relieving the muscles of this role and accounting for their myoelectric silence. Figure 5.4 shows the relative contribution of the muscles and the passive tissues (ligaments, disc, and gut) to the reaction moment throughout the movement, while table 5.5 documents the distribution of tissue forces and their moments and joint load consequence. Interestingly, the "relaxation" of the lumbar extensor muscles appeared to occur only in an electrical sense because they generated substantial force elastically during full spine flexion through stretching. Perhaps the term *flexion/relaxation* is inappropriate,

Table 5.4 Subject Averages of EMG Activation Normalized to That Activity Observed During a Maximal Effort (100%)—Mean and (Standard Deviation in Parentheses)

Psoas channels, external oblique, internal oblique, and transverse abdominis are intramuscular electrodes while rectus abdominis, rectus femoris, and erector spinae are surface electrodes.

Task	Psoas 1	Psoas 2	EOi	IOi	TAi	RAs	RFs	ESs
Straight-leg sit-up	15(±12)	24(±7)	44(±9)	15(±15)	11(±9)	48(±18)	16(±10)	4(±3)
Bent-knee sit-up	17(±10)	28(±7)	43("12)	16(±14)	10(±7)	55(±16)	14(±7)	6(±9)
Press-heel sit-up	28(±23)	34(±18)	51(±14)	22(±14)	20(±13)	51(±20)	15(±12)	4(±3)
Bent-knee curl-up	7(±8)	10(±14)	19(±14)	14(±10)	12(±9)	62(±22)	8(±12)	6(±10)
Bent-knee leg raise	24(±15)	25(±8)	22(±7)	8(±9)	7(±6)	32(±20)	8(±5)	6(±8)
Straight-leg raise	35(±20)	33(±8)	26(±9)	9(±8)	6(±4)	37(±24)	23(±12)	7(±11)
Isom. hand-to-knee (left hand → right knee,	16(±16)	16(±8)	68(±14)	30(±28)	28(±19)	69(±18)	8(±7)	6(±4)
right hand → left knee)	56(±28)	58(±16)	53(±12)	48(±23)	44(±18)	74(±25)	42(±29)	5(±4)
Cross curl-up (right shoulder → across,	5(±3)	4(±4)	23(±20)	24(±14)	20(±11)	57(±22)	10(±19)	5(±8)
left shoulder → across)	5(±3)	5(±5)	24(±17)	21(±16)	15(±13)	58(±24)	12(±24)	5(±8)
Isom. side support (left side down)	21(±17)	12(±8)	43(±13)	36(±29)	39(±24)	22(±13)	11(±11)	24(±15)
Dyn. side support (left side down)	26(±18)	13(±5)	44(±16)	42(±24)	44(±33)	41(±20)	9(±7)	29(±17)
Push-up from feet	24(±19)	12(±5)	29(±12)	10(±14)	9(±9)	29(±10)	10(±7)	3(±4)
Push-up from knees	14(±11)	10(±7)	19(±10)	7(±9)	8(±8)	19(±11)	5(±3)	3(±4)
Lift light load (20 kg)	9(±10)	3(±4)	3(±3)	6(±7)	6(±5)	14(±21)	6(±5)	37(±13)
Lift heavy load (50-110 kg)	16(±18)	5(±6)	5(±4)	10(±11)	10(±9)	17(±23)	6(±5)	62(±12)

(continued)

Table 5.4 *(continued)*

Task	Psoas 1	Psoas 2	EOi	IOi	TAi	RAs	RFs	ESs
Symmetric bucket hold, 20 kg (44 lb)	2(±4)	1(±1)	7(±4)	5(±3)	5(±1)	10(±7)	3(±3)	3(±6)
30 kg (66 lb)	3(±4)	1(±1)	9(±5)	6(±4)	6(±1)	10(±8)	3(±3)	4(±7)
40 kg (88 lb)	3(±5)	1(±1)	10(±6)	8(±6)	6(±2)	10(±8)	3(±3)	3(±2)
0 kg	1(±2)	0(±1)	2(±1)	2(±2)	2(±1)	10(±9)	2(±1)	2(±1)
Seated isom. twist CCW	30(±20)	17(±15)	18(±8)	43(±25)	49(±35)	17(±22)	7(±4)	14(±6)
Seated isom. twist CW	23(±20)	11(±8)	52(±13)	15(±11)	18(±19)	13(±10)	9(±10)	13(±8)
Standing hip internal rotation	21(±18)	10(±9)	18(±12)	24(±23)	33(±20)	13(±9)	9(±7)	18(±6)
Standing hip external rotation	27(±20)	22(±19)	17(±13)	21(±19)	31(±17)	13(±8)	19(±11)	17(±9)
Sitting hip internal rotation	19(±15)	21(±18)	36(±31)	30(±30)	31(±29)	18(±8)	20(±19)	12(±8)
Sitting hip external rotation	32(±25)	25(±20)	11(±9)	15(±17)	16(±13)	15(±9)	16(±13)	8(±8)
Sitting upright	12(±7)	7(±5)	3(±6)	3(±3)	4(±2)	17(±9)	4(±2)	5(±8)
Sitting slouched/relaxed	4(±4)	3(±3)	2(±5)	2(±2)	4(±3)	17(±11)	3(±2)	5(±8)
Quiet standing	2(±1)	1(±1)	3(±4)	5(±3)	4(±2)	5(±5)	3(±3)	11(±11)
Standing extended	3(±2)	2(±1)	12(±9)	6(±3)	5(±3)	11(±5)	4(±3)	7(±8)
Standing lateral bend, left	9(±10)	1(±2)	11(±8)	18(±14)	12(±7)	13(±7)	3(±2)	11(±13)
Standing lateral bend, right	6(±5)	1(±2)	19(±18)	18(±14)	25(±20)	14(±9)	3(±1)	8(±8)
Seated lateral bend moving left	2(±3)	1(±1)	21(±19)	7(±7)	7(±11)	13(±8)	4(±3)	6(±7)
Seated lateral bend, moving right	18(±12)	12(±2)	15(±26)	10(±7)	12(±7)	17(±20)	5(±4)	5(±8)
Upright	14(±9)	8(±4)	6(±4)	5(±3)	5(±5)	19(±23)	5(±3)	6(±8)

Reprinted, by permission, from D. Juker, S.M. McGill, S.M. Kropf, T. Steffen, 1998, "Quantitative intramuscular myoelectric activity of lumbar portions of psoas and the abdominal wall during a wide variety of tasks," *Medicine and Science in Sports and Exercise 30(2)*: 301-310. ©Lippincott, Williams, and Wilkins.

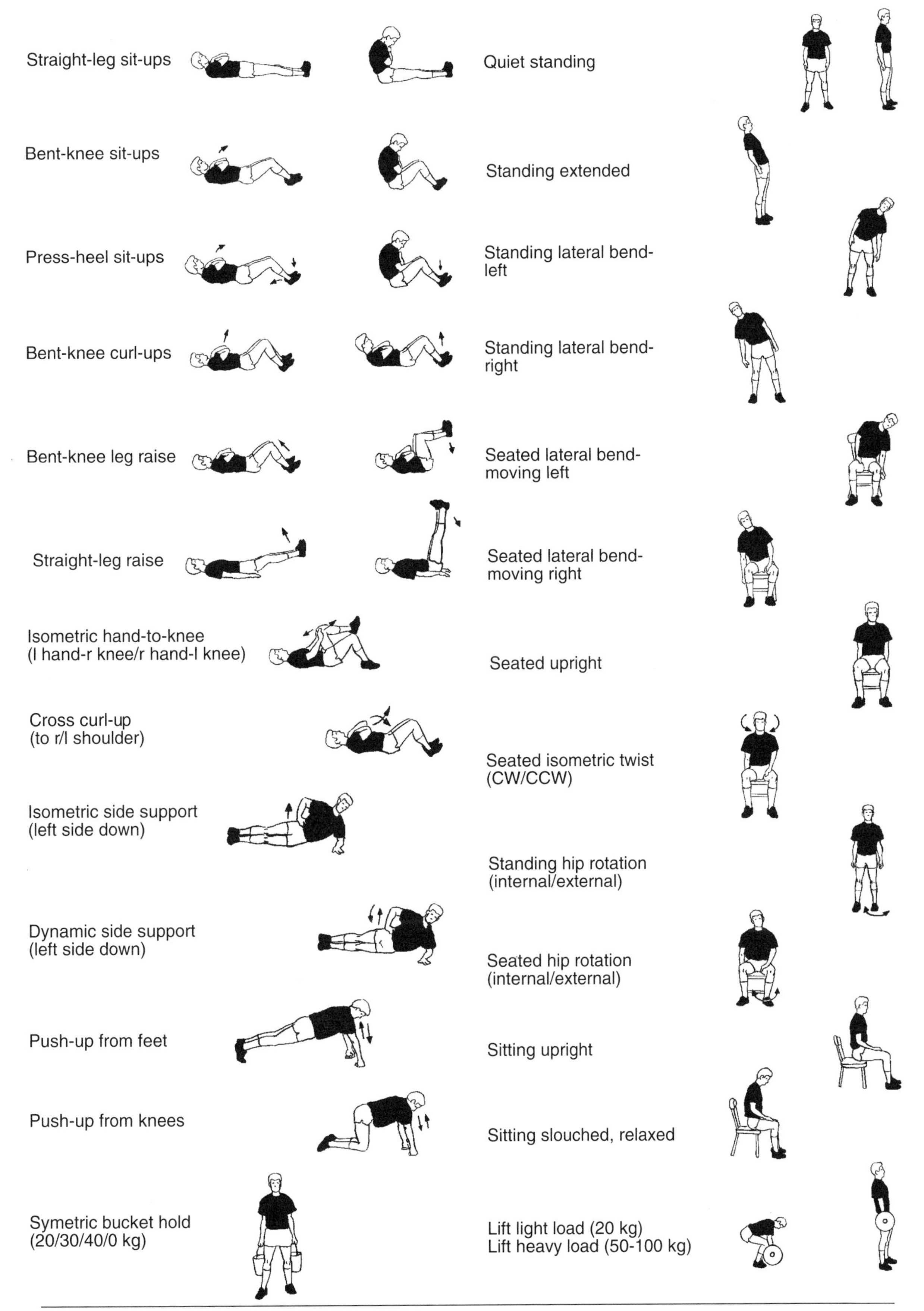

Figure 5.3 Schematic documenting various tasks during which EMG signals were obtained. They are listed in table 5.4.

Reprinted, by permission, from D. Juker, S.M. McGill, S.M. Kropf, T. Steffen, 1998, "Quantitative intramuscular myoelectric activity of lumbar portions of psoas and the abdominal wall during a wide variety of tasks," *Medicine and Science in Sports and Exercise* 30(2): 301-310. © Lippincott, Williams, and Wilkins.

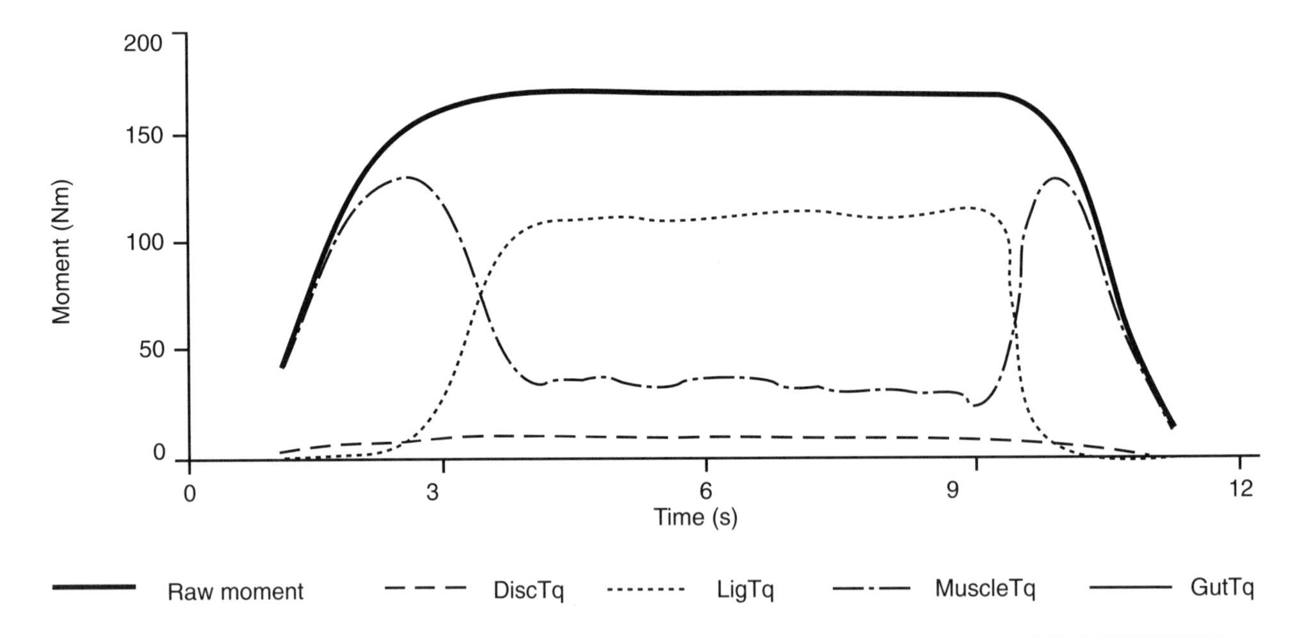

Figure 5.4 During standing to full forward flexion the extensor muscles eccentrically contract but transfer their moment supporting role (MuscleTq) to the passive tissues at full flexion—the disc (DiscTq), the buckled gut (GutTq), and the ligaments (LigTq). Note that some force remains in the muscles with passive stretching.

particularly for those who may be attempting to minimize forces in the muscle in clinical settings. Furthermore, the shear loading is substantial (see chapter 4 for a discussion of the ligament muscle vector directions and the loss of shear support in the extensors with full flexion) and would suggest caution for those with spondylolisthesis or other more subtle shear instabilities. Clearly, straight-leg toe touches or knees-to-chest stretches would cause similar concern.

Loads on the Low Back During Lifting

During lifting, muscle and ligament forces required to support the posture and facilitate movement impose mammoth loads on the spine. This is why lifting technique is so important to reduce low back moment demands and the risk of excessive loading. The following example demonstrates this concept.

A man is lifting 27 kg (59 lb) held in the hands using a squat lift style. This produces an extensor reaction moment in the low back of 450 Nm (332 ft lb). The forces in the various tissues that support this moment impose a compressive load on the lumbar spine of over 7000 N (1568 lb). Table 5.6 details the contributions to the total extension moment and to the forces from the muscular components. These forces and their effects are predicted using the sophisticated modeling approach, which uses biological signals obtained directly from the subject (see chapter 2). It should be noted here that 7000 N (1568 lb) of compression begins to cause damage in very weak spines, although the tolerance of the lumbar spine in an average healthy young man probably approaches 12-15 kN (2688-3360 lb) (Adams and Dolan, 1995). In extreme cases, compressive loads on the spines of competitive weightlifters have safely exceeded 20 kN (4480 lb) (Cholewicki, McGill, and Norman, 1991).

Understanding the individual muscle forces, their contribution to supporting the low back, and their components of compression and shear force that are imposed on the spine is very useful. In this particular example, the lifter avoided full spine flexion

Table 5.5 Individual Muscle and Passive Tissue Forces and Moments During Full Flexion

The moment was 171 Nm (38 Nm by muscle, 113 Nm by ligaments, and 20 Nm by passive tissues, such as the disc, skin, and buckled viscera). The joint compression was 3145 N, and shear was 1026 N.

	Force	Moment (Nm)			Compression	Shear (N)	
	N	Flexion	Lateral	Twist	N	Anteroposterior	Lateral
Muscle							
R rectus abdominis	16	−2	1	1	15	5	−4
L rectus abdominis	16	−2	−1	−1	15	5	4
R external oblique 1	10	−1	1	1	8	7	−3
L external oblique 1	10	−1	−1	−1	8	7	3
R external oblique 2	7	−1	1	0	6	2	−3
L external oblique 2	7	−1	−1	−0	6	2	3
R internal oblique 1	35	0	3	−2	21	−19	20
L internal oblique 1	35	0	−3	2	21	−19	−20
R internal oblique 2	29	−2	2	−3	8	−17	21
L internal oblique 2	29	−2	−2	3	8	−17	−21
R pars lumborum (L1)	21	2	1	0	21	6	2
L pars lumborum (L1)	21	2	−1	−0	21	6	−2
R pars lumborum (L2)	27	2	1	0	26	8	2
L pars lumborum (L2)	27	2	−1	−0	26	8	−2
R pars lumborum (L3)	31	1	1	0	29	−4	6
L pars lumborum (L3)	31	1	−1	−0	29	−4	−6
R pars lumborum (L4)	32	1	1	−0	30	−7	6
L pars lumborum (L4)	32	1	−1	0	30	−7	−6
R iliocostalis lumborum	58	5	4	1	57	14	−1
L iliocostalis lumborum	58	5	−4	−1	57	14	1
R longissimus thoracis	93	7	4	0	91	23	−6
L longissimus thoracis	93	7	−4	−0	91	23	6

(continued)

Table 5.5 *(continued)*

	Force	Moment (Nm)			Compression	Shear (N)	
	N	Flexion	Lateral	Twist	N	Anteroposterior	Lateral
Muscle *(continued)*							
R quadratus lumborum	25	1	2	−0	25	−1	1
L quadratus lumborum	25	1	−2	0	25	−1	−1
R latissimus dorsi (L5)	15	1	1	−0	14	−1	−6
L latissimus dorsi (L5)	15	1	−1	0	14	−1	6
R multifidus 1	28	1	1	1	26	6	9
L multifidus 1	28	1	−1	−1	26	6	−9
R multifidus 2	28	1	1	0	28	6	0
L multifidus 2	28	1	−1	−0	28	6	0
R psoas (L1)	25	1	2	0	24	9	6
L psoas (L1)	25	−1	−2	−0	24	9	−6
R psoas (L2)	25	−1	2	0	24	9	6
L psoas (L2)	25	−0	−2	−0	24	9	−6
R psoas (L3)	25	−0	1	0	24	9	7
L psoas (L3)	25	−0	−1	−0	24	9	−7
R psoas (L4)	25	−0	1	1	24	9	8
L psoas (L4)	25	−0	−1	−1	24	9	−8
Ligament							
Anterior longitudinal	0	0	0	0	0	0	–
Posterior longitudinal	86	2	0	0	261	44	–
Ligamentum flavum	21	1	0	0	21	2	–
R intertransverse	14	0	0	0	13	3	–
L intertransverse	14	0	−0	−0	13	3	–
R articular	74	2	1	1	65	40	–
L articular	74	2	−1	−1	65	40	–

Table 5.5 *(continued)*

	Force	Moment (Nm)			Compression	Shear (N)	
	N	Flexion	Lateral	Twist	N	Anteroposterior	Lateral
Ligament *(continued)*							
R articular 2	103	3	2	2	84	–3	–
L articular 2	103	3	–2	–2	84	–3	–
Interspinous 1	301	18	0	0	273	142	–
Interspinous 2	345	14	0	0	233	268	–
Interspinous 3	298	10	0	0	194	238	–
Supraspinous	592	41	0	0	591	79	–
R lumbodorsal fascia	122	8	1	–0	109	–1	–
L lumbodorsal fascia	122	8	–1	0	109	–1	–
Passive							
Disc	–	9	0	0	–	–	–
Gut, etc.	–	11	0	0	–	–	–

by flexing at the hip, minimizing ligament and other passive tissue tension and relegating the moment generation responsibility to the musculature. An example in which the spine is flexed is presented later in this chapter. As described in chapter 4, the pars thoracis extensors are very effective lumbar spine extensors, given their large moment arms. Also, since the lifter's upper body is flexed, large reaction shear forces on the spine are produced (the rib cage is trying to shear forward on the pelvis). These shear forces are supported to a very large degree by the pars lumborum extensor muscles. Furthermore, the abdominal muscles are activated but produce a negligible contribution to moment/posture support. Why are they active? These muscles are activated to stabilize the spinal column, although this mild abdominal activity imposes a compression penalty to the spine. A more robust explanation of stabilizing mechanics is described in chapter 6.

The preceding example demonstrates how diffusely the forces are distributed and illustrates how proper clinical interpretation requires anatomically detailed free body diagrams that represent reality (such as those incorporated into the virtual spine model). I believe that oversimplified free body diagrams have overlooked the important mechanical compressive and especially shear components of muscular force. This has compromised the assessment of injury mechanisms and the formulation of optimal therapeutic exercise.

Table 5.6 Musculature Components for Moment Generation of 450 Nm During Peak Loading for a Squat Lift of 27 kg (59.5 lb)

Muscle	Force (N)	Moment (Nm)	Compression (N)	Shear (N)
Rectus abdominis	25	-2	24	5
External oblique 1	45	1	39	24
External oblique 2	43	-2	30	31
Internal oblique 1	14	1	14	-2
Internal oblique 2	23	-1	17	-16
Longissimus thoracis pars lumborum L4	862	35	744	-436
Longissimus thoracis pars lumborum L3	1514	93	1422	-518
Longissimus thoracis pars lumborum L2	1342	121	1342	0
Longissimus thoracis pars lumborum L1	1302	110	1302	0
Iliocostalis lumborum pars thoracis	369	31	369	0
Longissimus thoracis pars thoracis	295	25	295	0
Quadratus lumborum	393	16	386	74
Latissimus dorsi L5	112	6	79	-2
Multifidus 1	136	8	134	18
Multifidus 2	226	8	189	124
Psoas L1	26	0	23	12
Psoas L2	28	0	27	8
Psoas L3	28	1	27	6
Psoas L4	28	1	27	5

Negative moments correspond to flexion, while negative shear corresponds to L4 shearing posteriorly on L5.

Loads on the Low Back During Walking

Thousands of low-level loading cycles are endured by the spine every day during walking. While the small loads in the low back during walking suggest it is a safe and tolerable activity, clinicians have found that walking provides relief to some individuals but is painful to others. Recent work has suggested that walking speed affects spine mechanics and may account for these individual differences. During walking, the compressive loads on the lumbar spine of approximately 2.5 times body weight,

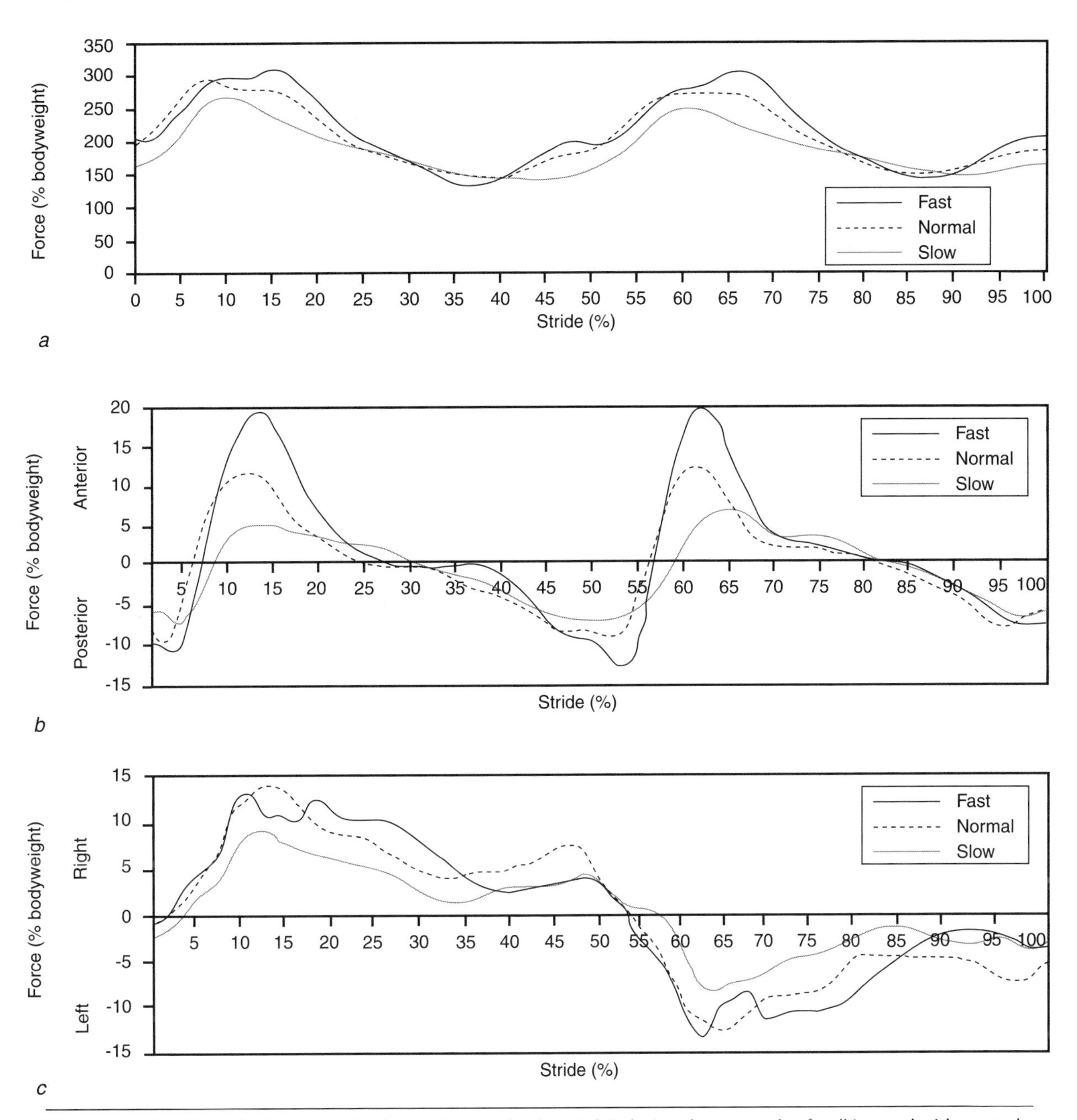

Figure 5.5 Loads on the lumbar spine (normalized to body weight) during three speeds of walking and with normal arm swing. The curves are normalized for one stride (right heel contact to right heel contact). *(a)* L4/L5 compression, *(b)* anterior/posterior shear forces in which positive indicates anterior shear of the superior vertebrae, *(c)* lateral shear force in which positive indicates right shear of the superior vertebrae.

Reprinted from *Clinical Biomechanics,* 14, J.P. Callaghan, A.E. Patla, and S.M. McGill, "Low back three-dimensional joint forces, kinematics and kinetics during walking," 203-216, 1999, with permission from Elsevier Science.

together with the very modest shear forces, are well below any known in-vitro failure load (see figure 5.5). Strolling reduces spine motion and produces almost static loading of tissues, however, while faster walking, with arms swinging, causes cyclic loading of tissues (Callaghan, Patla, and McGill, 1999) (see figure 5.6). This change in motion may begin to explain the relief experienced by some. Arm swinging while walking faster, with all other factors controlled, results in lower lumbar spine torques, muscle activity, and loading (see figure 5.7). In fact, we have observed up to 10% reduction in spine loads from arm swinging in some individuals. This may be because swinging the arms

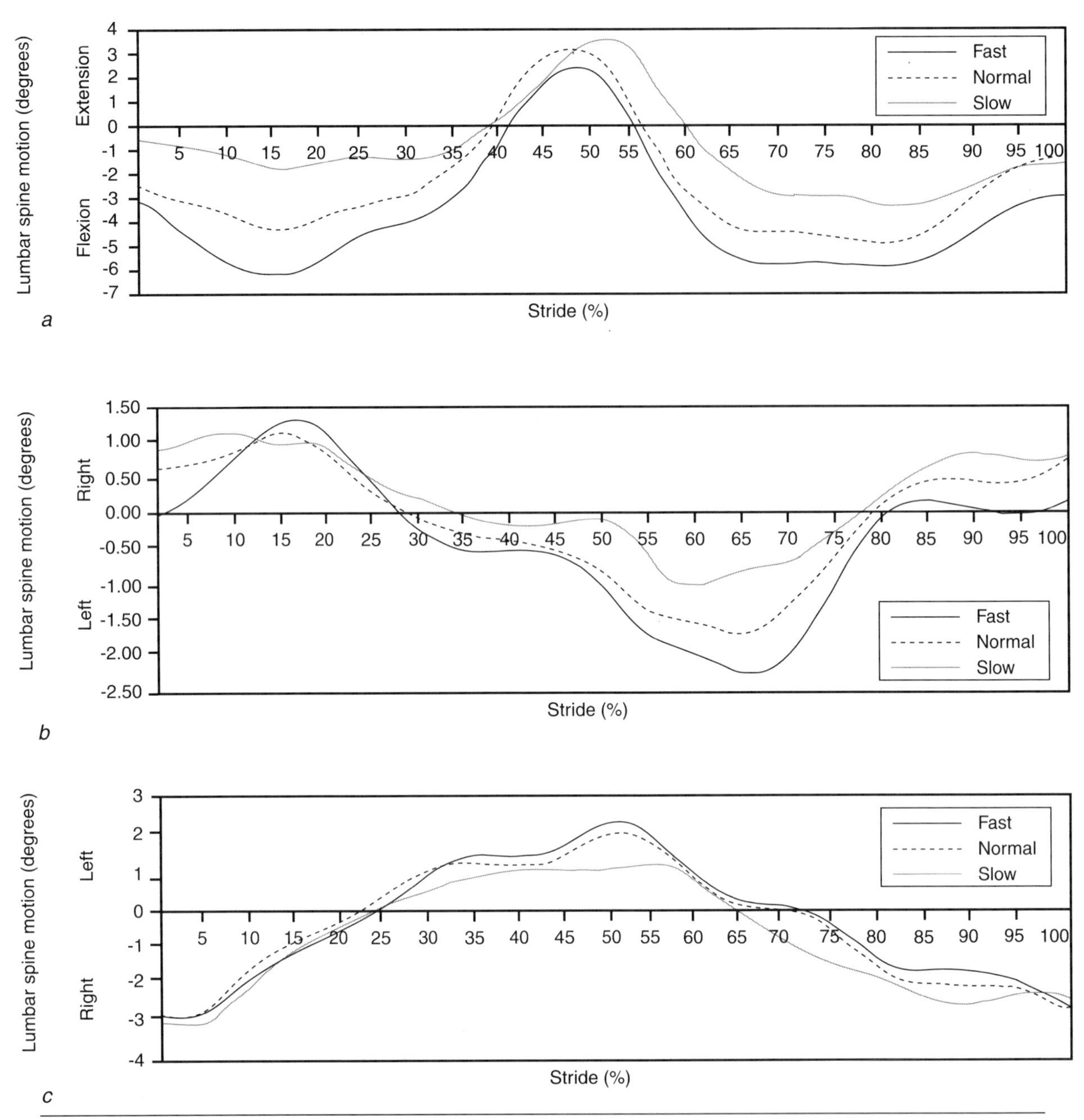

Figure 5.6 Lumbar motion during three speeds of walking for one normal subject (normalized to RHC to RHC): *(a)* Lumbar flexion/extension in which extension is positive, *(b)* lateral bend in which positive indicates bend to the right, *(c)* axial twist in which positive indicates the upper body twisting to the right.

facilitates efficient storage and recovery of elastic energy, reducing the need for concentric muscle contraction and the upper body accelerations associated with each step. Also interesting is the fact that fast walking has been shown to be a positive cofactor in prevention of, and more successful recovery from, low back troubles (Nutter, 1988).

CLINICAL RELEVANCE

Fast Walking

Fast walking is generally therapeutic (Nutter, 1988). Several mechanisms appear to account for this: reciprocal muscle activation and tissue load sharing, gentle motion, and reduced spine loads with energy conservancy from arm swing. In contrast, these benefits do not occur during slow walking or "mall strolling," which exacerbates symptoms in many because of the static loading that results.

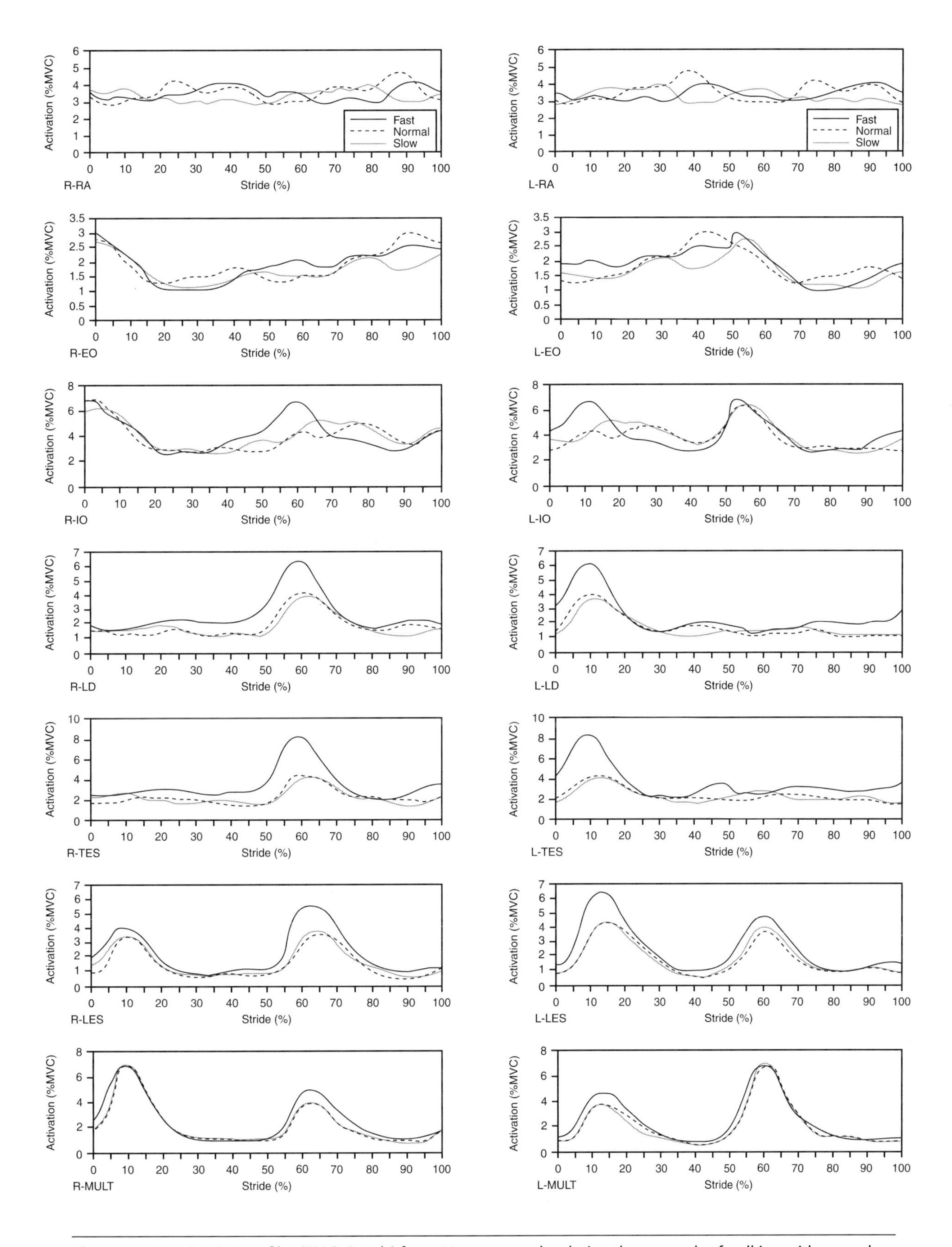

Figure 5.7 Activation profiles (EMG signals) from 14 torso muscles during three speeds of walking with normal arm swing and normalized to RHC to RHC. The muscle pairs are (RA) rectus abdominis, (EO) external oblique, (IO) internal oblique, (LD) latissimus dorsi, (TES) thoracic erector spinae, (LES) lumbar erector spinae, and (MULT) multifidus.

Loads on the Low Back During Sitting

Nachemson (1966), using intradiscal pressure measurements, documented the higher loads on the discs in various sitting postures compared to the standing posture. Normal sitting causes flexion in the lumbar spine, and people, if left alone, generally sit in a variety of flexed postures (Callaghan and McGill, 2001b). Sitting generally involves lower abdominal wall activity (particularly the deep abdominals) compared to standing and generally higher extensor activity with unsupported sitting (see Callaghan, Patla, and McGill, 1999 for walking and Callaghan and McGill, 2001b for sitting). Sitting slouched minimizes muscle activity, while sitting more upright requires higher activation of the psoas and the extensors (Juker, McGill, Kropf et al., 1998). Full flexion increases disc annulus stresses; this posture has produced disc herniations in the lab (e.g., Wilder, Pope, and Frymoyer, 1988). In fact, Kelsey (1975) discovered a specific link between prolonged sitting and the incidence of herniation. More upright sitting postures, and the concomitant psoas and other muscle activation, impose additional compressive loads on the spine. Changing lumbar postures causes a migration of the loads from one tissue to another. Callaghan and McGill (2001b) suggested that no single, ideal sitting posture exists; rather, a variable posture is recommended as a strategy to minimize the risk of tissue overload.

CLINICAL RELEVANCE

Sitting Posture

Despite the myth perpetuated in many ergonomic guidelines regarding a single "ideal" posture for sitting, the ideal sitting posture is one that continually changes, thus preventing any single tissue from accumulating too much strain. Sitting strategies are presented in chapter 8.

Loads on the Low Back During Flexion Exercises

Very few studies (only our own that we are aware of) have quantified the tissue loading on the low back tissues during various types of torso flexion exercises, although some have measured the EMG activity of selected muscles (e.g.. Flint, 1965; Halpern and Bleck, 1979; Jette, Sidney, and Cicutti, 1984). This type of information alone can provide insight into relative muscle challenge, but it is restrictive for guiding exercise prescription decisions because the resultant spine load is unknown. The goal is to challenge muscle at appropriate levels but in a way that spares the spine. Too many exercises are prescribed for back sufferers that exceed the tolerance of their compromised tissues. In fact, I believe that many commonly prescribed flexion exercises result in so much spine compression that they will ensure that the person remains a patient. For example, the traditional sit-up imposes approximately 3300 N (about 730 lb) of compression on the spine (Axler and McGill, 1997). (Figure 5.8 illustrates psoas and abdominal muscle activation levels in a variety of flexion tasks, while figure 5.9 illustrates activation patterns with bent-knee sit-ups, and figure 5.10 illustrates activation patterns with bent-knee curl-ups.) Note that muscle activation levels are expressed in normalized units (% MVC). This means that the activity is expressed as a percentage of what would be observed during a maximal voluntary contraction (MVC), thus quantifying activity in a physiological and functional context. Further, the spine is very flexed during the period of this load (McGill, 1998). The National Institute of Occupational Safety and Health (NIOSH) (1981) has set the action limit for low back compression at 3300 N; repetitive loading above this level is linked with higher injury rates in workers, yet this is imposed on the spine with each repetition of the sit-up! Table 5.7 shows the quantification of a variety of flexion exercises.

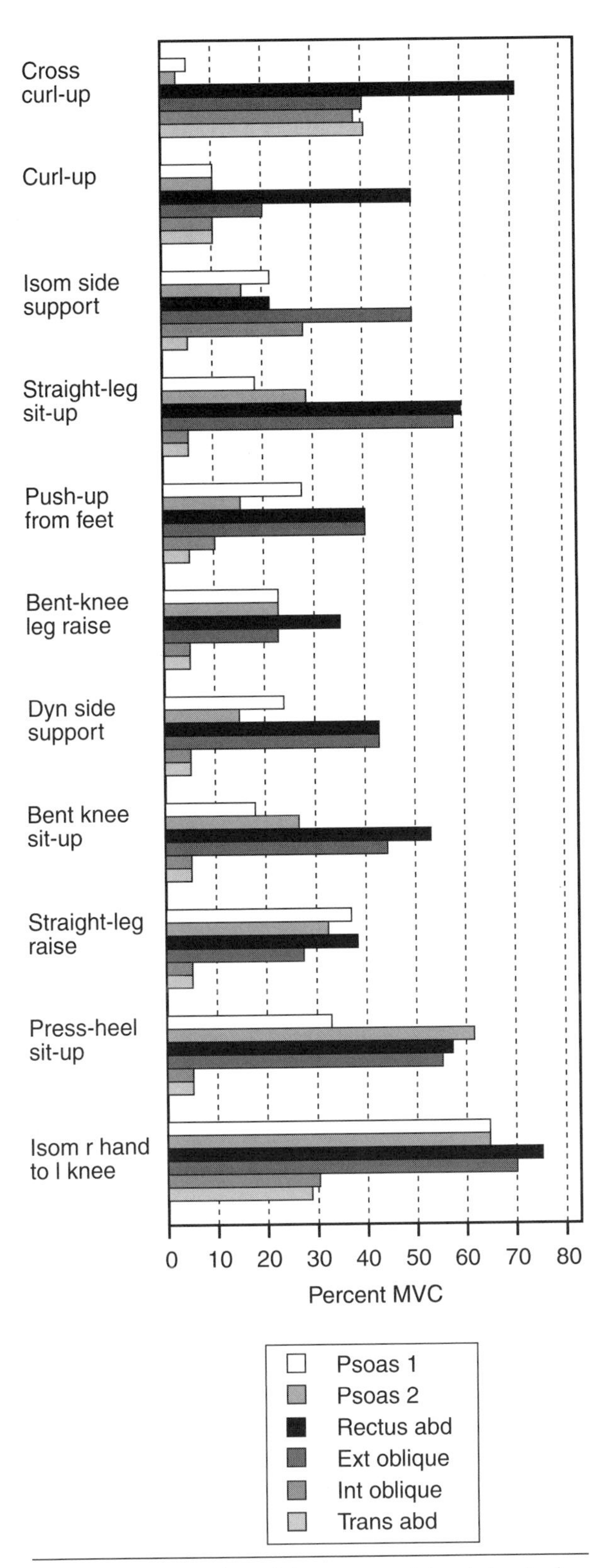

Figure 5.8 Activation of the psoas and the abdominal muscles in a variety of flexion tasks from a group of highly trained subjects (five men and three women).

Many recommend performing sit-ups with the knees bent, the theory being that the psoas is realigned to reduce compressive loading, or perhaps the psoas is shortened on the length–tension relationship so that the resulting forces are reduced. After examining both of these ideas, we found them to be untenable. We placed a group of women who were small enough to fit into an MRI scanner and varied the knee and hip angles while they were supine (Santaguida and McGill, 1995). The psoas did not change its line of action, nor could it since it is attached to each vertebral body and transverse process (as the lumbar spine increases lordosis, the psoas follows this curve). The psoas does not change its role from a flexor to an extensor as a function of lordosis—this interpretation error occurred from models in which the psoas was represented as a straight line puller. In fact, the psoas follows the lordotic curve as the lumbar spine flexes and extends. Further, it is true that the psoas is shortened with hip flexion, but its activation level is higher during bent-knee sit-ups (Juker, McGill, Kropf et al., 1998), not lower as has been previously thought. This is because the hip flexion torque must come from somewhere, and the shortened psoas must contract to higher levels of activation given its compromised length. Given that the sit-up imposes such a large compression load on the spine, regardless of the leg being bent or straight, the issue is not which type of sit-up should be recommended. Rather, sit-ups should not be performed at all by most people. Far better ways exist to preserve the abdominal muscle challenge while imposing lower spine loads. Those who are training for health never need to perform a sit-up; those training for performance may get better results by judiciously incorporating them into their routine.

While part III of this book (Low Back Rehabilitation) offers specific preferred exercises and challenges to specific muscles, a few flexion exercises will be reviewed here. First, hanging with the arms from an overhead bar and flexing the hips to raise the legs is often thought to impose low spine loads because the body is hanging in tension—not compression. This is faulty logic. This hanging exercise generates well over 100 Nm of abdominal torque (Axler and McGill, 1997). This produces almost maximal abdominal activation, which in turn imposes compressive forces on the spine (see table 5.7). Similar activation levels can be achieved with the side bridge (shown later and discussed in detail) with lower spine loads.

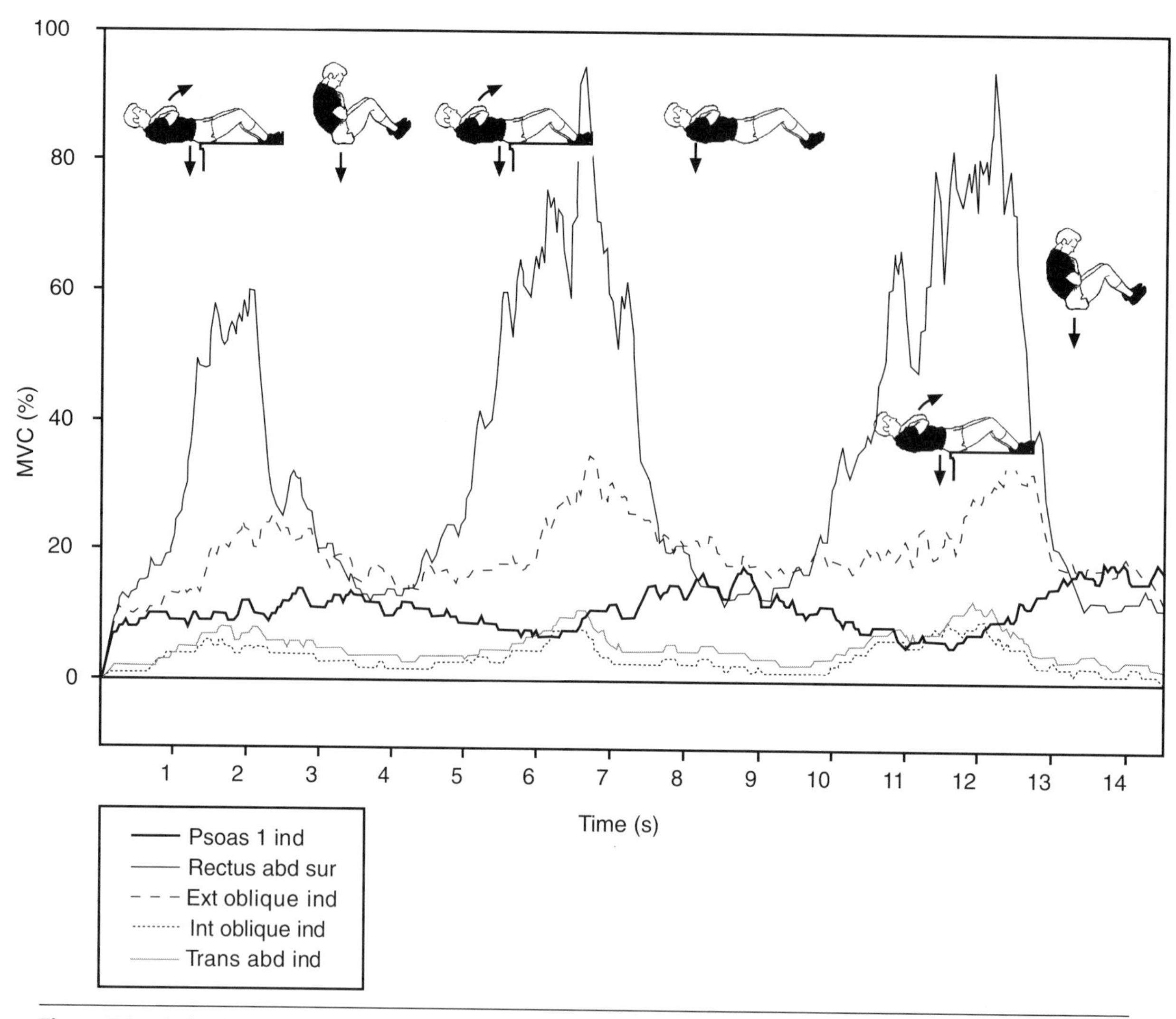

Figure 5.9 Activation-time histories of the same subjects as in figure 5.8 performing a bent-knee sit-up. Surface and indwelling electrodes are indicated.

Having stated this, those not interested in sparing their back and who are training with performance objectives may benefit from the high psoas challenge, together with rectus abdominis and oblique activity. Clearly, the curl-up primarily targets the rectus (both upper and lower), and generally other exercises should be performed to train the obliques. Some have suggested a twisting curl-up to engage the obliques, but this results in a poor ratio of oblique muscle challenge to spine compression compared to the side bridge exercise (Axler and McGill, 1997)—making the side bridge a preferred exercise.

Loads on the Low Back During Extension Exercises

As with the flexion exercises discussed in the previous paragraph, plenty of EMG-based studies have explored extension exercises, but only one attempted to quantify the resulting tissue loads. Exercise prescriptions will not be successful if the spine loading is not constrained for bad backs. Using the virtual spine approach, Callaghan,

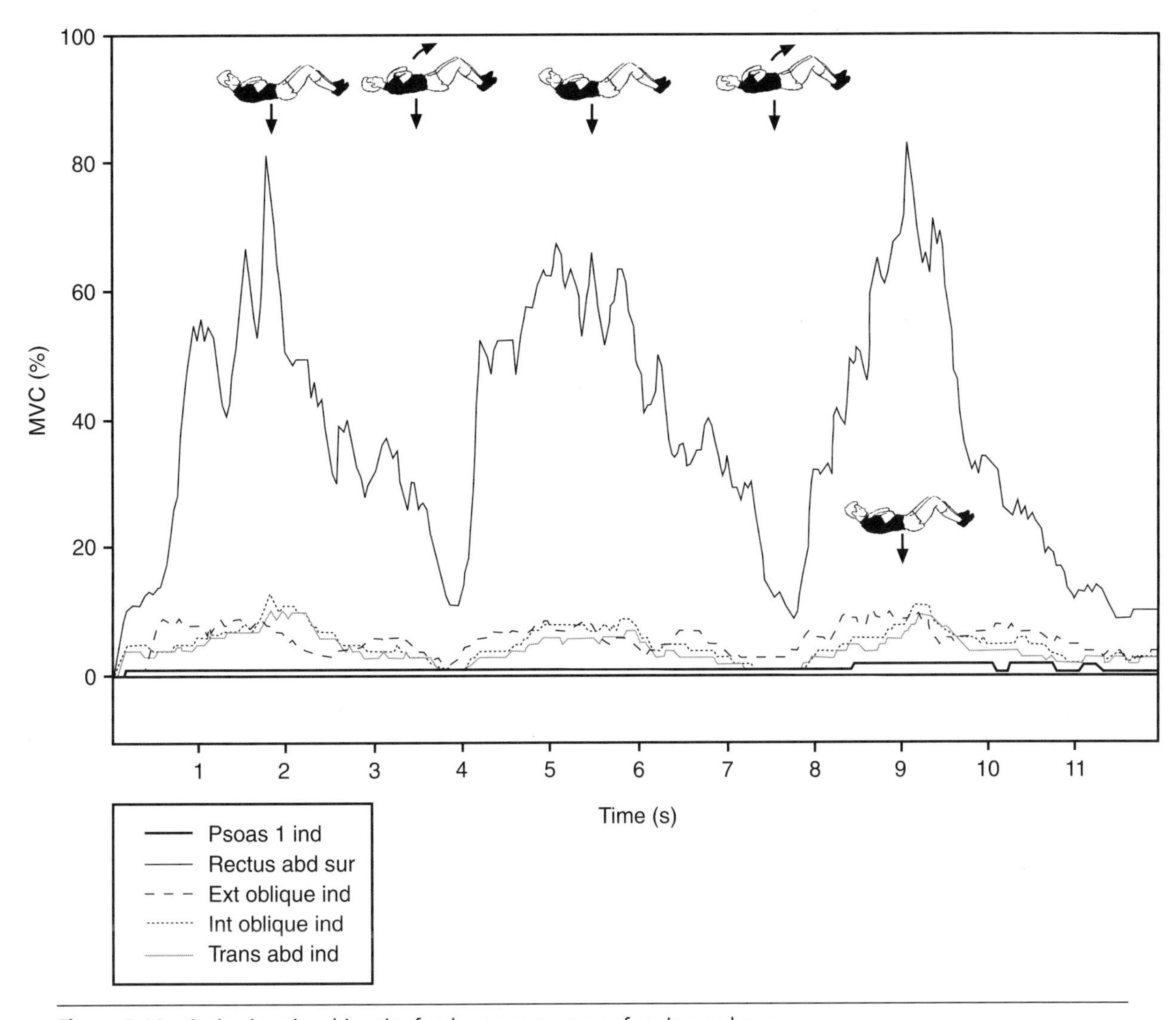

Figure 5.10 Activation-time histories for the same group performing curl-ups.

Gunning, and McGill (1998) attempted to rank extension exercises on the muscle challenge, the resultant spine load, and their optimal ratio. The key to preserving a therapeutic muscle activation level while minimizing the spine load is to activate only one side of the spine musculature at a time. The previous muscle anatomy section in chapter 4 describes the functional separation of the thoracic and lumbar portions of the longissimus and iliocostalis. For the purposes of this discussion, we can think of the extensors in four sections—right and left thoracic portions and right and left lumbar portions. The common extension task of performing torso extension with the legs braced and the cantilevered upper body extending over the end of a bench or Roman chair (figure 5.11a) activates all four extensor groups and typically imposes over 4000 N (about 890 lb) of compression on the spine. Even worse is the commonly prescribed back extension task in clinics, in which the patient lies prone and extends the legs and outstretched arms; this again activates all four extensor sections but imposes up to 6000 N (over 1300 lb) on a hyperextended spine (figure 5.11b). This is not justifiable for any patient! Several variations of exercise technique can preserve activation

Table 5.7 Low Back Moment, Abdominal Muscle Activity, and Lumbar Compressive Load During Several Types of Abdominal Exercises

		Muscle activation		
	Moment (Nm)	Rectus abdominis (% MVC)	External oblique	Compression (N)
Straight-leg sit-up	148	121	70	3506
Bent-leg sit-up	154	103	70	3350
Curl-up, feet anchored	92	87	45	2009
Curl-up, feet free	81	67	38	1991
Quarter sit-up	114	78	42	2392
Straight-leg raise	102	57	35	2525
Bent-leg raise	82	35	24	1767
Cross-knee curl-up	112	89	67	2964
Hanging, straight leg	107	112	90	2805
Hanging, bent leg	84	78	64	3313
Isometric side bridge	72	48	50	2585

MVC contractions were isometric. Activation values higher than 100% are often seen during dynamic exercise.

in portions of the extensors and greatly spare the spine of high load. For example, kneeling on all fours and extending one leg at the hip generally activates one side of the lumbar extensors to over 20% of maximum and imposes only 2000 N of compression (figure 5.11c). Performing the birddog, in which the opposite arm is extended at the shoulder while the leg is raised (figure 5.11d), adds activity to one side of the thoracic extensors (generally around 30-40% of maximum) and contains the spine load to about 3000 N. In addition, the special techniques shown for this exercise in the exercise prescription chapter (chapter 13) attempt to enhance the motor control system to groove stabilizing patterns. For data describing these exercises, see table 5.8.

Dubious Lifting Mechanisms

In the 1950s and 1960s spine biomechanists faced a paradox. The simple spine models of the day predicted that the spine would be crushed to the point of injury during certain lifting tasks, yet when people performed those tasks, they walked away uninjured. This motivated several research groups to theorize about mechanisms that could unload compressive stresses from the spine. Researchers proposed three major mechanisms: the intra-abdominal pressure mechanism, the lumbodorsal facia mechanism, and the hydraulic amplifier. Although none has survived scrutiny, clinical vestiges still remain. Nonetheless, some components provided insight for subsequent study and led to the understanding that we have today. For this reason they will be reviewed briefly here.

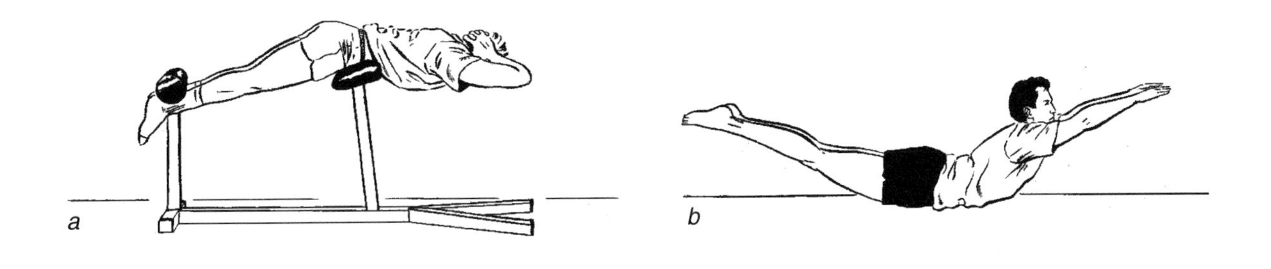

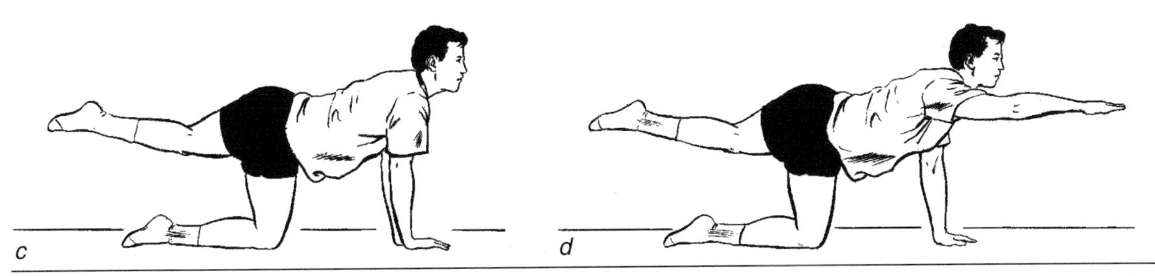

Figure 5.11 Specific extension exercises quantified for muscle activation and the resultant spine load (shown in table 5.8): *(a)* trunk extension, *(b)* prone leg and trunk extension, *(c)* single-leg extension, and *(d)* single-leg and contralateral arm extension (birddog).

Table 5.8 Mean Activation Levels (± 1 SD) of 14 EMG Channels for 13 Subjects Performing a Variety of Extensor-Dominant Exercises (Expressed As a Percentage of MVC)

	Extension						
Electromyographic channel[a]	**Right leg**	**Left leg**	**Right leg and left arm**	**Left leg and right arm**	**Trunk and legs**	**Trunk**	**Calibration posture**
Right RA							
X	3.3	2.7	4.0	3.5	4.7	3.1	1.4
SD	2.4	1.9	2.0	2.0	2.2	1.8	1.0
Right EO							
X	8.4	4.9	16.2	5.2	4.3	3.7	1.0
SD	4.9	1.5	6.0	2.3	2.5	1.7	0.6
Right IO							
X	12.0	8.2	15.6	12.0	12.1	12.7	1.9
SD	6.8	2.5	8.2	4.2	10.1	10.8	1.2
Right LD							
X	8.1	5.8	12.0	12.5	11.2	6.5	5.9
SD	5.4	3.5	9.6	6.2	4.3	4.0	8.5
Right TES							
X	5.7	13.7	11.5	46.8	66.1	45.4	21.0
SD	2.0	7.5	6.6	29.3	18.8	10.6	9.0

(continued)

Table 5.8 *(continued)*

Electromyographic channel[a]	Extension						Calibration posture
	Right leg	Left leg	Right leg and left arm	Left leg and right arm	Trunk and legs	Trunk	
Right LES							
X	19.7	11.7	28.4	19.4	59.2	57.8	21.3
SD	9.1	4.9	10.2	11.0	11.7	8.5	4.6
Right MF							
X	21.9	10.8	31.5	16.1	51.9	47.5	16.4
SD	6.3	6.0	8.2	12.0	14.7	12.3	5.6
Left RA							
X	4.3	3.6	4.4	4.2	6.5	3.7	2.2
SD	3.4	3.6	3.8	3.9	3.4	2.4	2.1
Left EO							
X	5.4	9.0	6.2	15.9	6.3	5.2	1.8
SD	2.0	3.8	2.5	6.6	3.2	5.2	1.0
Left IO							
X	16.0	11.3	22.6	15.2	11.0	12.5	1.6
SD	8.6	7.0	9.2	6.7	5.9	6.1	1.3
Left LD							
X	4.5	5.0	10.7	6.2	9.2	5.1	6.1
SD	4.3	4.5	18.2	4.4	5.1	4.1	8.5
Left TES							
X	15.0	4.5	42.9	10.5	63.6	41.6	21.2
SD	7.5	2.0	20.5	5.9	22.7	10.0	9.8
Left LES							
X	11.3	16.8	19.5	25.5	56.8	57.0	23.3
SD	6.6	4.5	7.4	7.3	14.5	14.7	8.4
Left MF							
X	11.9	22.3	16.6	33.8	57.3	53.3	18.7
SD	7.0	6.1	7.2	6.7	11.4	12.0	4.3

[a]Electromyographic channel: RA = rectus abdominis muscle, EO = external oblique muscle, IO = internal oblique muscle, LD = latissimus dorsi muscle, TES = thoracic erector spinae muscle, LES = lumbar erector spinae muscle, MF = multifidus muscle. Calibration posture: standing, trunk flexed 60°, neutral lumbar posture, 10 kg (22 lb) held in hands with arms hanging straight down.

Intra-Abdominal Pressure

Does intra-abdominal pressure (IAP) play an important role in the support of the lumbar spine, especially during strenuous lifting, as has been claimed for many years? Anatomical accuracy in representation of the involved tissues has been influential in this debate. Further, research on lifting mechanics has formed a cornerstone for the prescription of abdominal belts for industrial workers and has motivated the prescription of abdominal strengthening programs. Many researchers have advocated the use of intra-abdominal pressure as a mechanism to directly reduce lumbar spine compression (Bearn, 1961; Thomson, 1988). However, some researchers believe that the role of IAP in reducing spinal loads has been overemphasized (Grew, 1980; Krag et al., 1986).

Anatomical Consistency in Examining the Role of Intra-Abdominal Pressure

Morris, Lucas, and Bresler (1961) first operationalized the mechanics of the original proposal into a model and described it as follows. Pressurizing the abdomen by closing the glottis and bearing down during an exertion exerts hydraulic force down on the pelvic floor and up on the diaphragm, creating a tensile effect over the lumbar spine or at least alleviating some of the compression. Missing in the early calculations of this hydraulic potential was the full acknowledgment of the necessary abdominal activity (contacting the abdominal wall imposes extra compression on the spine). But evaluation of the trade-off between the extra abdominal muscle compression and the hydraulic relief depended on the geometrical assumptions made. Some of these assumptions appear to be outside of biological reality. In fact, experimental evidence suggests that somehow, in the process of building up IAP, the net compressive load on the spine is increased! Krag and coworkers (1986) observed increased low back EMG activity with increased IAP during voluntary Valsalva maneuvers. Nachemson and Morris (1964) and Nachemson, Andersson, and Schultz (1986) showed an increase in intradiscal pressure during a Valsalva maneuver, indicating a net increase in spine compression with an increase in IAP, presumably a result of abdominal wall musculature activity.

In our own investigation, in which we used our virtual spine model, we noted that net spine compression was increased from the necessary concomitant abdominal activity to increase IAP. Furthermore, the size of the cross-sectional area of the diaphragm and the moment arm used to estimate force and moment at the lower lumbar levels, produced by IAP, have a major effect on conclusions reached about the role of IAP (McGill and Norman, 1987). The diaphragm surface area was taken as 243 cm^2, and the centroid of this area was placed 3.8 cm anterior to the center of the T12 disc (compare these values with those used in other studies: 511 cm^2 for the pelvic floor, 465 cm^2 for the diaphragm, and moment arm distances of up to 11.4 cm, which is outside the chest in most people). During squat lifts, the net effect of the involvement of the abdominal musculature and IAP seems to be to increase compression rather than alleviate joint load. (A detailed description and analysis of the forces are in McGill and Norman, 1987). This theoretical finding agrees with the experimental evidence of Krag and colleagues (1986), who used EMG to evaluate the effect of reducing the need for the extensors to contract (they didn't), and of Nachemson and colleagues (1986), who documented increased intradiscal pressure with an increase in IAP.

Role of IAP During Lifting

The generation of appreciable IAP during load-handling tasks is well documented. The role of IAP is not. Farfan (1973) suggested that IAP creates a pressurized visceral cavity to maintain the hooplike geometry of the abdominals. In recent work in which they measured the distance of the abdominals to the spine (moment arms), McGill,

Juker, and Axler (1996) were unable to confirm substantial changes in abdominal geometry when activated in a standing posture. However, the compression penalty of abdominal activity cannot be discounted. The spine appears to be well suited to sustain increased compression loads if intrinsic stability is increased. An unstable spine buckles under extremely low compressive loads (e.g., approximately 20 N, or about 5 lb) (Lucas and Bresler, 1961). The geometry of the spinal musculature suggests that individual components exert lateral and anterior/posterior forces on the spine that can be thought of as guy wires on a mast to prevent bending and compressive buckling (Cholewicki and McGill, 1996). As well, activated abdominals create a rigid cylinder of the trunk, resulting in a stiffer structure. Recently, both Cholewicki and colleagues (1999) and Grenier and colleagues (in press) documented increased torso stiffness during elevated IAP even when accounting for similar abdominal wall contraction levels. Thus, although the increased IAP commonly observed during lifting and in people experiencing back pain does not have a direct role in reducing spinal compression or in adding to the extensor moment, it does stiffen the trunk and prevent tissue strain or failure from buckling.

Lumbodorsal Fascia

Recent studies have attributed various mechanical roles to the lumbodorsal fascia (LDF). In particular, some have suggested that the LDF reduces spine loads—solving the paradox noted earlier. In fact, some have recommended lifting postures based on various interpretations of the mechanics of the LDF. Gracovetsky, Farfan, and Lamy (1981) originally suggested that lateral forces generated by internal oblique and transverse abdominis are transmitted to the LDF via their attachments to the lateral border. They also claimed that the fascia could support substantial extensor moments. Further, lateral tension from abdominal wall attachments was hypothesized to increase longitudinal tension from **Poisson's effect**, causing the posterior spinous processes to move together resulting in lumbar extension. This sequence of events formed an attractive proposition because the LDF has the largest moment arm of all extensor tissues. As a result, any extensor forces within the LDF would impose the smallest compressive penalty to vertebral components of the spine.

Three independent studies, however, examined the mechanical role of the LDF and collectively questioned the idea that the LDF could support substantial extensor moments (Macintosh and Bogduk, 1987; McGill and Norman, 1988; Tesh, Dunn, and Evans, 1987). As previously noted, regardless of the choice of LDF activation strategy, the LDF contribution to the restorative extension moment was negligible compared with the much larger low back reaction moment required to support a load in the hands. Its function may be that of an extensor muscle retinaculum (Bogduk and Macintosh, 1984). Hukins, Aspden, and Hickey (1990) proposed on theoretical grounds that the LDF acts to increase the force per unit of cross-sectional area that muscle can produce by up to 30%. They suggested that this increase in force is achieved by constraining bulging of the muscles when they shorten. This contention remains to be proven. Tesh, Dunn, and Evans (1987) suggested that the LDF may be important for supporting lateral bending. Furthermore, there is no question that the LDF is involved in enhancing stability of the lumbar column (see chapter 6). No doubt, complete assessment of these notions will be pursued in the future.

Hydraulic Amplifier

The final mechanism hypothesized to unload compressive stresses from the spine was the hydraulic amplifier. This hybrid mechanism depends on three notions. First, the elevated IAP preserves the hooplike geometry of the abdominal wall during exertion.

The IAP must also exert hydraulic pressure posteriorly over the spine and presumably through to the underside of the lumbodorsal fascia. Finally, as the extensor muscle mass contacts, it was proposed to "bulk" upon shortening, again increasing the hydraulic pressure under the fascia. The biomechanical attraction of the pressure under the fascia is that any longitudinal forces generated in the fascia reduce the need for the underlying muscles to contribute extensor forces, thereby lowering the compressive load on the spine. Both of these proposals were dismissed. Given the size of the fascia, hydraulic pressures would have to reach levels of hundreds of mmHg. Pressures of this magnitude simply are not observed during recording (Carr et al., 1985). Moreover, the presence or absence of IAP makes little difference on the hooplike geometry of the abdominal wall (McGill, Juker, and Axler, 1996), as this is more modulated by the posture.

IAP, LDF, and Hydraulic Amplifier: A Summary

IAP, the mechanical role of the LDF, and the existence of the hydraulic amplifier were proposed to account for the paradox that people were able to perform lifts that the simple models suggested would crush their spines. Yet, although both IAP and the LDF appear to play some role in lifting, none of the three proposed mechanisms was a tenable explanation for the paradox of uncrushed spines under heavy loading, whether considered separately or combined with the other two. The problem lay in the simple models of three and four decades ago. Not only were the rather complex mechanics not represented with the necessary detail, but also the strength of the tissues to bear load was also quite underestimated in the early tests that used old cadaveric samples that were crushed in artificially stiff rams of materials-testing machines that caused failure too early.

Other Important Mechanisms of Normal Spine Mechanics

Several other features of spine mechanics influence function and ultimately underpin strategies for injury prevention and rehabilitation. The most important are presented here.

Biomechanics of Diurnal Spine Changes

Most people have experienced the ease of taking off their socks at night compared to putting them on in the morning. The diurnal variation in spine length (the spine being longer after a night's bed rest), together with the ability to flex forward, has been well documented. Reilly, Tynell, and Troup (1984) measured losses in sitting height over a day of up to 19 mm. They also noted that approximately 54% of this loss occurred in the first 30 minutes after rising. Over the course of a day, hydrostatic pressures cause a net outflow of fluid from the disc, resulting in narrowing of the space between the vertebrae, which in turn reduces tension in the ligaments. When a person lies down at night, osmotic pressures in the disc nucleus exceed the hydrostatic pressure, causing the disc to expand. Adams, Dolan, and Hutton (1987) noted that the range of lumbar flexion increased by 5-6° throughout the day. The increased fluid content after rising from bed caused the lumbar spine to be more resistant to bending, while the musculature did not appear to compensate by restricting the bending range. Adams and colleagues estimated that disc-bending stresses were increased by 300% and ligament stresses by 80% in the morning compared to the evening; they concluded that there is an increased risk of injury to these tissues when bending forward early in the morning. Recently, Snook and colleagues (1998) demonstrated that simply avoiding full lumbar flexion in the morning reduced back symptoms. We are beginning to understand the mechanism.

CLINICAL RELEVANCE

Early-Morning Exercise

People should not undertake spine exercises—particularly those that require full spine flexion or bending—just after rising from bed, given the elevated tissue stresses that result. This would hold true for any occupational task requiring full spine range of motion.

Spinal Memory

The function of the spine is modulated by certain previous activity. This occurs because the loading history determines disc hydration (and therefore the size of the disc space and disc geometry), which in turn modulates ligament rest length, joint mobility, stiffness, and load distribution. Consider the following scenario: McKenzie (1979) proposed that the nucleus within the annulus migrates anteriorly during spinal extension and posteriorly during flexion. McKenzie's program of passive extension of the lumbar spine (which is currently popular in physical therapy) was based on the supposition that an anterior movement of the nucleus would decrease pressure on the posterior portions of the annulus, which is the most problematic site of herniation. Because of the viscous properties of the nuclear material, such repositioning of the nucleus is not immediate after a postural change but rather takes time. Krag and coworkers (1987) observed anterior movement of the nucleus during lumbar extension, albeit quite minute, from an elaborate experiment that placed radio-opaque markers in the nucleus of cadaveric lumbar motion segments. Whether this observation was caused simply by a redistribution of the centroid of the wedge-shaped nuclear cavity moving forward with flexion or was a movement of the whole nucleus remains to be seen. Nonetheless, hydraulic theory would suggest lower bulging forces on the posterior annulus if the nuclear centroid moved anteriorly during extension. If compressive forces were applied to a disc in which the nuclear material were still posterior (as in lifting immediately after a prolonged period of flexion), a concentration of stress would occur on the posterior annulus.

While this specific area of research needs more development, a time constant seems to be associated with the redistribution of nuclear material. If this result is correct, it would be unwise to lift an object immediately following prolonged flexion, such as sitting or stooping (e.g., a stooped gardener should not stand erect and immediately lift a heavy object). Furthermore, Adams and Hutton (1988) suggested that prolonged full flexion may cause the posterior ligaments to creep, which may allow damaging flexion postures to go unchecked if lordosis is not controlled during subsequent lifts. In a study of posterior passive tissue creep while sitting in a slouched posture, McGill and Brown (1992) showed that over the 2 minutes following 20 minutes of full flexion, subjects regained only half of their intervertebral joint stiffness. Even after 30 minutes of rest, some residual joint laxity remained. This finding is of particular importance for individuals whose work is characterized by cyclic bouts of full end range of motion postures followed by exertion. Before lifting exertions following a stooped posture or after prolonged sitting, a case could be made for standing or even consciously extending the spine for a short period. Allowing the nuclear material to "equilibrate," or move anteriorly to a position associated with normal lordosis, may decrease forces on the posterior nucleus in a subsequent lifting task. Ligaments will regain some protective stiffness during a short period of lumbar extension. In conclusion, the anatomy and geometry of the spine are not static. Much research remains to be done to understand the importance of tissue loading history on subsequent biomechanics, rehabilitation therapies, and injury mechanics.

CLINICAL RELEVANCE

Functional Significance of Spinal Memory

It would appear protective to avoid loading immediately after a bout of prolonged flexion. In the occupational world this has relevance to ambulance drivers, for example, who drive to an accident scene without the luxury of time to warm up (or reset the passive tissues) before lifting. They would be wise to sit with a lumbar pad to avoid lumbar flexion and the associated creep. The athletic world provides good examples as well, such as sitting on the bench before engaging in play. Those with sensitive backs would do well to avoid siting on the bench with a flexed lumbar spine while waiting to perform. We recently quantified the loss of compliance in the lumbar spine with bench sitting between bouts of athletic performance (Green, Grenier, and McGill, in press) in elite volleyball players. **Viscosity** is also a consideration in prolonged postures since "internal friction" increases with prolonged static postures. Sitting in this way, and the associated changes in stiffness and viscosity, is detrimental to athletes' performance and increases their risk of injury. We will address this issue more completely in chapters 8 and 9.

Anatomical Flexible Beam and Truss: Muscle Cocontraction and Spine Stability

The osteoligamentous spine is somewhat of an anatomical paradox: it is a weight-bearing, upright, flexible rod. Observationally, the ability of the joints of the lumbar spine to bend in any direction is accomplished with large amounts of muscle coactivation. Such coactivation patterns are counterproductive to generating the torque necessary to support the applied load. Coactivation is also counterproductive to minimize the load penalty imposed on the spine from muscle contraction. Researchers have postulated several ideas to explain muscular coactivation: the abdominals are involved in the generation of intra-abdominal pressure (Davis, 1959) or in providing support forces to the lumbar spine via the lumbodorsal fascia (Gracovetsky, Farfan, and Lamy, 1981). These ideas have not been without opposition (see previous sections).

Another explanation for muscular coactivation is tenable. A ligamentous spine will fail under compressive loading in a buckling mode, at about 20 N (about 5 lb) (Lucas and Bresler, 1961). In other words, a bare spine is unable to bear compressive load! The spine can be likened to a flexible rod that will buckle under compressive loading. However, if the rod has guy wires connected to it, like the rigging on a ship's mast, although more compression is ultimately experienced by the rod, it is able to bear much more compressive load since it is stiffened and therefore more resistant to buckling. The cocontracting musculature of the lumbar spine (the flexible beam) can perform the role of stabilizing guy wires (the truss) to each lumbar vertebra, bracing it against buckling. Recent work by Crisco and Panjabi (1990), together with Cholewicki and McGill (1996), Cholewicki, Juluru, and McGill (1999), and Gardner-Morse, Stokes, and Laible (1995) has begun to quantify the influence of muscle architecture and the necessary coactivation on lumbar spine stability. The architecture of many torso muscles is especially suited for the role of stabilization (Macintosh and Bogduk, 1987; McGill and Norman, 1987).

CLINICAL RELEVANCE

Muscular Cocontraction

In order to invoke this antibuckling and stabilizing mechanism when lifting, one could justify lightly cocontracting the musculature to minimize the potential of spine buckling.

Injury Mechanisms

Many clinicians, engineers, and ergonomists believe that reducing the risk of low back injury involves the reduction of applied loads to the various anatomical components at risk of injury. Without question, reduction of excessive loads is beneficial, but this is an overly simplistic view. Optimal tissue health requires an envelope of loading, not too much or too little. While some occupations require lower loads to reduce the risk, in sedentary occupations, the risk can be better reduced with more loading, and varying the nature of the loading. To decide which is best, the clinician must have a thorough understanding of the biomechanics of injury, which comes in two parts. First is a brief review of the injury mechanisms of individual tissues; second is a description of the injury pathogenesis and several injury scenarios. Also needed is an understanding of generic situations for low back tissue damage described in the tissue injury primer at the end of chapter 1.

Summary of Specific Tissue Injury Mechanisms

This section provides a very brief description of damage from excessive load. All injuries noted are known to be accelerated with repetitive loading.

- **End plates.** Schmorl's nodes are thought to be healed end-plate fractures (Vernon-Roberts and Pirie, 1973) and pits that form from localized underlying trabecular bone collapse (Gunning, Callaghan, and McGill, 2001) and which are linked to trauma (Aggrawall et al., 1979). In fact, Kornberg (1988) documented (via MRI) traumatic Schmorl's node formation in a patient following forced lumbar flexion that resulted in an injury. People apparently are not born with Schmorl's nodes; their presence is associated with a more active lifestyle (Hardcastle, Annear, and Foster, 1992). Under excessive compressive loading of spinal units in the laboratory, the end plate appears to be the first structure to be injured (Brinckmann, Biggemann, and Hilweg, 1988; Callaghan and McGill, 2001a). Studies have revealed end-plate avulsion under excessive anterior/posterior shear loading.
- **Vertebrae.** Vertebral cancellous bone is damaged under compressive loading (Fyhrie and Schaffler, 1994) and often accompanies disc herniation and annular delamination (Gunning, Callaghan, and McGill, 2001).
- **Disc annulus.** Several types of damage to the disc annulus appear to occur. Classic disc herniation appears to be associated with repeated flexion motion with only moderate compressive loading required (Callaghan and McGill, 2001a) and with full flexion with lateral bending and twisting (Adams and Hutton, 1985; Gordon et al., 1991). Yingling and McGill (1999a and 1999b) documented avulsion of the lateral annulus under anterior/posterior shear loading.
- **Disc nucleus.** While Buckwalter (1995), when referring to the disc nucleus, stated that "no other musculoskeletal soft tissue structure undergoes more dramatic alterations with age," the relationships among loading, disc nutrition, decreasing concentration of viable cells, accumulation of degraded matrix molecules, and fatigue failure of the matrix remain obscure. However, recently Lotz and Chin (2000) documented that cell death (apoptosis) within the nucleus increases under excessive compressive load. Interestingly, these changes are generally not detectable or diagnosable in vivo.
- **Neural arch (posterior bony elements).** Spondylitic fractures are thought to occur from repeated stress–strain reversals associated with cyclic full flexion and extension (Burnett et al., 1996; Hardcastle, Annear, and Foster, 1992). Cripton and colleagues (1995) and Yingling and McGill (1999a) also documented that excessive shear forces can fracture parts of the arch.

• **Ligaments.** Ligaments seem to avulse at lower load rates but tear in their midsubstance at higher load rates (Noyes, DeLucas, and Torvik, 1994). McGill (1997) hypothesized that landing on the buttocks from a fall will rupture the interspinous complex given the documented forces (McGill and Callaghan, 1999) and joint tolerance. Falling on the behind increases the risk for prolonged disability (Troup, Martin, and Lloyd, 1981), which is consistent with the prolonged length of time it takes for ligamentous tissue to regain structural integrity, when compared with other tissues (Woo, Gomez, and Akeson, 1985).

CLINICAL RELEVANCE

Reducing Tissue Damage

Summarizing injury pathways from in-vitro testing, evidence suggests that reduction in specific tissue damage could be accomplished by doing the following:

1. Reducing peak (and cumulative) spine compressive loads to reduce the risk of end-plate fracture
2. Reducing repeated spine motion to full flexion to reduce the risk of disc herniation (reducing spine flexion in the morning reduces symptoms)
3. Reducing repeated full-range flexion to full-range extension to reduce the risk of pars (or neural arch) fracture
4. Reducing peak and cumulative shear forces to reduce the risk of facet and neural arch damage
5. Reducing slips and falls to reduce the risk of passive collagenous tissues such as ligaments
6. Reducing the length of time sitting, particularly exposure to seated vibration, to reduce the risk of disc herniation or accelerated degeneration

Injury Mechanics Involving the Lumbar Mechanism

Many researchers have established that too great a load placed on a tissue will result in injury. Epidemiological studies (Hilkka et al., 1990; Marras et al., 1993; Norman et al., 1998; Videman, Nurminen, and Troup, 1990) have proven this notion by identifying peak loading measures (i.e., shear, compression, trunk velocity, extensor moment, heavy work, etc.) as factors that explain the frequency and distribution of reporting of back pain or increased risk of back injury. However, the search for direct evidence that links spine load with occupational LBDs may have been hampered by focusing on too narrow a range of variables.

Researchers have paid a massive amount of attention, for example, to a single variable—namely, acute, or single maximum exposure to, lumbar compression. A few studies have suggested that higher levels of compression exposure increased the risk of LBD (e.g., Herrin et al., 1986), although the correlation was low. Yet some studies show that higher rates of LBDs occur when levels of lumbar compression are reasonably low.

What other mechanical variables modulate the risk of LBDs? As noted in chapter 1, they are as follows:

• Too many repetitions of force and motion and/or prolonged postures and loads have also been indicated as potential injury- or pain-causing mechanisms.

• Cumulative loading (i.e., compression, shear, or extensor moment) has been identified as a factor in the reporting of back pain (Kumar, 1990; Norman, Wells, and Neumann, 1998).

• Cumulative exposure to unchanging work has been linked to reporting of low back pain (Holmes, Hukins, and Freemont, 1993) and intervertebral disc injury (Videman, Nurminen, and Troup, 1990).

Many personal factors appear to affect spine tissue tolerance, for example, age and gender. In a compilation of the available literature on the tolerance of lumbar motion units to bear compressive load, Jager and colleagues (1991) noted that when males and females are matched for age, females are able to sustain only approximately two thirds of the compressive loads of males. Furthermore, Jager and colleagues' data showed that, within a given gender, a 60-year-old was able to tolerate only about two thirds of the load tolerated by a 20-year-old.

To complicate the picture, Holm and Nachemson (1983) showed that increased levels of motion are beneficial in providing nutrition to the structures of the intervertebral disc, while much of our lab's research has demonstrated that too many motion cycles (to full flexion) resulted in intervertebral disc herniation. Buckwalter (1995) associated intervertebral disc degeneration with decreased nutrition. Meanwhile, Videman and colleagues (1990) showed that too little motion from sedentary work resulted in intervertebral disc injury. While workers who performed heavy work were also at increased risk of developing a spine injury, workers who were involved in varied types of work (mixed work) had the lowest risk of developing a spine injury (Videman, Nurminen, and Troup, 1990). This presents the idea that too little motion or load, or too much motion or load can modulate the risk of spinal injury. In contrast, Hadler (1991) claimed that the incidence of back injury has not declined over the past 25 years, even after increased research and resources dedicated to the area over that time frame. Hadler suggested that the focus be turned from biomechanical causes of injury to developing more "comfortable" workplaces. However, the work described thus far in this text has clearly documented links to mechanical variables. Clearly, LBD causality is often extremely complex with all sorts of factors interacting. We will consider some of those factors in the following sections.

CLINICAL RELEVANCE

Putting Knowledge Into Practice

While the notion that too much of any activity (whether it be sitting at a desk or loading heavy boxes onto pallets) can be harmful is widely accepted, it is rarely taken into account in practice. For example, industrial ergonomics has not yet wholeheartedly embraced the idea that not all jobs need to be made less demanding, that some jobs need much more variety in the patterns of musculoskeletal loading, or that there is no such thing as "best posture" for sitting. It is time for the profession as a whole to remember that in any job the order and type of loading should be considered and the demand on tissues should be varied. All sedentary workers should be taught, for example, to adopt a variable posture that causes a migration of load from tissue to tissue, reducing the risk of troubles.

Stoop Versus Squat in Lifting Injury Risk

While the scientific method can prove that a phenomenon is possible if observed, failure to observe the expected result does not eliminate the possibility. One may only conclude that the experiment was insensitive to the particular phenomenon. The following discussion, as with many in this book, attempts to incorporate this limitation and temper it with clinical wisdom.

In a previous section, lifting with the torso flexing about the hips rather than flexing the spine was analyzed and described. Specifically, the lifter elected to maintain a neutral lumbar posture rather than allowing the lumbar region to flex. Here we will reexamine the lifting exercise, but the lifter will flex the spine sufficiently to cause the posterior ligaments to strain. This lifting strategy (spine flexion) has quite dramatic effects on shear loading of the intervertebral column and the resultant injury risk. The

Figure 5.12 This gardener appears to be adopting a fully flexed lumbar spine. Is this wise posture? The force analysis in figure 5.13 suggests it is not.

dominant direction of the pars lumborum fibers of the longissimus thoracis and iliocostalis lumborum muscles when the lumbar spine remained in neutral lordosis (page 64) caused these muscles to produce a posterior shear force on the superior vertebra. In contrast, with spine flexion the strained interspinous ligament complex generates forces with the opposite obliquity and therefore imposes an anterior shear force on the superior vertebra (see figures 5.12 and 5.13, a-d).

Let's examine the specific forces that result from flexing the lumbar spine. The recruited ligaments appear to contribute to the anterior shear force so that shear force levels are likely to exceed 1000 N (224 lb). Such large shear forces are of great concern from an injury risk viewpoint. However, when a more neutral lordotic posture is adopted, the extensor musculature is responsible for creating the extensor moment and at the same time provides a posterior shear force that supports the anterior shearing action of gravity on the upper body and handheld load. The joint shear forces are reduced to about 200 N (about 45 lb). Thus, using muscle to support the moment in a more neutral posture rather than being fully flexed with ligaments supporting the moment greatly reduces shear loading (see table 5.9).

Quantification of the risk of injury requires a comparison of the applied load to the tolerance of the tissue. Cripton and colleagues (1995) found the shear tolerance of the spine to be in the neighborhood of 2000-2800 N in adult cadavers, for one-time loading. Recent work by Yingling and McGill (1999a and 1999b) on pig spines has shown that load rate is not a major modulator of shear tolerance unless the load is very ballistic, such as what might occur during a slip and fall. This example

a

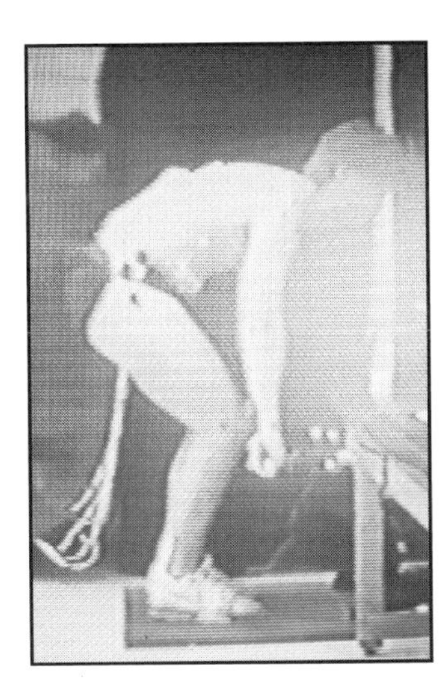

b

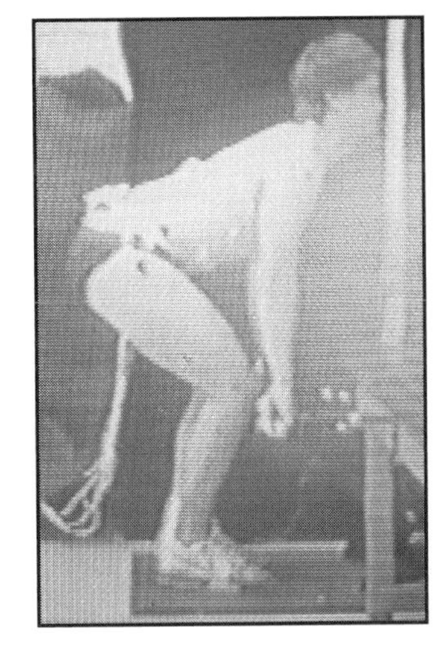

Figure 5.13 These original computer image bitmaps from the experiment conducted around 1987 illustrate the fully flexed spine *(a)* that is associated with myoelectric silence in the back extensors and strained posterior passive tissues and high shearing forces on the lumbar spine (from both reaction shear on the upper body and interspinous ligament strain (see figure 4.25). A more neutral spine *(b)* posture recruits the pars lumborum muscle groups . . .

(continued)

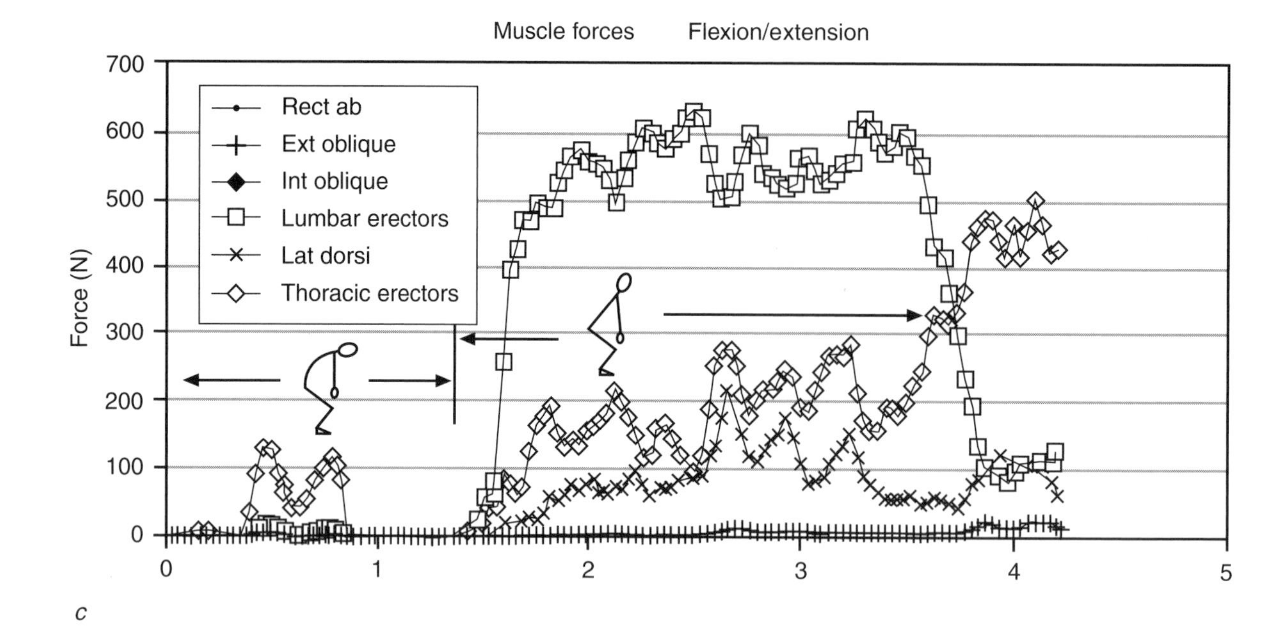

c

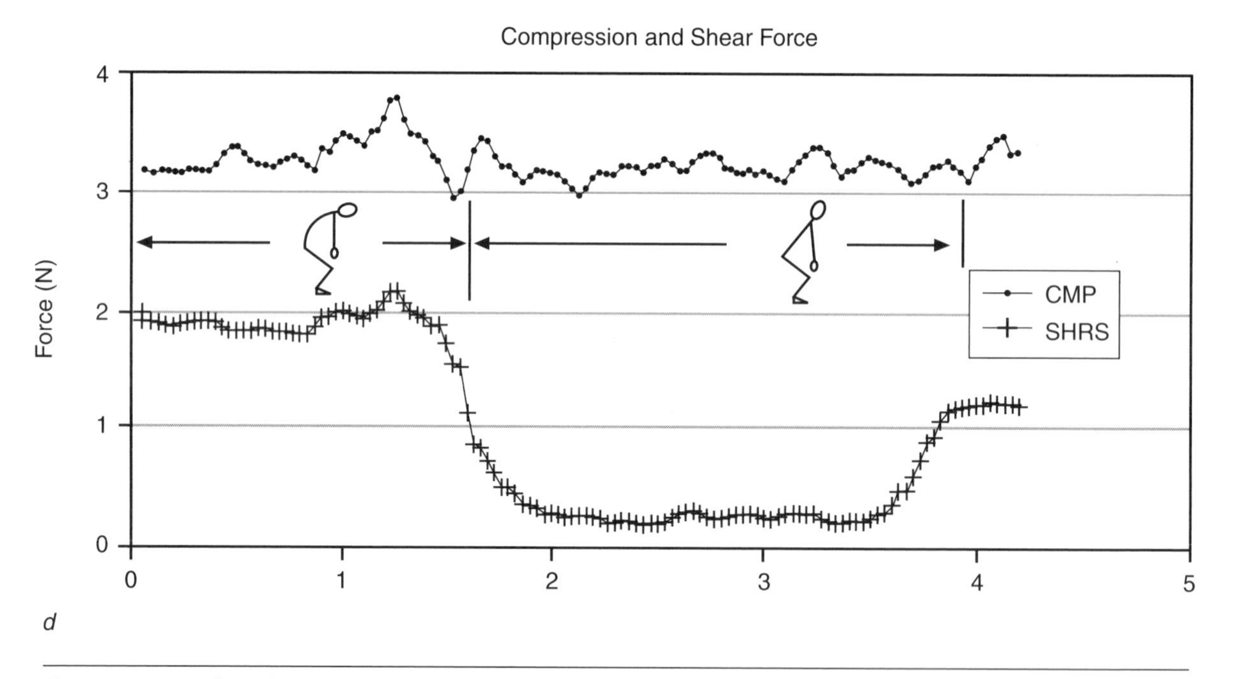

d

Figure 5.13 (continued) . . . *(c)* to support the reaction shear (see figure 4.19) and thus reduces total joint shear *(d)* (to approximately 200 N in this example).

Reprinted from *Journal of Biomechanics,* 30(5), S.M. McGill, "Invited paper: Biomechanics of low back injury: Implications on current practice and the clinic," 465-475, 1997, with permission from Elsevier Science.

Table 5.9 Individual Muscle and Passive Tissue Forces During Full Flexion and in a More Neutral Lumbar Posture Demonstrating the Shift from Muscle to Passive Tissue

The extensor moment with full lumbar flexion was 171 Nm producing 3145 N of compression and 954 N of anterior shear. The more neutral posture of 170 Nm produced 3490 N of compression and 269 N of shear.

	Fully flexed lumbar spine	Neutral lumbar spine
	Force (N)	Force (N)
Muscle		
R rectus abdominis	16	39
L rectus abdominis	16	62
R external oblique 1	10	68
L external oblique 1	10	40
R external oblique 2	7	62
L external oblique 2	7	31
R internal oblique 1	35	130
L internal oblique 1	35	102
R internal oblique 2	29	88
L internal oblique 2	29	116
R pars lumborum (L1)	21	253
L pars lumborum (L1)	21	285
R pars lumborum (L2)	27	281
L pars lumborum (L2)	27	317
R pars lumborum (L3)	31	327
L pars lumborum (L3)	31	333
R pars lumborum (L4)	32	402
L pars lumborum (L4)	32	355
R iliocostalis lumborum	58	100
L iliocostalis lumborum	58	137

(continued)

Table 5.9 *(continued)*

	Fully flexed lumbar spine	Neutral lumbar spine
	Force (N)	**Force (N)**
Muscle *(continued)*		
R longissimus thoracis	93	135
L longissimus thoracis	93	179
R quadratus lumborum	25	155
L quadratus lumborum	25	194
R latissimus dorsi (L5)	15	101
L latissimus dorsi (L5)	15	115
R multifidus 1	28	80
L multifidus 1	28	102
R multifidus 2	28	87
L multifidus 2	28	90
R psoas (L1)	25	61
L psoas (L1)	25	69
R psoas (L2)	25	62
L psoas (L2)	25	69
R psoas (L3)	25	62
L psoas (L3)	25	69
R psoas (L4)	25	61
L psoas (L4)	25	69
Ligament		
Anterior longitudinal	0	0
Posterior longitudinal	86	0
Ligamentum flavum	21	3

Table 5.9 *(continued)*

	Fully flexed lumbar spine	Neutral lumbar spine
	Force (N)	**Force (N)**
Ligament *(continued)*		
R intertransverse	14	0
L intertransverse	14	0
R articular	74	0
L articular	74	0
R articular 2	103	0
L articular 2	103	0
Interspinous 1	301	0
Interspinous 2	345	0
Interspinous 3	298	0
Supraspinous	592	0
R lumbodorsal fascia	122	0
L lumbodorsal fascia	122	0

demonstrates that the spine is at much greater risk of sustaining shear injury (>1000 N applied to the joint) (224 lb) than compressive injury (3000 N applied to the joint) (672 lb) simply because the spine is fully flexed. (For a more comprehensive discussion, see McGill, 1997.) The margin of safety is much larger in the compressive mode than in the shear mode since the spine can safely tolerate well over 10 kN in compression, but 1000 N of shear causes injury with cyclic loading. This example also illustrates the need for clinicians/ergonomists to consider other loading modes in addition to simple compression. In this example the real risk is anterior/posterior shear load. Interestingly, Norman and colleagues' 1998 study showed joint shear to be very important as a metric for risk of injury of auto plant workers, particularly cumulative shear (high repetitions of subfailure shear loads) over a workday.

Yet another consideration impinges on the interpretation of injury risk. The ability of the spine to bear load is a function of the curvature of the spine in vivo. For example, Adams and colleagues (1994) suggested that a fully flexed spine is weaker than one that is moderately flexed. In a recent study, Gunning, Callaghan, and McGill (2001), showed that a fully flexed spine (using a controlled porcine spine model) is 20-40% weaker than if it were in a neutral posture.

CLINICAL RELEVANCE

Stoop Versus Squat

So much has been written on the stoop style versus the squat style of lifting. Typically, conclusions were based on very simple analyses that measured only the low back moment. The reaction moment is a function of the size and position of the load in the hands and the position of the center of mass of the upper body. As the previous example demonstrated, the issue is actually much more complex—the lumbar spine curvature determines the sharing of the load between the muscles and the passive tissues. In addition, each individual elects the way in which agonists and antagonists are coactivated. Thus, the spine kinematic motion patterns, together with the muscle activation patterns, heavily influence the resulting spine load and the ability of the spine to bear load without damage. The risk of injury is the real issue that motivated most of these types of analyses, yet it was not addressed with sufficient detail as to reach a correct conclusion.

Motor Control Errors and Picking Up Pencils

Although clinicians often hear patients report injuries from seemingly benign tasks (such as when picking up a pencil from the floor), this phenomenon will not be found in the scientific literature. Because such an injury would not be deemed compensable in many jurisdictions, medical personnel rarely record this event. Instead they attribute the cause elsewhere. Moreover, while injury from large exertions is understandable, explaining injury that occurs when performing such light tasks is not. The following is worth considering.

A number of years ago, using video fluoroscopy for a sagittal view of the lumbar spine, we investigated the mechanics of power lifters' spines while they lifted extremely heavy loads (Cholewicki and McGill, 1992). The range of motion of the lifters' spines was calibrated and normalized to full flexion by first asking them to flex at the waist and support the upper body against gravity with no load in the hands. During the lifts, although the lifters appeared outwardly to have a very flexed spine, in fact, the lumbar joints were 2-3° per joint from full flexion (see figures 5.14 and 5.15). This explains how they could lift such magnificent loads (up to 210 kg, or approximately 462 lb) without sustaining the injuries that are suspected to be linked with full lumbar flexion.

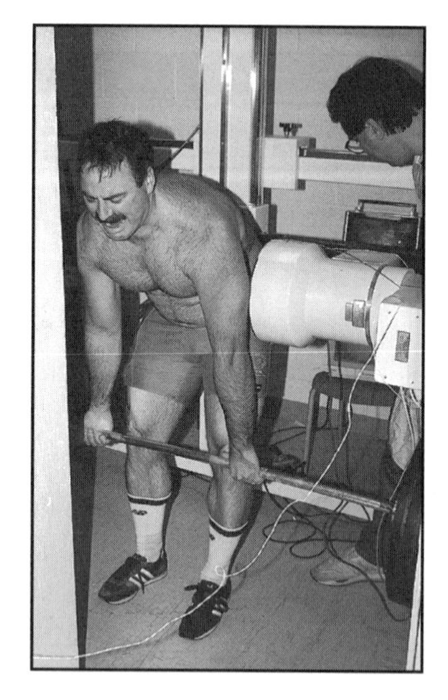

Figure 5.14 A group of power lifters lifted very heavy weights while their lumbar vertebrae motion patterns were quantified. Each joint was skillfully controlled to flex but not fully flex. Each joint was 2-3° away from the fully flexed, calibrated angle (see figure 5.15).

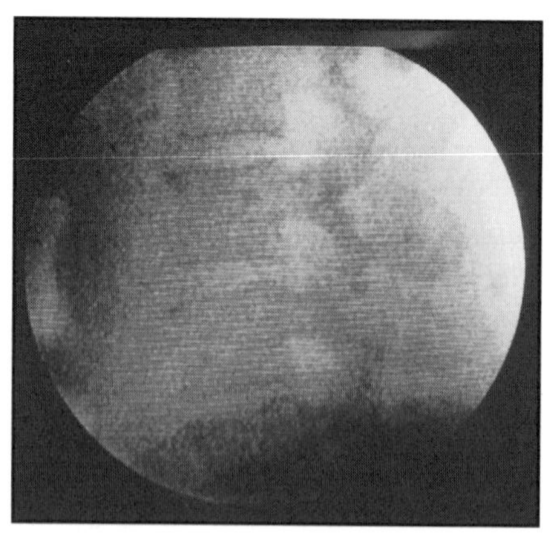

Figure 5.15 This fluoroscopy image of the power lifter's lumbar spine shows how individual lumbar joint motion can be quantified in vivo. While recording, an injury occurred in one lifter when the vertebra at just one joint went to its full flexion angle and surpassed it by about one-half degree.

However, during the execution of a lift, one lifter reported discomfort and pain. Upon examination of the video fluoroscopy records, one of the lumbar joints (specifically, the L2-L3 joint) reached the full flexion calibrated angle, while all other joints maintained their static position (2-3° short of full flexion). The spine buckled and caused injury (see figure 5.16). This is the first observation we know of reported in the scientific literature documenting proportionately more rotation occurring at a single lumbar joint; this unique occurrence appears to have been due to an inappropriate sequencing of muscle forces (or a temporary loss of motor control wisdom). This motivated the work of my colleague and former graduate student Professor Jacek Cholewicki to investigate and continuously quantify the stability of the lumbar spine throughout a wide variety of loading tasks (Cholewicki and McGill, 1996). Generally speaking, the occurrence of a motor control error that results in a temporary reduction in activation to one of the intersegmental muscles (perhaps, for example, a lamina of longissimus, iliocostalis, or multifidus) may allow rotation at just a single joint to the point at which passive or other tissues could become irritated or injured.

Cholewicki noted that the risk of such an event was greatest when high forces were developed by the large muscles and low forces by the small intersegmental muscles (a possibility with our power lifters) or when all muscle forces were low such as during a low-level exertion. Thus, injury from quite low intensity bending is possible. Adams and Dolan (1995) noted that passive tissues begin to experience damage with bending moments of 60 Nm. This can occur simply from a temporary loss of muscular support when bending over. This mechanism of motor control error resulting in temporary inappropriate neural activation explains how injury might occur during extremely low load situations, for example, picking up a pencil from the floor following a long day at work performing a very demanding job.

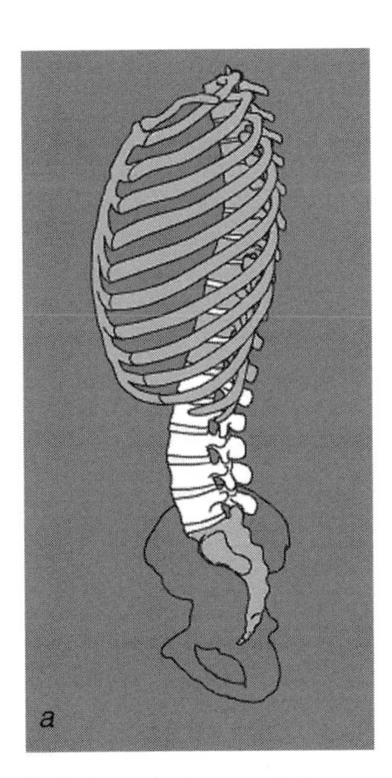

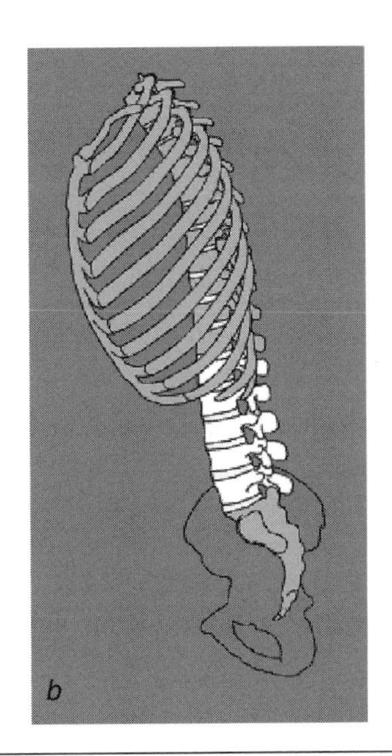

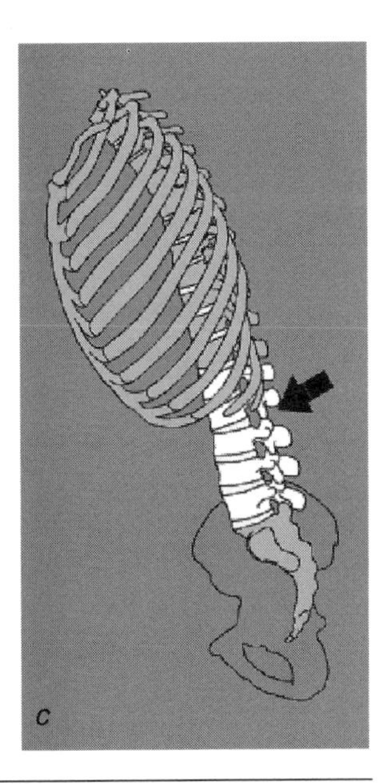

Figure 5.16 The power lifter flexed to grab the weight bar *(a)* and began to extend the spine during the lift *(b)*. As the weight was a few inches from the floor, a single joint (L2-L3) flexed to the full flexion angle, and the spine buckled *(c)*, indicated by the arrow.

CLINICAL RELEVANCE

Sources of Motor Control Errors

Another issue must be considered when dealing with the possibility of buckling that results from specific motor control errors. In our clinical testing we observed similarly inappropriate motor patterns in some men who were challenged by holding a load in the hands while breathing 10% CO_2 to elevate breathing. (Challenged breathing causes some of the spine-supporting musculature to drop to inappropriately low levels in some people; see McGill, Sharratt, and Seguin, 1995.) These deficient motor control mechanisms will heighten biomechanical susceptibility to injury or reinjury (Cholewicki and McGill, 1996). My lab is currently involved in a large-scale longitudinal investigation to assess the power of these motor control parameters together with some personal performance variables and their role in causing back injury over a multiyear period.

Sacroiliac Pain—Is It From the Joint?

Bogduk and colleagues (1996) proved that some low back pain is from the sacroiliac joint itself. The following discussion offers other considerations. The work that many have reported over recent years has demonstrated the extraordinary magnitudes of forces within the torso extensor musculature even during nonstrenuous tasks. While these forces have been interpreted for their mechanical role, clinicians have expressed interest in their potential to cause injury. One possibility worth considering is that the high muscular forces may damage the bony attachments of the corresponding muscle tendons. Such damage has perhaps been wrongfully attributed to alternate mechanisms. One such example follows.

Pain in the sacroiliac region is common and often attributed to disorders of the sarcoiliac (SI) joint itself or the iliolumbar ligament. For this reason, the role of the musculature may have been neglected. It is known that a large proportion of the extensor musculature obtains its origin in the SI and posterior superior iliac spine (PSIS) region (Bogduk, 1980). The area of tendon–periosteum attachment and extensor aponeurosis is relatively small in relation to the volume of muscle in series with the tendon complex. From this, a hypothesis evolved that the seeming mismatch of large muscle tissue to small attachment area for connective tissue places the connective tissue at high risk of sustaining microfailure, resulting in pain (McGill, 1987). Knowledge of the collective muscle force time histories enables speculation about one-time failure loads and cumulative trauma. For example, if the forces of muscles that originate in the SI region are tallied for the trial illustrated in table 5.4 (page 93), then the total force transmitted to the SI region during peak load would exceed 5.6 kN. Such a load would lift a small car off the ground!

The failure tolerance of these connective tissues is not known, which makes speculation over the potential for microfailure difficult. No doubt the risk of damage must increase with the increase in the extremely large loads observed in the extensor musculature and with the frequency of application. Industrial tasks comparable to lifting three containers in excess of 18 kg (40 lb) per minute over an eight-hour day are not unusual, suggesting that the potential for cumulative trauma is significant.

This mechanical explanation may account for local tenderness on palpation associated with many of the SI syndrome cases. As well, muscle strain and spasm often accompany SI pain. Nonetheless, treatment is often directed toward the articular joint despite the extreme difficulty in diagnosing the joint as the primary source of pain. While reduction of spasm through conventional techniques would reduce the sustained load on the damaged connective fibers, patients should be counseled on techniques to reduce internal muscle loads through effective lifting mechanics. This is a single example, of which there may be several, in which knowledge of individual muscle force time histories would suggest a mechanism for injury for which a specific—and possibly atypical—treatment modality could be prescribed.

Bed Rest and Back Pain

While bed rest has fallen from favor as therapy, given the generally poor patient outcome, the mechanism for poor outcome is just now starting to be understood. As noted on page 113, bed rest reduces the applied load (hydrostatic) below the disc osmotic pressure, resulting in a net inflow of fluid (McGill, van Wijk et al., 1996). McGill and Axler (1996) documented the growth in spine length over the usual 8 hours of bed rest and then over continued bed rest for another 32 hours. This is unusual sustained pressure and is suspected to cause backache. Oganov and colleagues (1991) documented increased mineral density in the vertebrae following prolonged space travel (weightlessness); this is notable because other bones lost mineral density. It indicates that the spine was stimulated to lay down bone in response to higher loads; in this case, the higher load was due to the swollen discs. The analog on earth is lying in bed for periods longer than eight hours—it actually stresses the spine!

CLINICAL RELEVANCE

Moderation in All Things

The phrase "moderation in all things" appears to encapsulate a lot of spine biomechanical wisdom; varying loading through constant change in activity is justifiable. Obviously, overload is problematic, but the previous example shows that even lying in bed too long is problematic—it increases annulus, end-plate, and vertebral stress!

Bending and Other Spine Motion

While some researchers have produced disc herniations under controlled conditions (e.g., Gordon et al., 1991), Callaghan and McGill (2001a) have been able to consistently produce disc herniations by mimicking the spine motion and load patterns seen in workers. Specifically, only a very modest amount of spine compression force seems to be required (only 800-1000 N, or 177-220 lb), but the spine specimen must be repeatedly flexed, mimicking repeated torso/spine flexion from continual bending to a fully flexed posture (see page 119). In these experiments, we documented the progressive tracking of disc nucleus material traveling posteriorly through the annulus of the disc with final sequestration of the nucleus material. This was noted to occur around 18,000-25,000 cycles of flexion with low levels of spine compression (about 1000 N, or 220 lb), but it would occur with much fewer cycles of bending with higher simultaneous compressive loads (about 5000 cycles with 3000 N, or 670 lb, of compression). Most interesting was the fact that the tracking of nuclear material initiated from the inside and slowly worked radially outward, suggesting that, had this been a living worker, there may have been little indication of cumulative damage until the culminating event.

CLINICAL RELEVANCE

Repeated Spine Flexion

While much ergonomic effort has been on reducing spine loads, it is becoming clearer that repeated spine flexion—even in the absence of moderate load—will lead to discogenic troubles. Ergonomic guidelines will be more effective once factors for flexion repetition are included.

Confusion Over Twisting

While twisting has been named in several studies as a risk factor for low back injury, the literature does not make the distinction between the kinematic variable of twisting and the kinetic variable of generating twisting torque. While many epidemiological surveillance studies link a higher risk of LBD with twisting, twisting with low twist moment demands results in lower muscle activity and lower spine load (McGill, 1991). Further, passive tissue loading is not substantial until the end of the twist range of motion (Duncan and Ahmed, 1991). However, developing twisting moment places very large compressive loads on the spine because of the enormous coactivation of the spine musculature (McGill, 1991). This can also occur when the spine is not twisted but in a neutral posture in which the ability to tolerate loads is higher. Either single variable (the kinematic act of twisting or generating the kinetic variable of twist torque while not twisting) seems less dangerous than epidemiological surveys suggest. However, elevated risk from very high tissue loading may occur when the spine is fully twisted and there is a need to generate high twisting torque (McGill, 1991). Several studies, then, have suggested that compression on the lumbar spine is not the sole risk factor. Both laboratory tissue tests and field surveillance suggest that shear loading of the spine, together with large twists coupled with twisting torque, increases the risk of tissue injury.

Is Viral Involvement Possible in Low Back Pain?

Research has suggested that the incidence of some musculoskeletal disorders of the upper extremity are elevated following exposure to a viral infection. Could viral infections also be responsible for low back disorders? While this scenario is possible in individual cases, it could only account for some. Since animal model studies use animals that are screened for disease, with only approved specimens being selected for testing, controlled mechanical loading would seem to account for the significant findings. Viral infections have been linked to an increased probability of developing such diseases as carpal tunnel syndrome or arthritis (Mody and Cassim, 1997; Phillips, 1997; Samii et al., 1996). However, Rossignol and colleagues (1997) found that work-related factors accounted for the majority of the causes attributed to developing carpal tunnel syndrome. Future research should examine the links between viral infections leading to spinal tissue degeneration and low back injury etiology. They seem to be involved in some select cases.

Section Summary

From this description of injury mechanisms with several modulating factors, it is clear that multifactorial links exist. Interpretation of these links will enhance injury prevention and rehabilitation efforts. In fact, effective interventions will not occur without an understanding of how the spine works and how it becomes injured.

Biomechanical and Physiological Changes Following Injury

While the foundation for good clinical practice requires an understanding of the mechanism for injury, an understanding of the lingering consequences is also helpful.

Tissue Damage Pathogenesis, Pain, and Performance

Many discuss spine injury as though it is a static entity, that is, by focusing on specific tissue damage. However, since tissue damage causes changes to the joint biomechanics, other tissues will be affected and drawn into the clinical picture. Understanding the links in this issue begins with the concept of pain.

Tissue damage causes pain. Some have said there is no proof that this statement is true since pain is a perception and no instrument can measure it directly. This notion ignores the large body of pain literature (a review of which is well outside the scope of this chapter), which has motivated a recent proposal from a diverse group of pain specialists to classify pain by mechanism—specifically, transient pain (which does not produce long-term sequelae) and tissue injury and nervous system injury pain, both of which have complex organic mechanisms (Woolf et al., 1998).

Siddall and Cousins (1997) wonderfully summarized the great advances made in the understanding of the neurobiology of pain—in particular, the long-term changes from noxious stimulation known as central sensitization. Briefly, tissue damage directly affects the response of the nociceptor to further stimulation; over longer periods of time, both increase the magnitude of the response, sensitivity to the stimulus, and size of the region served by the same afferents. Some clinicians have used this as a rationale for the prescription of analgesics early in acute LBP. Furthermore, some have based arguments on whether a certain tissue could initiate pain from the presence or absence of nociceptors or free nerve endings, requiring proof of damage to the specific tissue in question. But tissue damage (along a continuum from cell damage to macrostructural failure) changes the biomechanics of the spine. Not only may this change cause pain, but it appears to initiate a cascade of change that can cause disruptions to the joint and continual pain for years in leading to conditions such as facet arthritis, accelerated annular degeneration, and nerve root irritation, to name a few. (See Kirkaldy-Willis and Burton, 1992, for an excellent review of the cascade of spine degeneration initiated by damage.) Butler and colleagues (1990) documented that disc damage nearly always occurs before facet arthritis is observed. Injury and tissue damage initiate joint instabilities, causing the body to respond with arthritic activity to finally stabilize. This results in a loss of range of motion and, no doubt, pain (Brinckmann, 1985).

Injury Process: Motor Changes

It is conclusive that patients reporting debilitating low back pain suffer simultaneous changes in their motor control systems. Recognizing these changes is important since they affect the stabilizing system. Richardson and coworkers (1999) produced quite a comprehensive review of this literature and made a case for targeting specific muscle groups during rehabilitation. Specifically, their objective is to reeducate faulty motor control patterns postinjury. The challenge is to train the stabilizing system during steady-state activities and also during rapid voluntary motions and to withstand sudden surprise loads.

Among the wide variety of motor changes researchers have documented, they have paid particular attention to the transverse abdominis and multifidus muscles. For example, during anticipatory movements such as sudden shoulder flexion movements, the onset of transverse abdominis has been shown to be delayed in those with back troubles (Hodges and Richardson, 1996; Richardson et al., 1999). Such delayed onset of specific torso muscles during sudden events (Hodges and Richardson, 1996, 1999) would impair the ability of the spine to achieve protective stability during a situation such as a slip or fall. The Queensland group developed a rehabilitation protocol specifically intended to reeducate the motor control system for involvement of the transverse abdominis. A number of laboratories have also documented changes to the multifidus complex. Further, in a very nice study of 108 patients with histories of chronic LBD ranging from 4 months to 20 years, Sihvonen and colleagues (1997) noted that 50% had disturbed joint motion and 75% of those with radiating pain had abnormal electromyograms to the medial spine extensor muscles. Interestingly enough, in the many studies on the multifidus, EMG abnormalities in the more lateral longissimus seem to appear along with those in mulifidus in those with back troubles (Haig et al., 1995). Jorgensen and Nicolaisen (1987) associated lower endurance in the spine extensors in general, while Roy and colleagues (1995) established faster fatigue rates in the multifidus in those with low back troubles. In addition, changes in torso agonist/antagonist activity during gait (Arendt-Nielson et al., 1995) have been documented together with asymmetric extensor muscle output during isokinetic torso extensor efforts (Grabiner, Kah, and Ghazawi, 1992), which alters spine tissue loading.

Further, evidence indicates that the structure of the muscle itself experiences change following injury or pain episodes. Anatomical changes following low back injury include asymmetric atrophy in the multifidus (Hides, Richardson, and Jull, 1996) and fiber type changes in multifidus even five years after surgery (Rantanen et al., 1993). Long-term outcome was associated with certain composition characteristics. Specifically, good outcome was associated with normal fiber appearance, while poor outcome was associated with atrophy in the type II fibers and a "moth-eaten" appearance in type I fibers. Moreover, even after symptoms have resolved, Hides and colleagues (1996) documented a smaller multifidus and suggested impaired reflexes as a mechanism. This theory appears tenable, given documented evidence of this at other joints, particularly at the knee (Jayson and Dixon, 1970; Stokes and Young, 1984). Once again, the clinical emphasis on the multifidus may well be because the bulk of the research has been performed on this muscle. Yet researchers who have examined other muscles have observed similar changes in unilateral atrophy—for example, in the psoas (Dangaria and Naesh, 1998).

This collection of evidence strongly supports exercise prescription that promotes the patterns of muscular cocontraction observed in fit spines and quite powerfully documents patho-neural-mechanical changes associated with chronic LBD. Once again, these changes are lasting years—not 6 to 12 weeks!

Other Changes

Over the past two years we have extensively tested 72 workers. Of those workers, 26 had a history of back troubles sufficient to result in lost time, 24 had some back troubles but not severe enough to ever lose worktime, and the rest had never had any back complaints. Our subjects began by filling out detailed questionnaires that requested information about demographics, physical demands at work, job satisfaction and personal control, work organization, perceived fitness, and chronicity potential. This was followed with approximately four hours of lab testing. This included comprehensive testing for fitness parameters (body mass index, general strength, flexibility, $\dot{V}O_2$max, etc.) and a large inventory of spine-specific parameters (strength, muscle endurance, three-dimensional coupled motion, muscle usage during tasks requiring static balance, unexpected loading, anticipatory reaction, challenged breathing, unrestricted lifting from the floor, etc.). Throughout all tests, subjects were connected to instruments to document three-dimensional lumbar spine motion (3-SPACE, Polhemus, Colchester, Vermont) and torso musculature activity (14 channels of surface EMG).

Our findings are summarized here:

- Having a history of low back troubles was associated with a larger waist girth, a greater chronicity potential as predicted from psychosocial questionnaires, and perturbed flexion to extension strength and endurance ratios, among other factors.
- Those who had a history of back troubles had a lack of muscle endurance—specifically, a balance of endurance among torso muscle groups. Absolute torso strength was not related, although the ratio of flexor to extensor strength was important.
- Those with a history of low back trouble had diminished hip extension and internal rotation, suggesting psoas involvement.
- Those with a history of back troubles had a wide variety of motor control deficits including during challenged breathing (as would occur during challenging work), balancing, having to endure surprise loading, and so on. Generally, those who were poor in one motor task were likely to be poor in another.

CLINICAL RELEVANCE

Lingering Deficits Following Back Injury

Given that those workers who had missed work because of back troubles were measured 261 weeks, on average, after their last disabling episode, multiple deficits appear to remain for very lengthy periods. The broad implication of this work is that having a history of low back troubles, even when a substantial amount of time has elapsed, is associated with a variety of lingering deficits that would require a multidisciplinary intervention approach to diminish their presence. Whether these deficits were a cause of the troubles (perhaps they could have existed prior to the symptoms) or a consequence of the troubles remains to be seen. A longitudinal study based on these variables is under way.

References

Adams, M., and Dolan, P. (1995) Recent advances in lumbar spinal mechanics and their clinical significance. *Clinical Biomechanics,* 10 (1): 3.

Adams, M.A., Dolan, P., and Hutton, W.C. (1987) Diurnal variations in the stresses on the lumbar spine. *Spine*, 12 (2): 130.

Adams, M.A., and Hutton, W.C. (1985) Gradual disc prolapse. *Spine*, 10: 524.

Adams, M.A., and Hutton, W.C. (1988) Mechanics of the intervertebral disc. In: Ghosh, P. (Ed.), *The biology of the intervertebral disc*. Boca Raton, FL: CRC Press.

Adams, M.A., McNally, D.S., Chinn, H., and Dolan, P. (1994) Posture and the compressive strength of the lumbar spine. *Clinical Biomechanics*, 9: 5-14.

Aggrawall, N.D., Kavr, R., Kumar, S., and Mathur, D.N. (1979) A study of changes in the spine in weightlifters and other athletes. *British Journal of Sports Medicine,* 13: 58-61.

Arendt-Nielson, L., Graven-Neilson, T., Svarrer, H., and Svensson, P. (1995) The influence of low back pain on muscle activity and coordination during gait. *Pain,* 64: 231-240.

Ashton-Miller, J.A., and Schultz, A.B. (1988) Biomechanics of the human spine and trunk. In: Pandolf, K.B. (Ed.), *Exercise and sport science reviews* (Vol. 16), American College of Sports Medicine Series. New York: Macmillan.

Axler, C., and McGill, S.M. (1997). Low back loads over a variety of abdominal exercises: Searching for the safest abdominal challenge. *Medicine and Science in Sports and Exercise,* 29 (6): 804-811.

Bearn, J.G. (1961) The significance of the activity of the abdominal muscles in weight lifting. *Acta Anatomica*, 45: 83.

Berkson, M.H., Nachemson, A.L., and Shultz, A.B. (1979). Mechanical properties of human lumbar spine motion segments. Part II: Responses in compression and shear: Influence of gross morphology. *Journal of Biomechanical Engineering*, 101: 53.

Bogduk, N. (1980) A reappraisal of the anatomy of the human lumbar erector spinae. *Journal of Anatomy*, 131 (3): 525.

Bogduk, N., Derby, R., Aprill, C., Louis, S., and Schwartzer, R. (1996) Precision diagnosis in spinal pain. In: Campbell, J. (Ed.). *Pain 1996—An updated view* (pp. 313-323) Seattle: IASP Press.

Bogduk, N., and Macintosh, J.E. (1984) The applied anatomy of the thoracolumbar fascia. *Spine*, 9: 164.

Brinckmann, P. (1985) Pathology of the vertebral column. *Ergonomics*, 28: 235-244.

Brinckmann, P., Biggemann, M., and Hilweg, D. (1988) Prediction of the compressive strength of human lumbar vertebrae. *Clinical Biomechanics*, 4 (Suppl. 2).

Buckwalter, J.A. (1995) Spine update: Ageing and degeneration of the human intervertebral disc. *Spine*, 20: 1307-1314.

Burnett, A.F., Khangure, M., Elliot, B.C., Foster, D.H., Marshall, R.N., and Hardcastle, P. (1996) Thoracolumbar disc degeneration in young fast bowlers in cricket. A follow-up study. *Clinical Biomechanics*, 11: 305-310.

Butler, D., Trafimow, J.H., Andersson, G.B.J., McNiell, T.W., and Hackman, M.S. (1990) Discs degenerate before facets. *Spine*, 15: 111-113.

Carr, D., Gilbertson, L., Frymeyer, J., Krag, M., and Pope, M. (1985) Lumbar paraspinal compartment syndrome: A case report with physiologic and anatomic studies. *Spine*, 10: 816.

Callaghan, J.P., Gunning, J.L., and McGill, S.M. (1998) Relationship between lumbar spine load and muscle activity during extensor exercises. *Physical Therapy*, 78 (1): 8-18.

Callaghan, J.P., and McGill, S.M. (2001a) Intervertebral disc herniation: Studies on a porcine model exposed to highly repetitive flexion/extension motion with compressive force. *Clinical Biomechanics*, 16 (1): 28-37.

Callaghan, J.P., and McGill, S.M. (2001b) Low back joint loading and kinematics during standing and unsupported sitting. *Ergonomics*, 44 (4): 373-381.

Callaghan, J.P., Patla, A.E., and McGill, S.M. (1999) Low back three-dimensional joint forces, kinematics and kinetics during walking. *Clinical Biomechanics*, 14: 203-216.

Cholewicki, J., and McGill, S.M. (1992) Lumbar posterior ligament involvement during extremely heavy lifts estimated from fluoroscopic measurements. *Journal of Biomechanics*, 25 (1): 17.

Cholewicki, J., and McGill, S.M. (1996). Mechanical stability of the in vivo lumbar spine: Implications for injury and chronic low back pain. *Clinical Biomechanics*, 11 (1): 1-15.

Cholewicki, J., McGill, S.M., and Norman, R.W. (1991) Lumbar spine loads during lifting extremely heavy weights. *Medicine and Science in Sports and Exercise*, 23 (10): 1179-1186.

Cholewicki, J., Juluru, K., and McGill, S.M. (1999) The intra-abdominal pressure mechanism for stabilizing the lumbar spine. *Journal of Biomechanics*, 32 (1): 13-17.

Cholewicki, J., Juluru, K., Radebold, A., Panjabi, M.M., and McGill, S.M. (1999) Lumbar spine stability can be augmented with an abdominal belt and/or increased intra-abdominal pressure. *European Spine Journal*, 8: 388-395.

Crisco, J.J., and Panjabi, M.M. (1990) Postural biomechanical stability and gross muscular architecture in the spine (p. 438.). In: Winters, J., and Woo, S. (Eds.), *Multiple muscle systems*. New York: Springer-Verlag.

Cripton, P., Berlemen, U., Visarino, H., Begeman, P.C., Nolte, L.P., and Prasad, P. (1995) Response of the lumbar spine due to shear loading. In: *Injury prevention through biomechanics* (p. 111). Detroit: Wayne State University.

Dangaria, T.R., and Naesh, O. (1998) Changes in cross-sectional area of psoas major muscle in unilateral sciatica caused by disc herniation. *Spine*, 23 (8): 928-931.

Davis, P.R. (1959) The causation of herniae by weight-lifting. *Lancet*, 2: 155.

Duncan, N.A., and Ahmed, A.M. (1991) The role of axial rotation in the etiology of unilateral disc prolapse: An experimental and finite-element analysis. *Spine*, 16: 1089-1098.

Farfan, H.F. (1973) *Mechanical disorders of the low back*. Philadelphia: Lea and Febiger.

Flint, M.M. (1965) Abdominal muscle involvement during performance of various forms of sit-up exercises: Electromyographic study. *American Journal of Physical Medicine*, 44: 224-234.

Fyhrie, D.P., and Schaffler, M.B. (1994) Failure mechanisms in human vertebral cancellous bone. *Bone*, 15 (1): 105-109.

Gardner-Morse, M., Stokes, I.A.F., Laible, J.P. (1995) Role of the muscles in lumbar spine stability in maximum extension efforts. *Journal of Orthopaedic Research*, 13: 802-808.

Gordon, S.J., et al. (1991) Mechanism of disc rupture—A preliminary report. *Spine*, 16: 450.

Grabiner, M.D., Koh, T.J., and Ghazawi, A.E. (1992) Decoupling of bilateral excitation in subjects with low back pain. *Spine*, 17: 1219-1223.

Gracovetsky, S., Farfan, H.F., and Lamy, C. (1981) Mechanism of the lumbar spine. *Spine*, 6 (1): 249.

Green, J., Grenier, S., and McGill, S. (in press)

Grenier, S., Preuss, R., Russell, C., and McGill, S.M. (in press) Associations between low back disability sequellae and fitness variables.

Grew, N.D. (1980) Intra-abdominal pressure response to loads applied to the torso in normal subjects. *Spine*, 5 (2): 149.

Gunning, J.L., Callaghan, J.P., and McGill, S.M. (2001) The role of prior loading history and spinal posture on the compressive tolerance and type of failure in the spine using a porcine trauma model. *Clinical Biomechanics*, 16 (6): 471-480.

Hadler, N.M. (1991) Insuring against work capacity from spinal disorders. In: Frymoyer, J.W. (Ed.), *The adult spine* (pp. 77-83). New York: Raven Press.

Haig, A.J., LeBreck, D.B., and Powley, S.G. (1995) Paraspinal mapping: Quantified needle electromyography of the paraspinal muscles in persons without low back pain, *Spine*, 20 (6): 715-721.

Halpern, A.A., and Bleck, E.E. (1979) Sit-up exercises: An electromyographic study. *Clinical Orthopaedics and Related Research*, 145: 172-178.

Hardcastle, P., Annear, P., and Foster, D. (1992) Spinal abnormalities in young fast bowlers. *Journal of Bone and Joint Surgery*, 74B (3): 421.

Herrin, G.A., Jaraiedi, M., and Anderson, C.K. (1986) Prediction of overexertion injuries using biomechanical and psychophysical models. *American Industrial Hygiene Association Journal*, 47: 322-330.

Hides, J.A., Richardson, C.A., and Jull, G.A. (1996) Multifidus muscle recovery is not automatic following resolution of acute first episode low back pain. *Spine*, 21: 2763-2769.

Hilkka, R., Mattsson, T., Zitting, A., Wickstrom, G., Hanninen, K., and Waris, P. (1990) Radiographically detectable degenerative changes of lumbar spine among concrete reinforcement workers and house painters. *Spine*, 15, 114-119.

Hodges, P.W., and Richardson, C.A. (1996) Inefficient muscular stabilisation of the lumbar spine associated with low back pain: A motor control evaluation of transversus abdominis. *Spine*, 21: 2640-2650.

Hodges, P.W., and Richardson, C.A. (1999) Altered trunk muscle recruitment in people with low back pain with upper limb movement at different speeds. *Archives of Physical Medicine and Rehabilitation*, 80: 1005-1012.

Holm, S., and Nachemson, A. (1983) Variations in the nutrition of the canine intervertebral disc induced by motion. *Spine*, 8: 866-874.

Holmes, A.D., Hukins, D.W.L., and Freemont, A.J. (1993) End-plate displacement during compression of lumbar vertebra-disc-vertebra segments and the mechanism of failure. *Spine*, 18: 128-135.

Hukins, D.W.L., Aspden, R.M., and Hickey, D.S. (1990) Thoracolumbar fascia can increase the efficiency of the erector spinae muscles. *Clinical Biomechanics*, 5 (1): 30.

Jager, M., Luttman, A., and Laurig, W. (1991) Lumbar load during one-handed bricklaying. *International Journal of Industrial Ergonomics*, 8: 261-277.

Jayson, M., and Dixon, A. (1970) Intra-articular pressure in rheumatoid arthritis of the knee III, Pressure changes during joint use. *Annals of the Rheumatic Diseases*, 29: 401-408.

Jette, M., Sidney, K., and Cicutti, N. (1984) A critical analysis of sit-ups: A case for the partial curl-up as a test of muscular endurance. *Canadian Journal of Physical Education and Recreation*, Sept-Oct: 4-9.

Jorgensen, K., and Nicolaisen, T. (1987) Trunk extensor endurance: Determination and relation to low back trouble. *Ergonomics*, 30: 259-267.

Juker, D., McGill, S.M., and Kropf, P. (1998) Quantitative intramuscular myoelectric activity of lumbar portions of psoas and the abdominal wall during cycling. *Journal of Applied Biomechanics*, 14 (4): 428-438.

Juker, D., McGill, S.M., Kropf, P., and Steffen, T. (1998) Quantitative intramuscular myoelectric activity of lumbar portions of psoas and the abdominal wall during a wide variety of tasks. *Medicine and Science in Sports and Exercise*, 30 (2): 301-310.

Kelsey, J.L. (1975) An epidemiological study of the relationship between occupations and acute herniated lumbar intervertebral discs. *International Journal of Epidemiology*, 4: 197-205.

Kirkaldy-Willis, W.H., and Burton, C.V. (1992) *Managing low back pain* (3rd ed.). New York: Churchill Livingstone.

Kornberg, M. (1988) MRI diagnosis of traumatic Schmorles nodes. *Spine*, 13: 934-935.

Krag, M.H., Byrne, K.B., Gilbertson, L.G., and Haugh, L.D. (1986) Failure of intraabdominal pressurization to reduce erector spinae loads during lifting tasks (p. 87). In *Proceedings of the North American Congress on Biomechanics*, Montreal, August 25-27.

Krag, M.H., Seroussi, R.E., Wilder, D.G., and Pope, M.H. (1987) Internal displacement distribution from in vitro loading of human thoracic and lumbar spinal motion segments: Experimental results and theoretical predictions. *Spine*, 12 (10): 1001.

Kumar, S. (1990) Cumulative load as a risk factor for back pain. *Spine*, 15: 1311-1316.

Lotz, J.C., and Chin, J.R. (2000) Intervertebral disc cell death is dependent on the magnitude and duration of spinal loading. *Spine*, 25 (12): 1477-1483.

Lucas, D., and Bresler, B. (1961) Stability of the ligamentous spine. Tech. Report No. 40, Biomechanics Laboratory, University of California, San Francisco.

Macintosh, J.E., and Bogduk, N. (1987) The morphology of the lumbar erector spinae. *Spine*, 12 (7): 658.

Macintosh, J.E., Bogduk. N., and Gracovetsky, S. (1987) The biomechanics of the thoracolumbar fascia. *Clinical Biomechanics*, 2: 78.

Marras, W.S., Lavender, S.A., Leurgens, S.E., et al. (1993) The role of dynamic three-dimensional trunk motion in occupationally related low back disorders: The effects of workplace factors trunk position and trunk motion characteristics on risk of injury. *Spine,* 18: 617-628.

McGill, S.M. (1987). A biomechanical perspective of sacro-iliac pain. *Clinical Biomechanics*, 2 (3): 145-151.

McGill, S.M. (1991). The kinetic potential of the lumbar trunk musculature about three orthogonal orthopaedic axes in extreme postures. *Spine*, 16 (7): 809-815.

McGill, S.M. (1997) Invited paper: Biomechanics of low back injury: Implications on current practice and the clinic. *Journal of Biomechanics,* 30 (5): 465-475.

McGill, S.M. (1998) Invited paper: Low back exercises: Evidence for improving exercise regimens. *Physical Therapy*, 78 (7): 754-765.

McGill, S.M., and Axler, C.T. (1996) Changes in spine height throughout 32 hours of bedrest: Implications for bedrest and space travel on the low back. *Archives of Physical Medicine and Rehabilitation,* 38 (9): 925-927.

McGill, S.M., and Brown, S. (1992) Creep response of the lumbar spine to prolonged lumbar flexion. *Clinical Biomechanics*, 7: 43.

McGill, S.M., and Callaghan, J.P. (1999) Impact forces following the unexpected removal of a chair while sitting. *Accident Analysis and Prevention*, 31: 85-89.

McGill, S.M., Juker, D., and Axler, C. (1996). Correcting trunk muscle geometry obtained from MRI and CT scans of supine postures for use in standing postures. *Journal of Biomechanics,* 29 (5): 643-646.

McGill, S.M., van Wijk, M., Axler, C.T., Gletsu, M. (1996) Spinal shrinkage: Is it useful for evaluation of low back loads in the workplace? *Ergonomics,* 39 (1): 92-102.

McGill, S.M., Juker, D., and Kropf, P. (1996) Quantitative intramuscular myoelectric activity of quadratus lumborum during a wide variety of tasks. *Clinical Biomechanics*, 11 (3): 170.

McGill, S.M., and Kippers, V. (1994) Transfer of loads between lumbar tissues during the flexion relaxation phenomenon. *Spine*, 19 (19): 2190.

McGill, S.M., and Norman, R.W. (1987) Reassessment of the role of intraabdominal pressure in spinal compression. *Ergonomics*, 30 (11): 1565.

McGill, S.M., and Norman, R.W. (1988) The potential of lumbodorsal fascia forces to generate back extension moments during squat lifts. *Journal of Biomedical Engineering*, 10: 312.

McGill, S.M., Sharratt, M.T., and Seguin, J.P. (1995) Loads on spinal tissues during simultaneous lifting and ventilatory challenge. *Ergonomics,* 38: 1772-1792.

McGill, S.M., van Wijk, M., Axler, C.T., and Gletsu, M. (1996). Spinal shrinkage: Is it useful for evaluation of low back loads in the workplace? *Ergonomics*, 39 (1): 92-102.

McGill, S.M., Yingling, V.R., and Peach, J.P. (1999) Three dimensional kinematics and trunk muscle myoelectric activity in the elderly spine: A database compared to young people. *Clinical Biomechanics*, 14 (6): 389-395.

McGlashen, K.M., Miller, J.A.A., Shultz, A.B., and Anderson, G.B.J. (1987) Load displacement behaviour of the human lumbosacral joint. *Journal of Orthopedic Research*, 5: 488.

McKenzie, R.A. (1979) Prophylaxis in recurrent low back pain. *New Zealand Medical Journal,* 89: 22.

Mody, G.M., and Cassim, B. (1997) Rheumatologic manifestations of malignancy. *Current Opinions in Rheumatology,* 9: 75-79.

Morris, J.M., Lucas, D.B., Bresler, B. (1961) Role of the trunk in stability of the spine. *Journal of Bone and Joint Surgery,* 43A: 327-351.

Nachemson, A.L., and Morris, J.M. (1964) In vivo measurements of intradiscal pressure. *Journal of Bone and Joint Surgery,* 46A: 1077.

Nachemson, A. (1966) The load on lumbar discs in different positions of the body. *Clinical Orthopaedics and Related Research,* 45: 107.

Nachemson, A., Andersson, G.B.J., and Schultz, A.B. (1986) Valsalva manoeuvre biomechanics: Effects on lumbar trunk loads of elevated intra-abdominal pressure. *Spine,* 11 (5): 476.

Norman, R., Wells, R., Neumann, P., Frank, P., Shannon, H., and Kerr, M. (1998) A comparison of peak vs cumulative physical work exposure risk factors for the reporting of low back pain in the automotive industry. *Clinical Biomechanics,* 13: 561-573.

Noyes, F.R., De Lucas, J.L., and Torvik, P.J. (1994) Biomechanics of ligament failure: An analysis of strain-rate sensitivity and mechanisms of failure in primates. *Journal of Bone and Joint Surgery,* 56A: 236.

Nutter, P. (1988) Aerobic exercise in the treatment and prevention of low back pain. *State of the Art Review of Occupational Medicine,* 3: 137.

Oganov, V.S., Rakhmanov, A.S., Novikov, V.E., Zatsepin, S.T., Rodionova, S.S., and Cann, C. (1991) The state of human bone tissue during space flight. *Acta Astronomy,* 213: 129-133.

Pearcy, M.J., Portek, J., and Shepherd, J. (1984) Three dimensional x-ray analysis of normal measurement in the lumbar spine. *Spine,* 9: 294.

Pearcy, M.J., and Tibrewal, S.B. (1984) Axial rotation and lateral bending in the normal lumbar spine measured by three-dimensional radiography. *Spine,* 9: 582.

Phillips, P.E. (1997) Viral arthritis. *Current Opinions in Rheumatology,* 9: 337-344.

Rantanen, J., Hurme, M., Falck, B., et al. (1993) The lumbar multifidus muscle five years after surgery for a lumbar intervertebral disc herniation. *Spine,* 18: 568-574.

Reilly, T., Tynell, A., and Troup, J.D.G. (1984) Circadian variation in human stature. *Chronobiology International,* 1: 121.

Richardson, C., Jull, G., Hodges, P., and Hides, J. (1999) Therapeutic exercise for spinal segmental stabilization in low back pain. Edinburgh, Scotland: Churchill Livingstone.

Roaf, R. (1960) A study of the mechanics of spinal injuries. *Journal of Joint and Bone Surgery,* 42B: 810.

Ross, J.K., Bereznik, D., and McGill, S.M. (1999) Atlas-axis facet asymmetry: Implications for manual palpation. *Spine,* 24 (12): 1203-1209.

Rossignol, M., Stock, S., Patry, L., and Armstrong, B. (1997) Carpal tunnel syndrome: What is attributable to work? The Montreal study. *Occupational and Environmental Medicine,* 54: 519-523.

Roy, S.H., De Luca, C.J., Emley, M., and Buijs, R.J.C. (1995) Spectral electromyographic assessment of back muscles in patients with low back pain undergoing rehabilitation. *Spine,* 20: 38-48.

Samii, K., Cassinotti, P., de Freudenreich, J., Gallopin, Y., Le Fort, D., and Stalder, H. (1996) Acute bilateral carpal tunnel syndrome associated with the human parvovirus B19 infection. *Clinical Infectious Diseases: An Official Publication of the Infectious Diseases Society of America,* 22: 162-164.

Santaguida, P., and McGill, S.M. (1995) The psoas major muscle: A three dimensional mechanical modelling study with respect to the spine based on MRI measurement. *Journal of Biomechanics,* 28: 339-345.

Schultz, A.B., Warwick, D.N., Berkson, M.H., and Nachemson, A.L. (1979) Mechanical properties of human lumbar spine motion segments. Part I: Response in flexion, extension, lateral bending and torsion. *Journal of Biomechanical Engineering,* 101: 46.

Siddall, P.J., and Cousins, M.J. (1997) Spine pain mechanisms. *Spine,* 22: 98-104.

Sihvonen, T., Lindgren, K., Airaksinen, O., and Manninen, H. (1997) Movement disturbances of the lumbar spine and abnormal back muscle electromyographic findings in recurrent low back pain. *Spine,* 22: 289-295.

Snook, S.H., Webster, B.S., McGorry, R.W., Fogleman, M.T., and McCann, K.B. (1998) The reduction of chronic nonspecific low back pain through the control of early morning lumbar flexion. *Spine,* 23: 2601-2607.

Spencer, D.L., Miller, J.A.A., and Schultz, A.B. (1985) The effects of chemonucleolysis on the mechanical properties of the canine lumbar disc. *Spine,* 10: 555.

Stokes, M., and Young, A. (1984) The contribution of reflex inhibition to arthrogenous muscle weakness. *Clinical Science,* 67: 7-14.

Tanaka, N., An, H.S., Lim, T-H., Fujiwara, A., Jeon, C., and Haughton, V.M. (2001) The relationship between disc degeneration and flexibility of the lumbar spine. *Spine,* 1: 47-56.

Tesh, K.M., Dunn, J., and Evans, J.H. (1987) The abdominal muscles and vertebral stability. *Spine*, 12 (5): 501.

Thomson, K.D. (1988) On the bending moment capability of the pressurized abdominal cavity during human lifting activity. *Ergonomics*, 31 (5): 817.

Troup, J.D.G., Martin, J.W., and Lloyd, D.C.E.F. (1981) Back pain in industry—A prospective study. *Spine,* 6: 61-69.

Vernon-Roberts, B., and Pirie, C.J. (1973) Healing trabecular microfractures in the bodies of lumbar vertebrae. *Annals of the Rheumatic Diseases,* 32: 406-412.

Videman, T., Nurminen, M., and Troup, J.D.G. (1990) Lumbar spinal pathology in cadaveric material in relation to history of back pain, occupation and physical loading. *Spine*, 15 (8): 728.

White, A.A., and Panjabi, M.M. (1978) *Clinical biomechanics of the spine.* Philadelphia: J.B. Lippincott.

Wilder, D.G., Pope, M.H., and Frymoyer, J.W. (1988) The biomechanics of lumbar disc herniation and the effect of overload and instability. *Journal of Spinal Disorders,* 1 (1): 16.

Woo, S.L.-Y., Gomez, M.A., and Akeson, W.H. (1985) Mechanical behaviors of soft tissues: Measurements, modifications, injuries, and treatment. In: Nahum, H.M., and Melvin, J. (Eds.), *Biomechanics of trauma* (pp. 109-133). Norwalk, CT: Appleton Century Crofts.

Woolf, C.J., Bennett, G.J., Doherty, M., Dubner, R., Kidd, B., Koltzenburg, M., Lipton, R., Loeser, J.D., Payne, R., and Torebjork, E. (1998) Towards a mechanism-based classification of pain? *Pain*, 77: 227-229.

Yingling, V.R., and McGill, S.M. (1999a) Anterior shear of spinal motion segments: Kinematics, kinetics and resulting injuries observed in a porcine model. *Spine,* 24 (18): 1882-1889.

Yingling, V.R., and McGill, S.M. (1999b) Mechanical properties and failure mechanics of the spine under posterior shear load: Observations from a porcine model. *Journal of Spinal Disorders,* 12 (6): 501-508.

CHAPTER 6

Lumbar Spine Stability: Myths and Realities

Stability is a popular term when discussing the low back, but it may be widely misunderstood and inappropriately used. In previous chapters we established several relevant facts. First, all sorts of tissue damage results in joint laxity, which in turn can lead to instability. For example, strained or failed ligaments cause joint laxity and unstable motion under load. End-plate fractures with loss of disc height are another example of tissue damage that results in unstable joint behavior. Clearly, joint instability is a consequence of tissue damage (this is nicely summarized by Oxland et al., 1991). A fundamental tenet is that lost mechanical integrity in any load-bearing tissue will result in stiffness losses and an increased risk of unstable behavior. We also saw in chapter 5 that during an event in which instability was observed (the buckling power lifter's spine), injury resulted. So, instability can both cause and be the result of injury. Finally, overlaying the tissue-based aspects of stability are the motor control aspects since coordinated contraction stiffens the joints and ultimately determines joint stability.

The purpose of this chapter is to provide a definition of stability and an understanding of how it is increased or decreased. This is a critical foundation for those prescribing stabilization exercise or recommending strategies to prevent injury.

Attempts to enhance stability and prevent instability are compromised without an understanding of the influencing factors. To quantify those factors, however, we must agree on definitions. What exactly do we mean when we use the terms *spine stability*, *core stability*, and *stabilization exercise*? Often they depend on the background of the individual: To the biomechanist they pertain to a mechanical structure that can become unstable when a "critical point" is reached; a surgeon may view abnormal joint motion patterns as unstable but correctable by changing the anatomy; and the manual medicine practitioner may interpret patterns of muscle coordination and posture as indicative of instability and attempt to alter one, or a few, muscle activation profiles. Several groups have made contributions to the stability issue, but only a very few have attempted to actually quantify stability. This critical issue is addressed here.

Upon completion of this chapter, you will understand stability and its importance in injury prevention and rehabilitation. Furthermore, you will understand why certain approaches are preferable for achieving sufficient stability.

Stability: A Qualitative Analogy

The following demonstration of structural stability illustrates key issues. Suppose a fishing rod is placed upright and vertical with the butt on the ground. If the rod were to have a small load placed in its tip, perhaps a pound or two, it would soon bend and buckle. Now suppose the same rod has guy wires attached at different levels along its length and those wires are also attached to the ground in a circular pattern (see figure 6.1, a-b). Each guy wire is pulled to the same tension (this is critical). Now if the tip of the rod is loaded as before, the rod can sustain the compressive forces successfully. If you reduce the tension in just one of the wires, the rod will buckle at a reduced load; we could actually predict the node, or locus, of the buckle.

Compressive loading similar to this has been performed on human lumbar spines. Typically, an osteoligamentous lumbar spine from a cadaver with muscles removed (and no guy wires) will buckle under approximately 90 N (about 20 lb) of compressive load (first noted by Lucas and Bresler, 1961). This is all that a spine can withstand!

This analogy demonstrates the critical role of the muscles (the guy wires) to first ensure sufficient stability of the spine so that it is prepared to withstand loading and sustain postures and movement. Also demonstrated with this example is the role of the motor control system, which ensures that the tensions in the cables are proportional so as to not create a nodal point where buckling will occur. Revisiting the buckling injury that we observed fluoroscopically in the power lifter (page 125), we would hypothesize that it was caused by a motor control error in which possibly one muscle reduced its activation or, from the previous analogy, lost its stiffness. The synchrony of balanced stiffness produced by the motor control system is absolutely critical. Now we can address how stability is quantified and modulated.

Figure 6.1 The spine is analogous to a fishing rod placed upright with the butt on the ground. When compressive load is applied downward to the tip, it will buckle quickly *(a)*. Attaching guy wires at different levels and in different directions and, most important, tensioning each guy wire to the same tension will ensure stability even with massive compressive loads *(b)*. Note that the guy wires need not have high tension forces but that the tensile forces must be of roughly equal magnitude. This is the role of the musculature in ensuring sufficient spine stability.

Quantitative Foundation of Stability

This section quantifies the notion of stability from a spine perspective. During the 1980s, Professor Anders Bergmark of Sweden very elegantly formalized stability in a spine model with joint stiffness and 40 muscles (Bergmark, 1987). In this classic work he was able to represent mathematically the concepts of "energy wells," stiffness, stability, and instability. For the most part, this seminal work went unrecognized largely because the engineers who understood the mechanics did not have the biological/clinical perspective and the clinicians were hindered in the interpretation and implications of the engineering mechanics. This section synthesizes Bergmark's pioneering effort and its continued evolution in the work of several others and attempts to encapsulate the critical notions without mathematical complexity. If you are mathematically inclined, see Bergmark's original work or its formalization by Cholewicki and McGill (1996).

Potential Energy as a Function of Height

The concept of stability begins with potential energy, which for our purposes here is of two basic forms. In the first form, objects have potential energy (PE) by virtue of their height above a datum.

$$PE = mass \cdot gravity \cdot height$$

Critical to measuring stability are the notions of energy wells and minimum potential energy. A ball in a bowl is considered stable because if a force (or a perturbation) were applied to it, it would rise up the side of the bowl but then come to rest again in the position of least potential energy at the bottom of the bowl (the energy well) (see figure 6.2, a-d). As noted by Bergmark, "stable equilibrium prevails when the potential energy of the system is minimum." The system is made more stable by deepening the bowl and/or by increasing the steepness of the sides of the bowl (see figure 6.3). Thus, the notion of stability encompasses

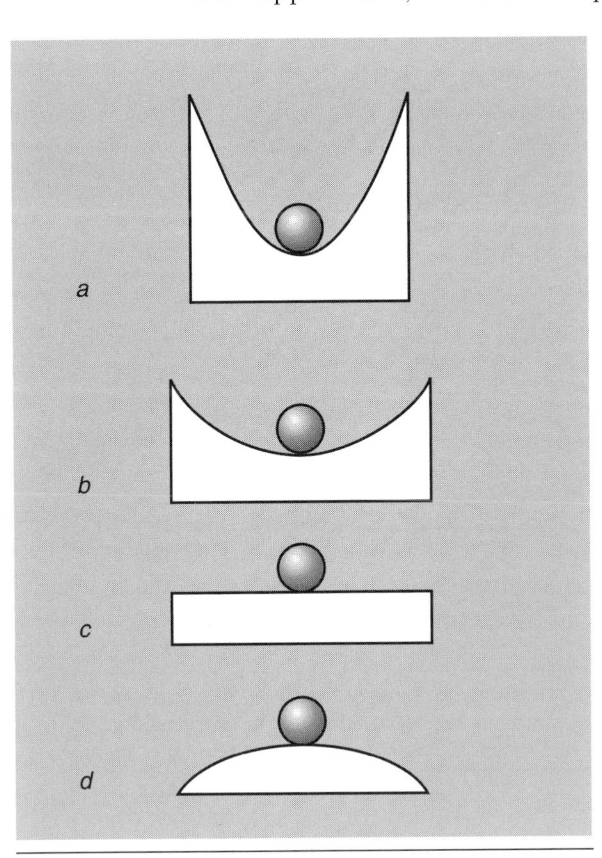

Figure 6.2 The continuum of stability. The deepest bowl *(a)* is most stable, and the hump *(d)* is least stable. The ball in the bowl seeks the energy well or position of minimum potential energy ($m \cdot g \cdot h$). Deepening the bowl or increasing the steepness of the sides increases the ability to survive perturbation. This increases stability.

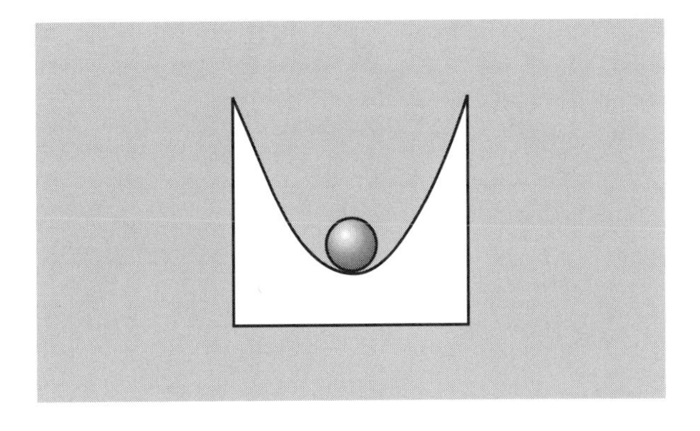

PE = m* g* h
Slope = joint stiffness
Width = joint laxity

Figure 6.3 The steepness of the sides of the bowl corresponds to the stiffness of the passive tissues of the joint, which create the mechanical stop to motion. The width of the bottom of the bowl corresponds to joint laxity. For example, a positive "drawer test" on the knee would be represented by a flattened bottom of the curve in which small applied forces produce large unopposed motion.

the unperturbed energy state of a system and a study of the system following perturbation—if the joules of work done by the perturbation are less than the joules of potential energy inherent to the system, then the system will remain stable (i.e., the ball will not roll out of the bowl). The corollary is that the mechanical system will become unstable and possibly collapse if the applied load exceeds a critical value (determined by potential energy and stiffness).

The previous ball analogy is a two-dimensional example. This would be analogous to a hinged skeletal joint that has the capacity only for flexion/extension. Spinal joints can rotate in three planes and translate along three axes, requiring a six-dimensional bowl for each joint. Mathematics enables us to examine a 36-dimensional bowl (6 lumbar joints with 6 degrees of freedom each) representing the whole lumbar spine. If the height of the bowl were decreased in any one of these 36 dimensions, the ball could roll out. In clinical terms, a single muscle having an inappropriate force (and thus stiffness) or a damaged passive tissue that has lost stiffness can cause instability that is both predictable and quantifiable.

Some clinicians have confused spine stability with whole body balance and stability, which involves the center of mass and base of support in the context of falling over; this is quite different from spine stability. Figure 6.4, a and b illustrate the mechanics of whole body balance and stability. Figure 6.5, a and b show the clinical practice of prodding patients. This approach is misguided for enhancing spine stability.

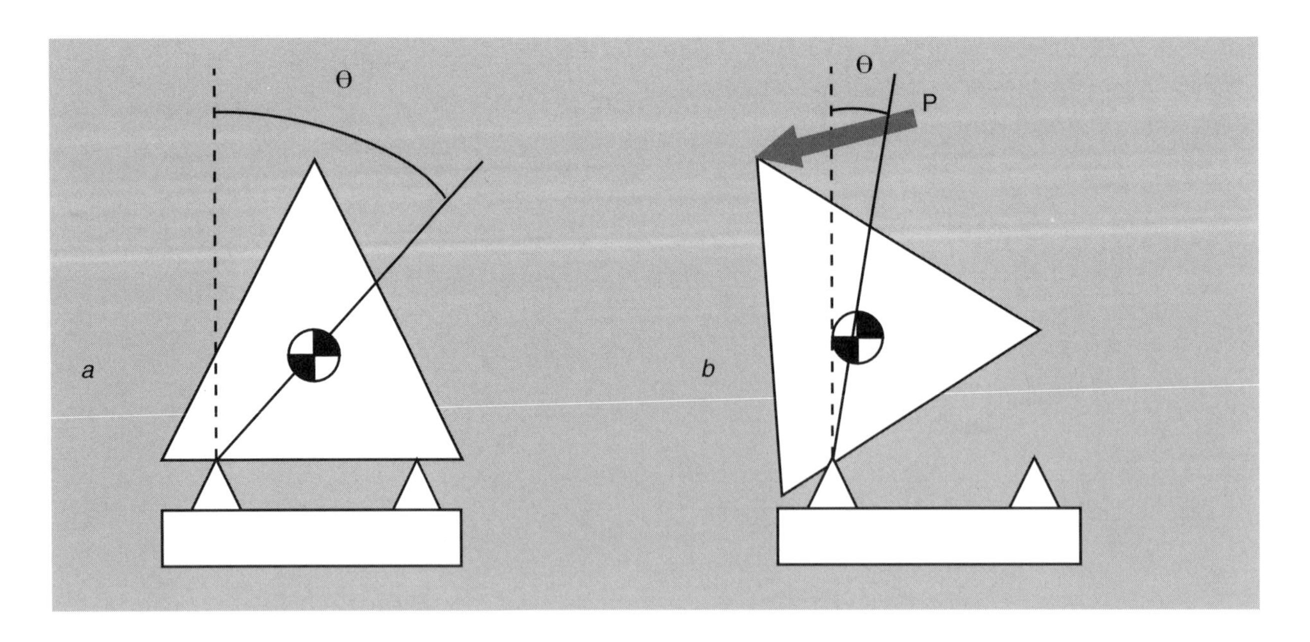

Figure 6.4 Another type of stability, often confused with spine column stability, involves the center of mass and base of support in the context of falling over. The triangle in *(a)* is stable because a small perturbation to its top would not cause it to fall. The size of this stability can be quantified, in part, by the size of the angle theta: as the center of the mass approaches a vertical line drawn from the base of support, it would require a smaller perturbing force to fall, demonstrating that it is stable but less so *(b)*.

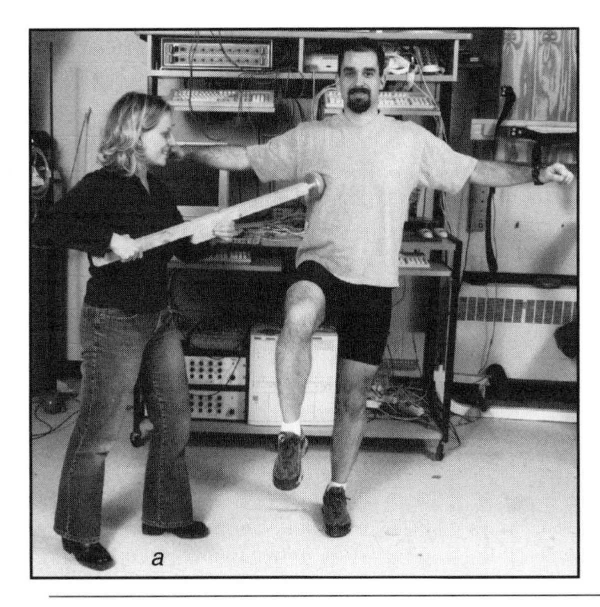

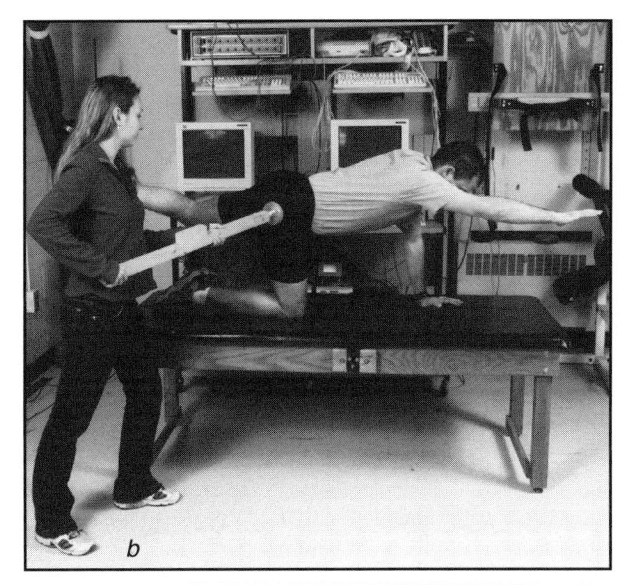

Figure 6.5 Misunderstanding stability, illustrated in the example given in figure 6.4, has led clinicians to try to enhance spine stability by prodding patients and attempting to knock them off balance *(a)* and *(b)*. This is not spine stability but whole body stability.

Potential Energy as a Function of Stiffness and Elastic Energy Storage

While potential energy by virtue of height is useful for illustrating the concept, potential energy as a function of stiffness and storage of elastic energy is actually used for musculoskeletal application. Elastic potential energy is calculated from stiffness (k) and deformation (x) in the elastic element, as shown here:

$$PE = 1/2 \cdot k \cdot x^2$$

We will use an elastic band as an example. Stretching the band with a stiffness (k) a distance x, it will store energy (PE). The stretched band has the potential to do work when stretched. In other words, the greater the stiffness (k), the greater the steepness of the sides of the bowl (from the previous analogy), and the more stable the structure. Thus, stiffness creates stability (see figure 6.6). More specifically, symmetric stiffness creates

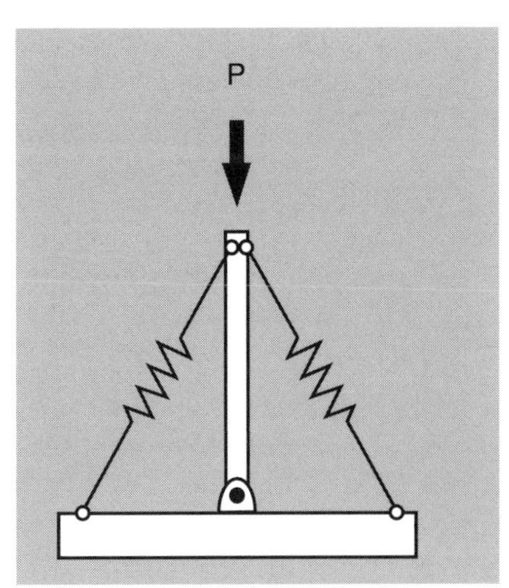

PE = 1/2* k* x^2
S: PE–P
S<0 : unstable
S>0 : stable

Figure 6.6 Increasing the stiffness of the cables (muscles) increases the stability (or deepens the bowl) and increases the ability to support larger applied loads (*P*) without falling. But most important, ensuring that the stiffness is balanced on each side enables the column to survive perturbation from either side. Increasing stiffness of just one spring will actually decrease the ability to bear compressive load (lowering PE in one direction and mode), illustrating the need for "balancing" stiffness for a given posture and moment demand.

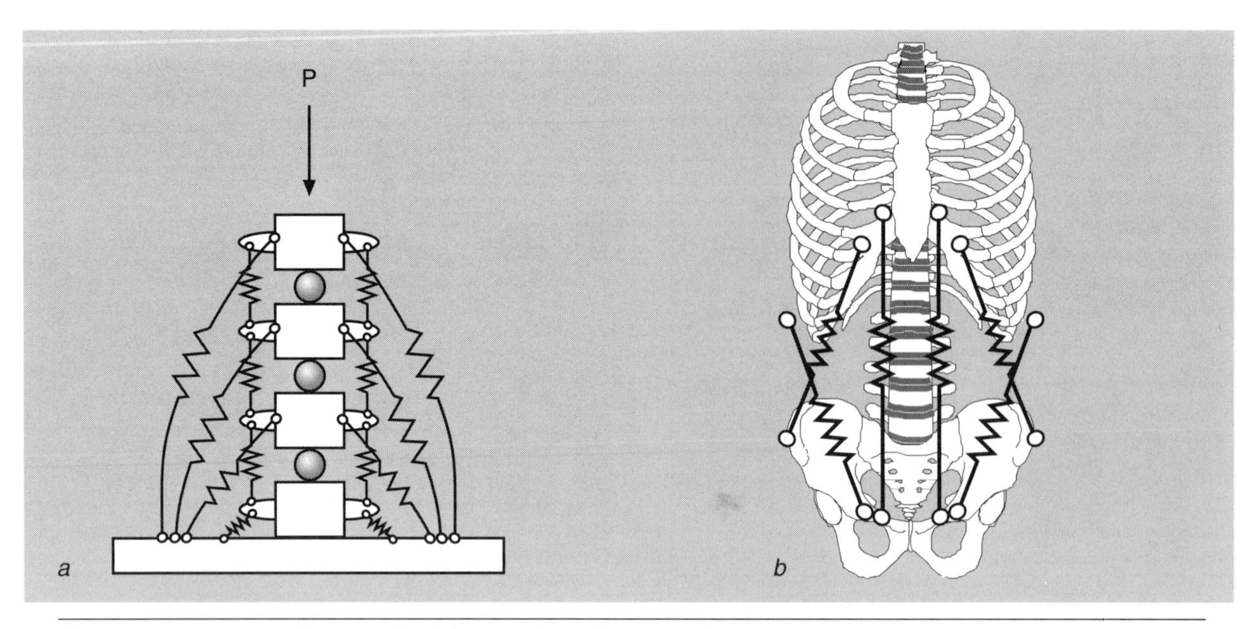

Figure 6.7 *(a)* Spine stiffness (and stability) is achieved by a complex interaction of stiffening structures along the spine and *(b)* those forming the torso wall. Balancing stiffness on all sides of the spine is more critical to ensuring stability than having high forces on a single side.

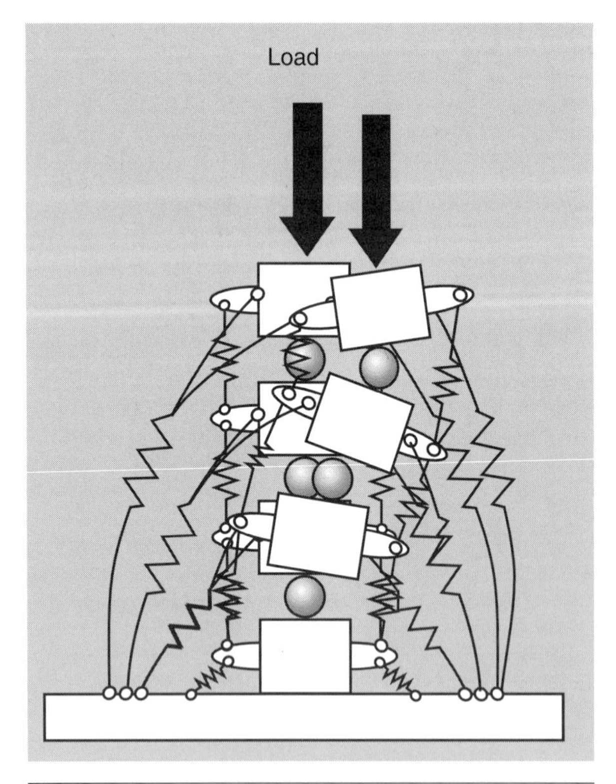

Figure 6.8 If more joules of work are performed on the spine than joules of potential energy due to stiffness, the spine will become unstable—in this case, the instability mode is buckling.

even more stability. (Symmetry in stiffness is achieved by virtually all muscles of the torso; see figure 6.7, a-b.) If more joules of work are performed on the spine than joules of potential energy due to stiffness, the spine will become unstable (see figure 6.8).

Active muscle acts like a stiff spring, and in fact, the greater the activation of the muscle, the greater this stiffness. Hoffer and Andreassen (1981) showed that joint stiffness increases rapidly and nonlinearly with muscle activation such that only very modest levels of muscle activity create sufficiently stiff and stable joints. Furthermore, joints possess inherent joint stiffness as the passive capsules and ligaments contribute stiffness, particularly at the end range of motion. The motor control system is able to control stability of the joints through coordinated muscle coactivation and to a lesser degree by placing joints in positions that modulate passive stiffness contribution. However, a faulty motor control system can lead to inappropriate magnitudes of muscle force and stiffness, allowing a valley for the ball to roll out or, clinically, for a joint to buckle or undergo shear translation. However, we are limited in our ability to analyze the local stability of mechanical systems and particularly musculoskeletal linkages, since the energy wells are not infinitely deep and the many anatomical components contribute force and stiffness in synchrony to create surfaces of potential energy in which many local wells exist. Thus, we locate local minima by examining the derivative of the energy surface. (See Bergmark, 1987; Cholewicki and McGill, 1996, for mathematical detail.)

Spine stability, then, is quantified by forming a matrix in which the total "stiffness energy" for each degree of freedom of joint motion is represented by a number (or **eigenvalue**), and the magnitude of that number represents its contribution to forming the height of the bowl in that particular dimension. Eigenvalues less than zero indicate the potential for instability. The **eigenvector** (different from the eigenvalue) can then identify the mode in which the instability occurred, while sensitivity analysis is used to reveal the possible contributors to unstable behavior. Gardner-Morse and colleagues (1995) initiated interesting investigations into eigenvectors by predicting patterns of spine deformation due to impaired muscular intersegmental control. Their question was, Which muscular pattern would have prevented the instability? Crisco and Panjabi (1992) began investigations into the contributions of the various passive tissues. Our group has been investigating the eigenvector by systematically adjusting the stiffness of each muscle and assessing stability in a variety of tasks and exercises. The contributions of individual muscles to stability are shown later in this chapter.

Activating a group of muscle **synergists** and **antagonists** in the optimal way now becomes a critical issue. In clinical terms, the full complement of the stabilizing musculature must work harmoniously to ensure stability, generation of the required moment, and desired joint movement. But only one muscle with inappropriate activation amplitude may produce instability, or at least unstable behavior could result from inappropriate activation at lower applied loads.

Sufficient Stability

How much stability is necessary? Obviously, insufficient stiffness renders the joint unstable, but too much stiffness and coactivation imposes massive load penalties on the joints and prevents motion. We can define sufficient stability as the muscular stiffness necessary for stability, with a modest amount of extra stability to form a margin of safety. Interestingly, given the rapid increase in joint stiffness with modest muscle force, large muscular forces are rarely required.

Cholewicki's work (Cholewicki and McGill, 1996; Cholewicki, Simons, and Radebold, 2000) demonstrated that, in most people with an undeviated spine, modest levels of coactivation of the paraspinal and abdominal wall muscles will result in sufficient stability of the lumbar spine. This means that people, from patients to athletes, must be able to maintain sufficient stability in all activities—with low but continuous muscle activation. Thus, maintaining a stability margin of safety when performing tasks, particularly the tasks of daily living, is not compromised by insufficient strength but probably by insufficient endurance. We are now beginning to understand the mechanistic pathway of those studies showing the efficacy of endurance training for the muscles that stabilize the spine. Having strong abdominals does not necessarily provide the prophylactic effect that many hoped for. However, several works suggest that muscular endurance reduces the risk of future back troubles (Biering-Sorensen, 1984). Finally, the Queensland group (e.g., Richardson et al., 1999) and several others (e.g., O'Sullivan, Twomey, and Allison, 1997) noted the disturbances in the motor control system following injury (detailed in chapter 5). These disturbances compromise the ability to maintain sufficient stability. In summary, stability comes from stiffness, passive stiffness is lost with tissue damage, and active stiffness throughout the range of motion is lost with perturbed motor patterns following injury.

Stability Myths, Facts, and Clinical Implications

Having explored some of the issues of spine stability, we will now consider a few crucial questions that will help enhance clinical decisions.

• **How much muscle activation is needed to ensure sufficient stability?** The amount of muscle activation needed to ensure sufficient stability depends on the task. Generally, for most tasks of daily living, very modest levels of abdominal wall cocontraction (activation of about 10% of MVC or even less) are sufficient. Again, depending on the task, cocontraction with the extensors (this will also include the quadratus) will ensure stability. However, if a joint has lost stiffness because of damage, more cocontraction is needed. A specific example is shown in chapter 13.

• **Is any single muscle most important?** Several clinical groups have suggested focusing on one or two muscles to enhance stability. This would be similar to emphasizing a single guy wire in the fishing rod example. Rarely would it help but rather would be detrimental for achieving the balance in stiffness needed to ensure stability throughout the changing task demands. In particular, clinical groups have emphasized the multifidus and transverse abdominis. The Queensland group performed some of the original work emphasizing these two muscles. This was based on their research, which noted motor disturbances in these muscles following injury. In fact, they developed a tissue damage model that suggested that chronically poor motor control (and motion patterns) initiates microtrauma in tissues that accumulates, leading to symptomatic injury (Richardson et al., 1999). Further, according to the Queensland model, injury leads to further deleterious change in motor patterns such that chronicity can only be broken with specific techniques to reeducate the local muscle–motor control system. The intentions of the Queensland group were to address the documented motor deficits and attempt to reduce the risk of aberrant motor patterns that could lead to the pathology-inducing patterns in their damage model. In other words, their recommendations appear to be directed toward reeducating the faulty motor patterns. However, many clinical groups have interpreted this approach to mean that these two muscles should be the specific targets when teaching stability maintenance over all sorts of tasks. Hopefully the understanding obtained from the stability explanation provided earlier will underscore the folly of this unidimensional emphasis. In fact, the multifidus or any other muscle can be the most important stabilizer at an instance in time. The transverse abdominis, when quantified for its influence on stability, acts the same way. Four exercises are shown as an example (see figure 6.9) where the contributors to stability are shown. You should realize that they will change and adjust with subtle changes in relative muscle activation. In these examples entire muscles are adjusted to evaluate their role. Obviously, adjustments in single laminae of some muscles would result in yet a different distribution of contributions to stability. Evaluating a wide variety of tasks and exercises is most revealing.The muscular and motor control system must satisfy requirements to sustain postures, create movements, brace against sudden motion or unexpected forces, build pressure, and assist challenged breathing, all the while ensuring sufficient stability. Virtually all muscles play a role in ensuring stability, but their importance at any point in time is determined by the unique combination of the demands just listed.

• **Do local/global or intrinsic/extrinsic stabilizers exist?** From the classic definition and quantification of stability, the answer would be no. Stability results from the stiffness at each joint in a particular degree of freedom. The relative contribution from every muscle source is dynamically changing depending on its need to contract for other purposes. The way the various contributors to stiffness add up, however, is important. In some instances, removing a muscle from the analysis has very little effect on joint stability. Once again, it depends on the demands and constraints unique to the task at that instant. The point is that all contributors are important for some tasks and should be recognized as potentially important in any prevention or rehabilitation

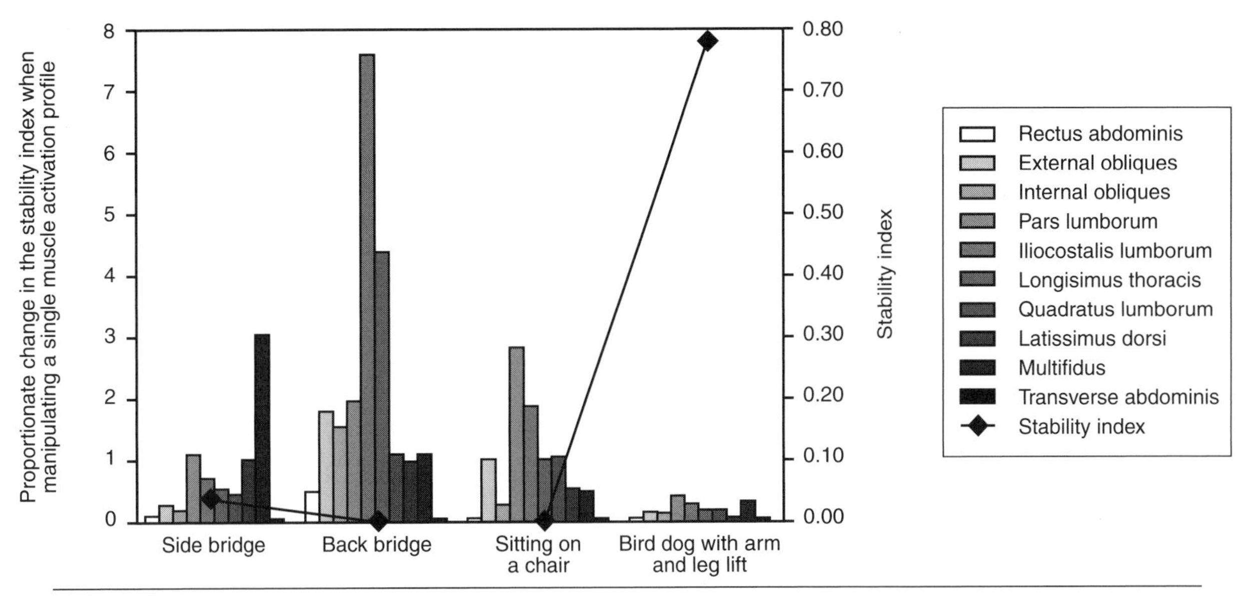

Figure 6.9 Measuring individual muscle influence on lumbar spine stability while performing four different "stabilization" exercises demonstrates that all muscles are important and that their relative importance changes with the task. Total stability is also indicated by the determinant demonstrating how stability can change between tasks. (This is an example from a single subject. Typically, group means indicate a different stability ranking of exercises.)

Courtesy of N. Kavcic and S.McGill.

program. Conversely, conceiving stabilizers as "intrinsic" or "extrinsic" may offer no benefit for clinical decision making.

- **What are stabilization exercises?** Stability exercises and a stable core are often discussed in exercise forums. What are stabilization exercises? The fact is that any exercise can be a stabilization exercise depending on how it is performed. Sufficient joint stiffness is achieved by creating specific motor patterns. To adapt a phrase popular in motor control circles, "Practice does not make perfect; it makes permanent." Ideally, good stabilization exercises that are performed properly groove motor and motion patterns that ensure stability while satisfying all other demands. Thus, an exercise, repeated in a way that grooves motor patterns and ensures a stable spine, constitutes a stabilization exercise (McGill, 2001). However, some stabilization exercises are better than others; again, it depends on the objectives. For example, the resultant load on the spine is rarely considered. One key to improving bad backs is to select stabilization exercises that impose the lowest load on the damaged spine. Chapter 5 provides a ranking of such exercises. On the other hand, the best stabilization exercise for a high-performance athlete will involve the grooving of a dynamic and complex motion pattern, all the while ensuring sufficient spine stability. An important requirement for many athletes is to ensure a stable spine while breathing hard, such as when playing high-intensity sports. This will be discussed in chapter 14.

- **Can clinicians identify those who are poor stabilizers?** From a qualitative perspective, the Queensland group observed aberrant motor patterns that they hypothesized to compromise the ability of the affected person to stabilize efficiently. From a quantitative perspective, we have used modeling analysis to try to find ways to identify those who compromise their lumbar stability from specific motor control errors. We observed such inappropriate muscle sequencing in men who were challenged by holding a load in the hands while breathing 10% CO_2 to elevate breathing. On one hand, the muscles must cocontract to ensure sufficient spine stability, but on the other, challenged breathing is often characterized by a rhythmic contraction/relaxation of the abdominal wall (McGill, Sharratt, and Seguin, 1995). Thus, the motor system is presented with a conflict: Should the torso muscles remain active isometrically to maintain spine stability, or will they rhythmically relax and contract to assist with active expiration (but sacrifice spine stability)? Fit motor systems appear to meet the simultaneous breathing and spine support challenge; unfit ones may not. All of these deficient motor control mechanisms will heighten biomechanical susceptibility to injury or reinjury (Cholewicki and McGill, 1996). We are currently using this paradigm in a longitudinal study to see if those who bordered on instability during this challenge will be the ones who develop low back troubles. Some other tests to detect functional instability are described in chapter 12.

In summary, achieving stability is not just a matter of activating a few targeted muscles, be they the multifidus, transverse abdominis, or any other. Sufficient stability is a moving target that continually changes as a function of the three-dimensional torques needed to support postures. It involves achieving the stiffness needed to endure unexpected loads, preparing for moving quickly, and ensuring sufficient stiffness in any degree of freedom of the joint that may be compromised from injury. Motor control fitness is essential for achieving the stability target under all possible conditions for performance and injury avoidance.

References

Bergmark, A. (1987) Mechanical stability of the human lumbar spine. Doctoral dissertation, Department of Solid Mechanics, Lund University, Sweden.

Biering-Sorensen, F. (1984) Physical measurements as risk indicators for low back trouble over a one year period. *Spine*, 9: 106-119.

Cholewicki, J., and McGill, S.M. (1996) Mechanical stability of the in vivo lumbar spine: Implications for injury and chronic low back pain. *Clinical Biomechanics,* 11 (1): 1-15.

Cholewicki, J., Simons, A.P.D., and Radebold, A. (2000) Effects of external trunk loads on lumbar spine stability. *Journal of Biomechanics,* 33 (11): 1377-1385.

Crisco, J.J., and Panjabi, M.M. (1992) Euler stability of the human ligamentous lumbar spine. Part I: Theory and Part II: Experiment. *Clinical Biomechanics,* 7: 27-32.

Gardner-Morse, M., Stokes, I.A.F., and Laible, J.P. (1995) Role of the muscles in lumbar spine stability in maximum extension efforts. *Journal of Orthopaedic Research,* 13: 802-808.

Hoffer, J., and Andreassen, S. (1981) Regulation of soleus muscle stiffness in premamillary cats. *Journal of Neurophysiology,* 45: 267-285.

Lucas, D., and Bresler, B. (1961) Stability of the ligamentous spine. In: Tech report No. 40, Biomechanics Laboratory, University of California, San Francisco.

McGill, S.M. (2001) Invited review. Low back stability: From formal description to issues for performance and rehabilitation. *Exercise and Sports Science Reviews*, 29 (1): 26-31.

McGill, S.M., Sharratt, M.T., and Seguin, J.P. (1995) Loads on the spinal tissues during simultaneous lifting and ventilatory challenge. *Ergonomics,* 38 (9): 1772-1792.

O'Sullivan, P., Twomey, L.T., and Allison, G.T. (1997) Altered pattern of abdominal muscle activation in chronic back pain patients. *Australian Journal of Physiotherapy,* 43: 91-98.

Oxland, T.R., Panjabi, M.M., Southern, E.P., and Duranceau, J.S. (1991) An anatomic basis for spinal instability: A porcine trauma model. *Journal of Orthopaedic Research,* 9: 452-462.

Richardson, C., Jull, G., Hodges, P., and Hides, J. (1999) Therapeutic exercise for spinal segmental stabilization in low back pain. Edinburgh, Scotland: Churchill Livingstone.

PART II

Injury Prevention

In part I we went back to school for an update on lumbar function. In part II this foundation is used to justify the best injury prevention approaches.

LBD Risk Assessment

Two types of physical risk factors predispose people to developing low back troubles: those linked to the person (for example, muscle endurance or the lack of) and those linked to the demands of performing a certain task (for example, applied loads). This chapter describes how to assess the risk of back troubles that result from the task demands.

Upon completion of this chapter, you will understand the variables that are important to include for risk assessment and be familiar with different tools to assess the risk. These tools include the NIOSH approach, the Snook psychophysical approach, the lumbar motion monitor (LMM) approach, and the 4D WATBAK approach.

Tissue overload causes damage and subsequent low back troubles. Assessing the risk involves comparing applied loads with some form of load-bearing tolerance value. Chapter 4 included examples of applied loads, tissue loads, and injury scenarios. Although direct measurement of tissue loads would be ideal, this remains impractical for large numbers of people performing a wide variety of occupational tasks and activities of daily living. Researchers have developed various modeling approaches to predict these loads, but they remain methodologically complex. For this reason, surrogates for tissue load that are linked with posture, applied load, and motion have been proposed. These were introduced in chapter 3, Epidemiological Studies on Low Back Disorders (LBDs).

See "Brief Review of the Risk Factors for LBD" (page 150) for a list of known risk factors that also describes and critiques several approaches to assessing them.

Four risk assessment approaches are presented more or less from least complex to most complex. The simplest approaches use metrics that are the easiest and cheapest to employ for injury risk assessment. However, these are compromised by a lack of accuracy and specificity and are not sensitive to individual worker/person technique. The more complex approaches involve more sophisticated methods and are more expensive to conduct, but they are more robust in their sensitivity to specific tasks and the individual ways in which people elect to perform them.

CLINICAL RELEVANCE

Brief Review of the Risk Factors for LBD

Risk Factors Identified From Epidemiological Approaches

1. Static work postures, specifically prolonged trunk flexion and a twisted or laterally bent trunk posture
2. Seated working postures
3. Frequent torso motion, higher spine rotational velocity, and spine rotational deviations
4. Frequent lifting, pushing, and pulling
5. Vibration exposure, particularly seated whole-body vibration
6. Peak and cumulative low back shear force, compressive force, and extensor moment
7. Incidence of slips and falls

Risk Factors Identified From Tissue-Based Studies

8. Repeated full lumbar flexion
9. Time of day (or time after rising from bed)
10. Excessive magnitude and repetition of compressive loads, shear loads, and torsional displacement and moments
11. Insufficient loading so that tissue strength is compromised
12. Rapid ballistic loading (such as results during landing from a fall)

Personal Variables Identified As Risk Factors

13. Increased spine mobility (range of motion)
14. Lower torso muscle endurance
15. Perturbed motor control patterns
16. Age
17. Gender
18. Abdominal/torso girth

NIOSH Approach to Risk Assessment

For about the past 40 years both field surveillance and laboratory studies have focused on the relationship between back injury and compressive forces experienced by the lumbar spine. For example, in 1981 the National Institute for Safety and Health (NIOSH) published guidelines for maximum compressive loading of the lumbar spine (both an action limit of 3400 N [about 750 lb] and a maximum permissible limit of 6300 N [about 1400 lb]) based on some earlier evidence. In 1994 Leamon questioned the use of compression when he quite correctly stated that the supporting evidence for compression as a metric, or index of risk, was sparse. However, since that time, several good data sets have shown that restricting the amount of low back compression (both peak and cumulative) is one valid approach to reducing the risk of low back injury (for example, data sets in the previously discussed Norman et al., 1998, study and the Marras et al., 1995, study). In addition, NIOSH has also proposed lifting guidelines to limit the amount of load lifted in the hands. These limits are described in the following sections.

The influence of NIOSH has been far reaching; many groups have used these values in an attempt to reduce the risk in workers. In fact, the NIOSH approach continues to be widely used today primarily because of its ease of use.

1981 Guideline

The first lifting guide (NIOSH, 1981) was restricted to lifts in the sagittal plane and lifts that involved only slow and smooth motions. Predictions of the load lifted in the hands was based on some rudimentary distances to characterize the kinematics of the lift and load moment at the low back. Two limits were defined: the action limit (AL), which, if exceeded, triggered action to apply engineering and administrative controls, and the maximum permissible limit (MPL) for the load lifted, above which the risk is too high and not permissible. Weights lighter than the AL are considered safe. Experts with biomechanical, physiological, and psychophysical expertise chose the magnitude of these limits and the variables needed to compute them. The general form of the formula used to compute the AL for weight in the hands is as follows:

$$AL\ (kg) = 40\ (HF)\ (VF)\ (DF)\ (FF)$$

where

HF = horizontal factor, or the horizontal distance (H) from a point bisecting the ankles to the center of gravity of the load at the lift origin. This was defined as (15/H).

VF = vertical factor, or the height (V) of the load at lift origin. This was defined as (0.004)/(V-75).

DF = distance factor, or the vertical travel distance (D) of the load. This was defined as .7 + 7.5/D.

FF = frequency factor, or the lifting rate defined as 1 – F/Fmax.

F = average frequency of the lift, while Fmax is obtained from tabulated data.

The logic of the equation is to set a maximal load of 40 kg (88 lb) and multiply this value against variables that act as discount factors. Thus, the maximal lift is 40 kg (88 lb) under optimal conditions and when all discount variables equal unity (1). Suboptimal lifting conditions cause the discount multipliers to drop to smaller values, reducing the "safe" load.

The MPL is computed as three times the AL for a particular set of lifting circumstances:

$$MPL = 3(AL)$$

1993 Guideline

The revised NIOSH equation (Waters et al., 1993) was introduced to address those lifting tasks that violated the sagittally symmetric lifting task restriction of the earlier equation. In addition, the concept of the AL and MPL were replaced with a recommended weight limit (RWL) for a particular situation. If the actual load to be lifted exceeds the RWL, the risk of developing an LBD is elevated. While the RWL equation is similar to the 1981 equation, two additional factors were incorporated. These include an asymmetric variable for nonsagittal lifts and a factor for whether or not the object has handles. Note that the specific variable weightings have also changed from the 1981 equation form.

$$\text{RWL (kg)} = 23\,(25/H)\,[1 - (0.003\,|V - 75|)]\,[.82 + 4.5/D])\,(FM)\,[1 - (0.0032A)]\,(CM)$$

where

H = horizontal location forward of the midpoint between the ankles at the origin of the lift. If significant control is required at the destination, then H should be measured both at the origin and destination of the lift.

V = vertical location at the origin of the lift

D = vertical travel distance between the origin and destination of the lift

FM = frequency multiplier is obtained from a table supplied by NIOSH

A = angle between the midpoint of the ankles and the midpoint between the hands at the origin of the lift

CM = coupling multiplier ranked as either good, fair, or poor. It is obtained from a table.

The 1993 equation predicts smaller loads that can be lifted safely, when compared to the 1981 equation, and is thus more conservative. Interestingly, NIOSH offered no provision for the difference capacities of men and women; they are treated similarly for largely political reasons. (I discussed this issue of discrimination and the impact on protecting vulnerable workers in chapter 1.) Further, the approach ignored any individual differences in body mechanics used while lifting. In addition, some have suggested that the handle factors may not be consistent with the subsequent forces endured by the body. For example, while NIOSH assumed that having handles on the lifted object is better, our work has shown that handles allow the lifter to apply even more force, resulting in subsequently higher back loads (Honsa et al., 1998)! Nonetheless, the equations form a rudimentary approach to risk assessment that is easy to conduct. Marras and colleagues (1999) concluded that the 1981 NIOSH guide identified low-risk jobs well (specificity of 91%) but did not predict the high-risk jobs well. The 1993 guide correctly identified 73% of the high-risk jobs but did not identify the low- and medium-risk jobs well. Overall, both NIOSH guides predicted those jobs resulting in LBD with odds ratios between 3.1 and 4.6. Marras and colleagues (1999) noted that the most powerful individual variable was average low back moment (load in hands and the horizontal distance between the anteriorly placed load from the spine).

Snook Psychophysical Approach

The psychophysical approach to setting load, or work limits, is based on peoples' perceptions as to what they feel is a tolerable work rate. The main criticism of this approach is that most people perceive physiologically related discomfort and not the internal loading on tissues that actually causes damage. For example, Dul, Douwes, and Smitt (1994) made a case that muscle discomfort, which is readily perceived, may not be correlated with actual tissue loads. Workers may not be cognizant of the tissue loads as they reach damaging levels (McGill, 1997). Furthermore, Karwowski and colleagues (Karwowski and Pongpatanasuegsa, 1989; Karwowski, 1991, 1992) noted in a series of studies that the general psychophysical approach is very dependent on such factors as the instructions provided to the experimental subjects and the color of the object being lifted, to name a couple. For example, the Snook studies (1978) advised the workers to determine a work rate and load to be moved based on the instruction of not to become unduly fatigued. The workers selected the loads they perceived not to be fatiguing. Yet when Karwowski repeated the experiments but changed the instruction to lift loads so that they would not become injured, the acceptable loads changed (they chose smaller loads). Karwowski and Pongpatanasuegsa (1989), when evaluating the effect of the color of the boxes being lifted, noted that workers would lift more when the boxes were white, as they perceived the black boxes as being more dense. The psychophysical approach appears to depend on many factors that modulate perception.

The work of Snook (Snook, 1978; Snook and Ciriello, 1991) is probably the most recognized in the psychophysics area. They experimentally controlled variables such as object size, height of the lift, and movement distance in lifting and pushing and pulling tasks for both men and women. They constructed tables compiling the acceptable loads for both men and women over a variety of tasks. On one hand, this approach inherently incorporates the dynamics of the tasks, while on the other hand, the concerns raised in the previous paragraph are worth considering. Nonetheless, the Snook approach remains popular for its ease of implementation and because it is one of the few assessment methods for pushing and pulling tasks.

Lumbar Motion Monitor (LMM)

Figure 7.1 Wearing the lumbar motion monitor to record the three-dimensional kinematics of the lumbar spine while performing industrial tasks. Photo courtesy of my good friend Professor Bill Marras.

Several studies have documented the links between spine motion and the development of LBD but none more thoroughly than those of Marras and colleagues (1993, 1995). They developed several types of regression models that demonstrated that spine motion and, in particular, the velocity of motion and the range of motion are important predictors of those jobs with high rates of disorders (in fact, some odds ratios exceeded 10). The lumbar motion monitor is a three-dimensional goniometer that measures the three-dimensional kinematics of the lumbar spine (see figure 7.1). Marras and colleagues showed that the kinematic variables obtained from workers while wearing the LMM, when combined with simple measurements of lift frequency or motion duty cycle and load moment (the load magnitude multiplied by the distance to the low back), provide impressive risk predictions. Using the LMM on people while they are performing real on-site tasks and not laboratory mock-ups is relatively easy to do. In addition, the LMM captures the individual ways in which people use and move their spines. See chapter 5 for a discussion of the mechanisms of injury that are dependent on spine posture and motion.

4D WATBAK

Biomechanical models have been used to estimate loads in the low back tissues and identify high-risk jobs for approximately three decades. Some models were intended as a simple tool for health and safety personnel to provide an approximate index of injury risk on the plant floor. Other models were designed to be more robust in illustrating injury mechanisms and being sensitive to worker variance. Decisions as to which model to use boil down to the purpose and necessary complexity of the model.

The better "simple" models need the "complex" models to assess the many simplifying assumptions that affect accuracy and validity of output, which in turn depend on the type of application. Further, in many workplaces, the most blatant or overt ergonomic injury risks have been addressed, and only the more subtle risks remain. Ergonomists need simple models but also must be conversant with the more complex models that will assist in rectifying the more subtle injury risks and assist in developing more effective intervention strategies.

Although simple biomechanically based models to obtain quick estimates of low back compression exist, risk assessment for most jobs requires more complex metrics and analysis approaches. In an effort to optimize biofidelity and industrial utility, Bob Norman, Mardy Frazer, Rich Wells, and I developed the software package 4D WATBAK. The "4D" corresponds to the 3D moments computed at joints while workers perform 3D postures/tasks plus the fourth dimension of time and the effect of repetition on determining the safe load. (It is available by contacting Professor Frazer.) The package uses the detailed output of the virtual spine (described in chapter 2) as a foundation but makes assumptions to simplify data collection and to facilitate routine analysis. In this way individual worker behavior and spine mechanics are quantified but simplified into "average" muscular responses seen in workers performing similar tasks. This greatly simplifies data collection and preserves the benefits of better anatomical representation and more valid predictions of low back loads. Furthermore, by incorporating the injury data obtained from samples of workers, 4D WATBAK quantifies the risk of back injury during fully three-dimensional tasks and postures. The fourth dimension is time, in this case a variable needed to take into account the repetition of tasks over a work shift.

When using the 4D WATBAK, the joint coordinate data is entered manually or the user manipulates an on-screen mannequin into the work posture of interest. The software executes the difficult task of determining the three-dimensional joint moments by computing the Euler angles and transforming them into moments about the orthopedic axes of each joint. In this way, the package is capable of calculating joint loads in any posture and for any combination of lift, lower, push, or pull task. It also incorporates algorithms that capture the average muscular response measured from workers to support three-dimensional spine moments of force together with strength data for both men and women. As well, the model is sensitive to spine posture in that ligament and passive tissue forces are invoked during fully flexed spinal postures. Thus, the model more accurately predicts low back shear forces supported by the spine (as compared to reaction shear forces of other models) together with compression forces that result from the load and the many torso muscles contracting to support a particular set of three-dimensional moments (see figure 7.2, a-b). Finally, the risk is also calculated from accumulated loads during repetitive work by comparing the load–time integrals with epidemiologic data obtained from a large surveillance study conducted in an automotive assembly plant (Norman et al., 1998). (See figure 7.3, a-d.)

Euler Angles and Orthopedic Moments

Estimations of three-dimensional joint moments are typically performed using three orthogonal axes (XYZ). But as a person moves, the orthopedic axes of the joints (for example, in the lumbar spine, axis 1—flexion/extension, axis 2—lateral bend, axis 3—axial twist) also move so that they no longer align with the inertial XYZ axes. The difference between the orthopedic joint axes and the XYZ axes is described with Euler angles. The orthopedic moments are then obtained from the E-vector approach described by Grood and Suntay (1983). This involves taking the cross-product of the long axis (the primary axis, usually twist) of two adjacent segments to form the secondary axis (usually flexion/extension). The cross-product of the primary and secondary axes forms the E-vector, which in turn becomes the third joint orthopedic axis (usually lateral bend). Moments computed about axes XYZ can now be transformed into joint-specific orthopedic axes. This technique ensures that the joint axes convention "stay with the joint" as the person moves within the inertial coordinate system.

The 4D WATBAK approach represents a compromise between simple models that are easy to implement but are not very robust in their risk assessment and more complex approaches. 4D WATBAK is still relatively easy to use, and it provides some degree of sensitivity to the way individual workers perform complex tasks. It uses many risk indexes to quantify the subsequent risk from a specific task performed over a work shift.

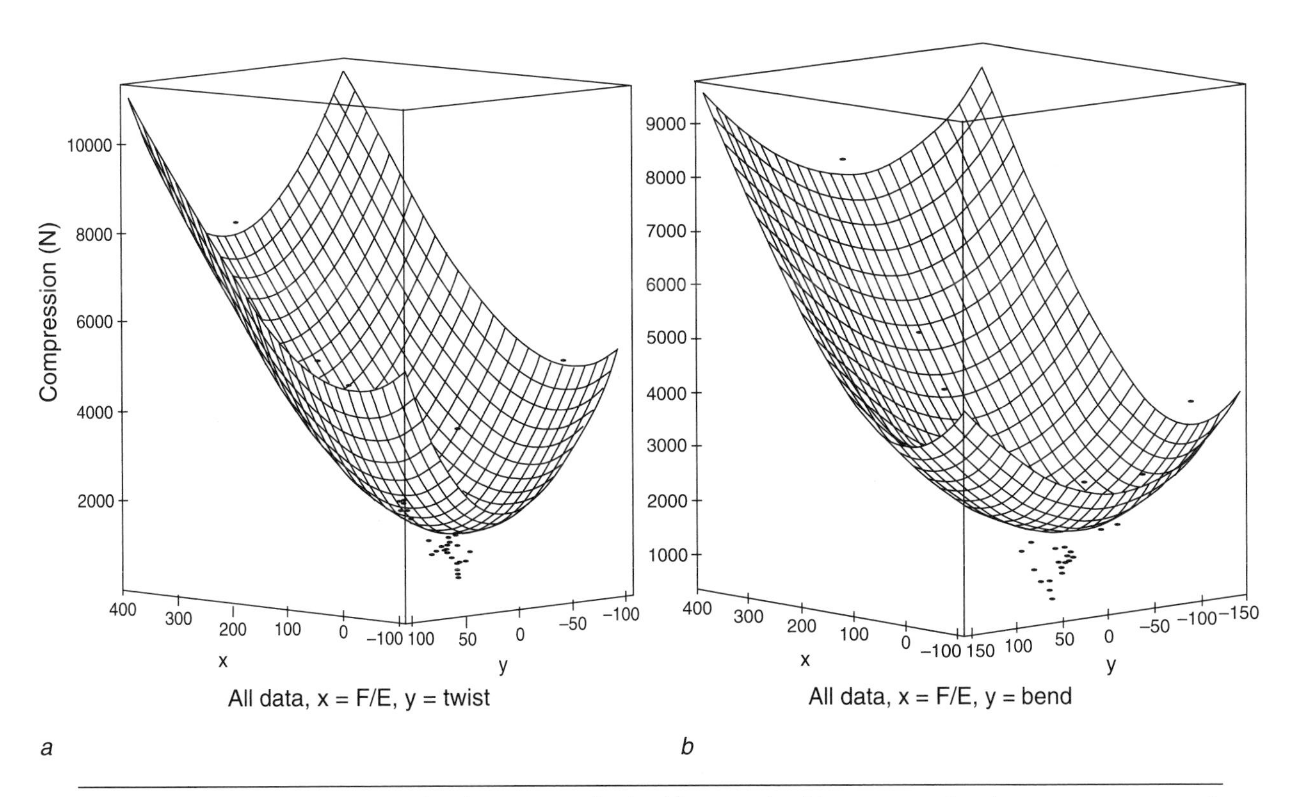

Figure 7.2 4D WATBAK computes lumbar compressive load from an algorithm representing the average response of the many muscles, measured from real workers, that combine to support the infinite combinations of three low back moments: flexion/extension, lateral bend, and axial twist. Because only three dimensions can be graphed at one time, *(a)* shows the surface of lumbar compression from combinations of flexion/extension and twist moments, while *(b)* shows the surface of compression from flexion/extension and lateral bend.

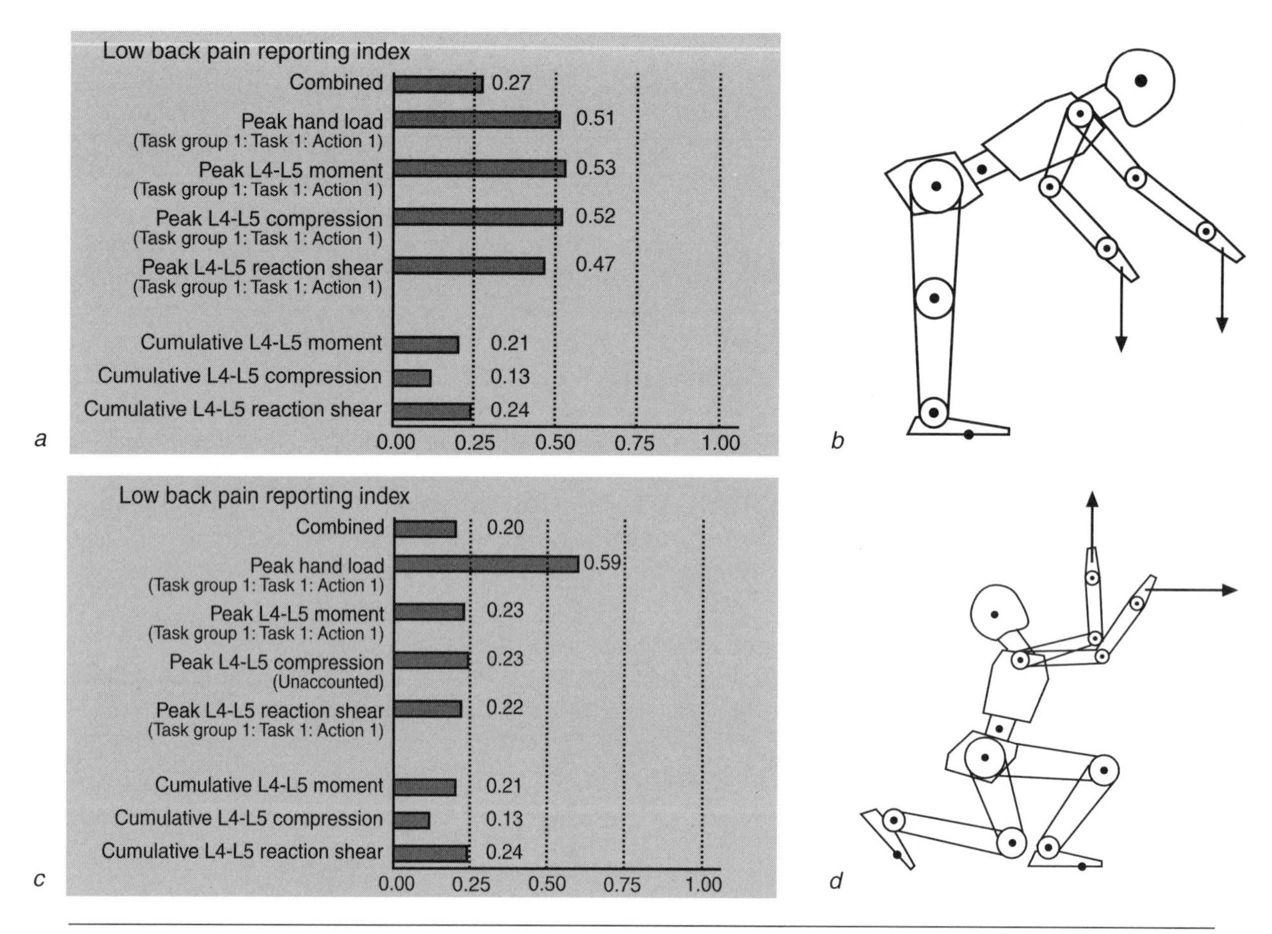

Figure 7.3 4D WATBAK computes the joint-specific, three-dimensional orthopedic moments at each joint and uses several metrics to calculate risk. For example, the low back risk includes compression and shear load threshold values along with accumulated loads during repetitive work by comparing the load–time integrals with epidemiologic data obtained from a large surveillance study. Two examples are shown: a reaching lift *(a)* and *(b)*, and an overhead assembly task *(c)* and *(d)*.

Biological Signal–Driven Model Approaches

The final approach for risk assessment is to measure biological signals from each subject in an attempt to capture the individual ways people perform their jobs and then use sophisticated anatomical, biomechanical, and physiological relationships to assign forces to the various tissues.

The Marras Model and the McGill Model

The Marras model (Marras and Sommerich, 1991a, 1991b) measures EMGs from several muscles and, using known physiological relationships, assigns forces to the muscles during virtually any industrial task. As noted earlier, this approach revealed powerful evidence linking the physical demands of specific occupational tasks with the incidence of LBD.

The McGill model (McGill, 1992; McGill and Norman, 1986) uses the same philosophical approach to assign forces to the muscles but attempts to include the highest level of anatomical accuracy possible. (This model was introduced in chapter 2.) For example, it also measures three-dimensional spine curvature to assess forces to the various passive tissues, including the intervertebral discs and the various ligaments. By assigning forces to muscles and passive tissues throughout the full range of spine motion, it captures the different ways people perform their jobs and even how they change with repetitions of the same job. The obvious liability of the approach is its enormous computational requirements, postprocessing of data, and difficulty of collecting such comprehensive data from a wide number of workers in the field.

The question of the validity of this type of model must be addressed along with other models based on the biological approach. Some have argued that since these models contain known biomechanical and physiological relationships, they contain a certain amount of content validity. Moreover, both the Marras model and the McGill model have been quite successful in estimating the various passive tissue and muscle forces that sum together to produce flexion and extension, lateral bend, and axial twisting moments. (These moments have been well predicted, with the exception of the twisting moment.) In summary, if twisting is not a dominant moment of force in a particular job, these models appear to be predicting accurate distributions of forces among the support tissues.

EMG-Assisted Optimization

Perhaps the current pinnacle of model evolution is a hybrid modeling approach known as EMG-assisted optimization (developed by Cholewicki and McGill, 1994) with stability analysis and tissue load prediction (Cholewicki and McGill, 1996). This approach exploits the asset of the biological EMG approach to distribute loads among the tissues based on the biological behavior of the subject. It then uses optimization to fine-tune the satisfaction of joint torques about several low back joints. The optimization takes the "biologically" predicted forces and adjusts muscle forces the minimal amount possible to satisfy the three-dimensional moment axes at every joint over the length of the lumbar spine. Once the three-dimensional moments have been assigned and tissue forces determined, an analysis of spine stability is performed by comparing the joules of work imposed on the spine through perturbation with the joules of potential energy existing in the stiffened column.

This most highly evolved of spine models provides the biomechanist/ergonomist with insight into injury mechanisms caused by instability (as witnessed by Cholewicki and McGill, 1992) such as when picking up a very light load from the floor. Even though patients have reported back injury from incidents such as bending over to tie a shoe, previous modeling approaches were sensitive only to tissue damage and injury scenarios produced from large loads and moments. Now a biomechanical explanation is available to explain the subsequent tissue damage, and a method is available to detect the risk of its happening. Although the routine use of this type of sophisticated model by ergonomists is not feasible, it is useful (by trained scientists) for analyzing individual workers and identifying those who are at elevated risk of injury because of faulty personal motor patterns.

Simple or Complex Models?

In summary, the complex models provide a tool to investigate the mechanisms of injury and the effects of technique during material handling on the risk of injury. The most complex and highest evolved models provide insight as to how injury occurs with all types of heavy and light loads. On the other hand, the simpler models, while sacrificing accuracy, can be a powerful tool for routine surveillance of physical demands in the workplace but must be wisely interpreted in each case for their limitations and constraints. Biomechanics and ergonomics need the full continuum of sophistication in models. The choice of which one to use depends on the issue in question.

The Challenge Before Us

In this chapter I have summarized and critiqued several approaches for their assets and liabilities. The choice of the most appropriate tool depends on the objective at hand. Yet many ergonomists and industrial engineers believe that minimal tissue loading is best for all jobs, motivating them to direct their interventions toward making all jobs easier. This concept is faulty. Biological tissues require repeated loading and stress to be healthy. Virtually all risk assessment tools consider only the risk of too much load. Risk assessment tools that survey too little loading for optimal health are just starting to emerge. The challenge is to develop a wise rest break strategy to facilitate optimal tissue adaptation. More robust assessment tools will be developed in the future based on the most healthy combination of work and rest. This will be achieved by first understanding the biomechanical, physiological, and psychological parameters of injury and human performance and then applying this wisdom thoughtfully.

References

Cholewicki, J., and McGill, S.M. (1992) Lumbar posterior ligament involvement during extremely heavy lifts estimated from fluoroscopic measurements. *Journal of Biomechanics,* 25 (1): 17-28.

Cholewicki, J., and McGill, S.M. (1994) EMG assisted optimization: A hybrid approach for estimating muscle forces in an indeterminate biomechanical model. *Journal of Biomechanics,* 27: 1287-1289.

Cholewicki, J., and McGill, S.M. (1996) Mechanical stability of the in vivo lumbar spine: Implications for injury and chronic low back pain. *Clinical Biomechanics,* 11 (1): 1-15.

Dul, J., Douwes, M., and Smitt, P. (1994) Ergonomic guidelines for the prevention of discomfort of static postures can be based on endurance data. *Ergonomics,* 37: 807-815.

Grood, E.S., and Suntay, W.J. (1983) A joint coordinate system for the clinical description of three-dimensional motions: Application to the knee. *Journal of Biomechanical Engineering,* 105: 136-144.

Honsa, K., Vennettelli, M., Mott, N., Silvera, D., Niechwiej, E., Wagar, S., Howard, M., Zettel, J., and McGill, S.M. (1998) The efficacy of the NIOSH hand-to-container coupling factor. Proceedings of the 30th Annual Conference of the Human Factors Association of Canada, p. 253. (Winner of HFAC/ACE Best Undergraduate Presentation and HFAC Ontario Chapter Award)

Karwowski, W. (1991) Psychophysical acceptability and perception of load heaviness by females. *Ergonomics*, 34 (4): 487-496.

Karwowski, W. (1992) Comments on the assumption of multiplicity of risk factors in the draft revisions to NIOSH lifting guide. In: Kumar, S. (Ed.), *Advances in industrial ergonomics and safety.* London: Taylor and Francis.

Karwowski, W., and Pongpatanasuegsa, N. (1989) The effect of color on human perception of load heaviness. In: Mital, A. (Ed.), *Advances in industrial ergonomics and safety* (pp. 673-678). London: Taylor and Francis.

Leamon, T.B. (1994) Research to reality in a critical review of the validity of various criteria to the prevention of occupationally induced low back pain disability. *Ergonomics,* 37 (12): 1959-1974.

Marras, W.S., Fine, L.J., Ferguson, S.A., Waters, T.R. (1999) The effectiveness of commonly used lifting assessment methods to identify industrial jobs associated with elevated risk of low-back disorders. *Ergonomics,* 42 (1): 229-245.

Marras, W.S., Lavender, S.A., Leurgens, S.E., et al. (1993) The role of dynamic three-dimensional trunk motion in occupationally related low back disorders: The effects of workplace factors trunk position and trunk motion characteristics on risk of injury. *Spine,* 18: 617-628.

Marras, W.S., Lavender, S.A., Leurgans, S.E., Fathallah, F.A., Ferguson, S.A., Allread, W.G., and Rajulu, S.L. (1995) Biomechanical risk factors for occupationally related low back disorders. *Ergonomics,* 38: 377-410.

Marras, W.S., and Sommerich, C.M. (1991a) A three-dimensional motion model of loads on the lumbar spine. I: Model structure. *Human Factors*, 32: 123-137.

Marras, W.S., and Sommerich, C.M. (1991b) A three-dimensional motion model of loads on the lumbar spine. II: Model structure. *Human Factors*, 32: 139-149.

McGill, S.M. (1992) A myoelectrically based dynamic three-dimensional model to predict loads on lumbar spine tissues during lateral bending. *Journal of Biomechanics,* 25: 395-414.

McGill, S.M. (1997) The biomechanics of low back injury: Implications on current practice in industry and the clinic. *Journal of Biomechanics,* 30: 465-475.

McGill, S.M., and Norman, R.W. (1986) Partitioning of the L4/L5 dynamic moment into disc, ligamentous and muscular components during lifting. *Spine,* 11: 666-677.

National Institute for Occupational Safety and Health (NIOSH) (1981) *Work practices guide for manual lifting.* Department of Health and Human Services (DHHS), NIOSH Publication No. 81-122.

Norman, R., Wells, R., Neumann, P., Frank, P., Shannon, H., and Kerr, M. (1998) A comparison of peak vs cumulative physical work exposure risk factors for the reporting of low back pain in the automotive industry. *Clinical Biomechanics,* 13: 561-573.

Snook, S.H. (1978) The ergonomics society—The Society's Lecture 1978. *Ergonomics*, 21 (12): 963-985.

Snook, S.H., and Ciriello, V.M. (1991) The design of manual handling tasks: Revised tables of maximum acceptable weights and forces. *Ergonomics*, 34 (9): 1197-1213.

Waters, T.R., Putz-Anderson, V., Garg, A., and Fine, L.J. (1993) Revised NIOSH equation for the design and evaluation of manual lifting tasks. *Ergonomics*, 36 (7): 749-776.

Reducing the Risk at Work

As I have suggested in previous chapters, many of the classic instructions from "experts" about lifting (for example, bend the knees and not the back) are in most cases nonsense. In fact, very few occupational lifting tasks can be performed this way. The resulting tragedy is that clinicians or other experts lose credibility in the eyes of the worker; the worker knows she cannot do her job that way. She lifts objects from the floor, out of parts bins, over tables, and so forth. The science shows that squatting can have an increased physiological cost and that, depending on the characteristics of the load, it may not even reduce loads. Squatting is not always the best choice of position; it depends on the specifics of the task (size, weight, and density of the object; pick and place location; number of repetitions, etc.). For example, the golfer's lift (see figure 8.6 later in this chapter) may be more preferable for reducing spine loads than bending the knees and keeping the back straight for someone repeatedly lifting light objects from the floor.

Reducing pain and improving function for patients with low back pain involve two components:

- Removing the stressors that create or exacerbate damage
- Enhancing activities that build healthy supportive tissues

This chapter addresses the issue of prevention, specifically, reducing the overloading stressors that cause occupational LBD. After reviewing lessons from the literature and presenting the scientific issues, I will provide two sets of guidelines: one for workers and another for management. Finally, I will offer a few notes to enhance the effectiveness of consultants. The lists provided in this chapter will result in more successful injury prevention programs. The practitioner or consultant who employs them will stand out from his peers.

Upon completion of this chapter, you will be able to formulate scientifically based guidelines effective for any activity that reduce the risk of occupational low back troubles. Further, you will be aware of, and so will be able to avoid, the pitfalls that cause ergonomic approaches to fail. Finally, you will be a more effective consultant by harnessing the real expertise—that of the worker.

Lessons From the Literature

Why does industry care about the backs of workers? Competitiveness in the new economy requires being profitable, and for many the only way to enhance profit is to maximize loss control. A major source of losses to North American industry is worker injury, which results in direct costs for compensation and indirect costs of hiring and training replacement workers and reduced productivity due to lower speed and increased errors. For this reason, preventing injury and promoting the rapid return to work of injured workers have become a major focus for industry. Because back injury represents an enormous cost in both real dollars and suffering, companies are realizing the benefits of supporting injury prevention and rehabilitation programs. Following are a number of studies that have addressed the issues that must be examined if such programs are to be successful.

Ergonomic Studies

If LBDs are associated with loading, then changes in loading should change injury and absenteeism rates. But surprisingly, the literature does not provide clear guidance for reaching this objective; it needs interpretation. The traditional ergonomic approach is to establish a criterion that, if applied in the job design, would lead to a reduction in incidence rate. In a review in 1996, Frank and colleagues suggested that modified work with lower demands can be successful in reducing the number of injuries but that other changes such as organizational parameters made it difficult to conclude that physical demand changes accounted for any reported differences. This is because very few studies have simply altered job demands. Furthermore, Winkel and Westgaard (1996) noted that the implementation of ergonomics can lead to what they called an ergonomic pitfall. That is, the new ergonomic consciousness causes many workers to report concerns that they did not previously realize were a result of their job situations. Thus, many newly implemented ergonomic intervention programs have resulted in a temporary rise in musculoskeletal disorders. This effect appears to have confounded any study of ergonomic efficacy of insufficient duration. The number of good investigations documenting the effects of job change, or the implementation of ergonomic principles, is also low because such research is very time consuming and difficult to perform on site.

Nonetheless, several nice studies qualify by documenting only the effect of ergonomic job redesign. For example, in a report on a series of studies on working women in Norway, Aaras (1994) noted a collectively documented reduction in sick leave due to musculoskeletal troubles from job redesign (specifically, from 5.3% to 3.1%) and a reduction in employee turnover from 30.1% to 7.6%.

Rehab and Prevention Studies

Loisel and colleagues (1997) conducted a very interesting and important study in which they used a randomized control trial design with four groups. One group of people with back injuries received "clinical" intervention consisting of a visit to a back pain specialist, back school, and functional rehabilitation therapy. Another group received an occupational intervention consisting of a visit to an occupational physician and then an ergonomist to arrive at ergonomic solutions. A third group, the "full intervention group," received both of these approaches, and the fourth group received "usual care." The group receiving full intervention returned to regular work 2.41 times faster than the usual care group, although the specific effect of occupational intervention (ergonomics) accounted for the largest proportion of this result with a rate ratio of return to regular work of 1.91.

Those paying for injury (government agencies and compensation boards and the insurance industry) could reasonably argue from this evidence that, to reduce costs, care for the injured back should be removed from medical hands (once the medics have ruled out red flag conditions such as tumors) and given to ergonomists! Tongue-in-cheek as this statement may be, its point is worthy of deep consideration. Hopefully, the last section of this book will provide clues for more efficacious medical intervention. Krause, Dasinger, and Neubraser (1998) reviewed the literature on the role of modified work in the return to work and rated as high quality the Loisel study as well as the studies of Baldwin, Johnson, and Butler (1996). This latter review concluded that modified work (involving the modification of musculoskeletal loading) is effective in facilitating the return to work of disabled workers.

In addition to workplace design modifications, workplace interventions may benefit from addressing the personal movement strategies of each worker. In a fascinating study Snook and colleagues (1998) demonstrated that of 85 patients randomly assigned to a group that controlled the amount of early-morning lumbar flexion, the experimental group had a significant reduction in pain intensity compared to a control group. When the control group received the experimental treatment, they responded with similar reductions. This is yet another example of how personal spine motion patterns and loading posture can influence whether the person will become injured. The next section blends notions of job design with personal movement strategies to reduce loading and the risk of LBD.

Studies on the Connection Between Fitness and Injury Disability

It is fruitful to discuss briefly the role of fitness in the link between injury and disability. Although several studies have shown links between various fitness factors and the incidence of LBD (e.g., Suni et al., 1998, who showed that higher $\dot{V}O_2max$ scores were linked to LBDs), these cross-sectional studies cannot infer causation.

Probably the most widely cited longitudinal study was reported by Cady and colleagues (1979), who assessed the fitness of Los Angeles firefighters and noted that those who were rated "more fit" had fewer subsequent back injuries. However, what is not widely quoted by those citing this study is that when the more fit did become injured, the injury was more severe. Perhaps the more fit were willing to experience higher physical loads. Several have suggested that a psychological profile is associated with being fit (e.g., Farmer et al., 1988; Hughes, 1984; Ross and Hayes, 1988; Young, 1979) and that the unfit may complain more about the more minor aches. Along those lines, some athletes have demonstrated the ability to compete despite injury. Burnett and colleagues (1996) reported cricket bowlers with pars fractures who were still able to compete. Is this due to their supreme fitness and ability to achieve spine stability or their mental toughness? Perhaps it is both. The issue remains unresolved.

It is also interesting to note that personal fitness factors appear to play some role in first-time occurrence. Biering-Sorensen (1984) tested 449 men and 479 women for a variety of physical characteristics and showed that those with larger amounts of spine mobility and lower extensor muscle endurance (independent factors) had an increased occurrence of subsequent first-time back troubles. Luoto and colleagues (1995) reached similar conclusions. It would appear that muscle endurance, and not anthropometric variables, are protective.

LBD Prevention for Workers

This section addresses a list of issues that are scientifically justifiable to reduce the risk of occupational LBDs.

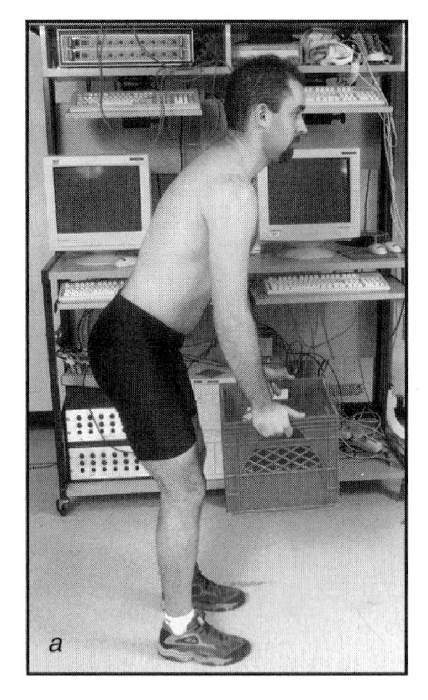

Figure 8.1 Flexing the torso involves either hip flexion or spine flexion, or both. *(a)* A neutral spine with hip flexion. *(b)* Spine flexion.

Should Workers Avoid End Range of Spine Motion During Exertion?

Generally, the answer to whether end range of motion should be avoided is yes—for several reasons. Maintaining a more neutrally lordotic spine will maximize shear support, ensure a high tolerance of the joint to withstand compressive forces, eliminate the risk of ligamentous damage since the ligaments remain unstrained, eliminate the risk of disc herniation since this is associated with a fully flexed spine, and qualitatively emulate the spine postures that Olympic lifters adopt to avoid injury (see figure 8.1, a-b for an illustration of a flexed and a neutral spine). Unfortunately, this issue has become confused with issues such as whether it is better to stoop or squat, for example. Another source of confusion has evolved from the common recommendation to perform a pelvic tilt when lifting; the scientific base for this clinically popular notion is nonexistent! Performing a pelvic tilt increases tissue stresses and the risk of injury!

What spine and hip posture best minimizes the risk of injury? We do know that very few lifting tasks in industry can be accomplished by bending the knees and not the back. Furthermore, most workers rarely adhere to this technique when repetitive lifts are required—a fact that is quite probably due to the increased physiological cost of squatting compared with stooping (Garg and Herrin, 1979). However, a case can be made for preserving neutral lumbar spine curvature while lifting (specifically, avoiding end range limits of spine motion about any of the three axes). This is a different concept from trunk angle, as the posture of the lumbar spine can be maintained independent of thigh and trunk angles.

The literature is confused between trunk angle or inclination and the amount of flexion in the lumbar spine. Bending over is accomplished by either hip flexion or spine flexion or both. It is the issue of specific lumbar spine flexion that is of importance here. Normal lordosis can be considered to be the curvature of the lumbar spine associated with the upright standing posture. (To be precise, the lumbar spine is slightly extended from elastic equilibrium when standing; see Scannell and McGill, in press.) In figure 8.2 a warehouse worker is successfully sparing his spine by avoiding end range of spine motion even though he is not bending the knees; he has accomplished torso flexion by rotating about

Figure 8.2 This warehouse worker is not bending the knees, yet he is sparing the spine by electing to bend and rotate about the hips; the lumbar spine is not flexed.

Figure 8.3 This firefighter *(a)*, who is flexing his lumbar region, is loading the passive tissues and increasing his risk of back troubles. *(b)* Avoiding full lumbar flexion and rotating about the hips spares the spine.

the hips. In figure 8.3, a firefighter is demonstrating a spine-sparing technique (figure 8.3b) versus a spine-damaging technique (figure 8.3a) for lifting extinguishers. Chapter 11 offers further occupation-specific examples (e.g., the construction workers demonstrating spine-sparing postures in figure 11.4).

In chapter 5, I explained the load distribution among the tissues of the low back. Of interest here are the dramatic effects on shear loading as a function of spine curvature. Recall the following facts:

- A spine that is not fully flexed ensures that the pars lumborum fibers of the longissimus thoracis and iliocostalis lumborum are able to provide a supporting posterior shear force on the superior vertebra, while full flexion causes the interspinous ligament complex to strain, imposing an anterior shear force on the superior vertebra. For these reasons, avoiding full flexion not only ensures a lower shear load but also eliminates ligament damage.
- A fully flexed spine is significantly compromised in its ability to withstand compressive load.
- Because herniation of the nucleus through the annulus is caused by repeated or prolonged full flexion, avoidance of the posture minimizes the risk of herniation (a cumulative trauma problem) and minimizes the stresses on any developing annular bulges.

Olympic lifters provide a convincing example of the efficacy of avoiding lumbar flexion in lifting. They lock their spines in a neutral posture and emphasize rotation about the hips (see figure 8.4), lifting enormous weights without injury. The ambulance attendants in figure 8.5 are also sparing their backs by locking the lumbar regions to avoid flexion.

Figure 8.4 A gold-medal lift by Naim Suleymanoglu in Olympic weightlifting. His lumbar spine is locked in neutral, and the motion takes place about the hips and knees.

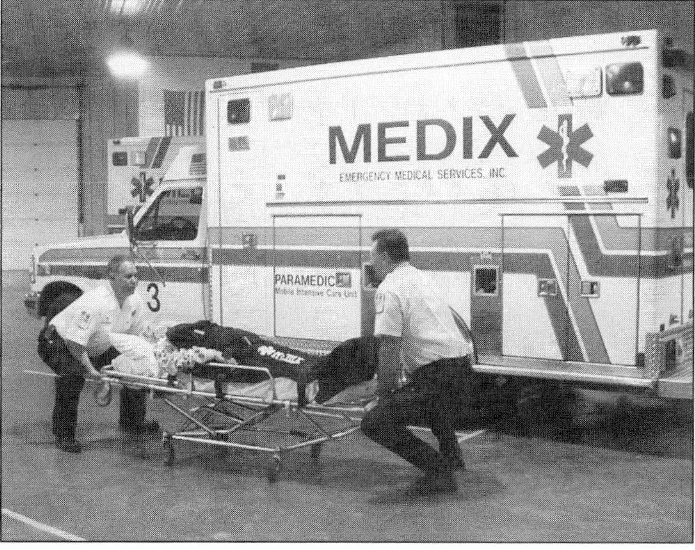

Figure 8.5 These ambulance attendants are attempting to spare their backs by avoiding lumbar flexion and lifting together to share the load. They have also been taught to lightly contract the stabilizing abdominal musculature.

Figure 8.6 The golfer's lift has been documented to minimize low back motion and reduce the loads on the lumbar tissues by using the leg, which is cantilevered behind, as a counterweight; the hips act as a fulcrum to raise the torso to upright. This is an effective technique for repeated lifting of light objects from floor level. Most still adhere to the general instruction to bend the knees and keep the back straight *(a)*, not considering the spine-conserving benefits of the golfer's lift *(b)*.

Thus, the important issue is not whether it is better to stoop lift or to squat lift; rather, the emphasis should be to place the load close to the body to reduce the reaction moment (and the subsequent extensor forces and resultant compressive joint loading) and to avoid a fully flexed spine. Sometimes it may be better to squat to achieve this; in cases in which the object is too large to fit between the knees, however, it may be better to stoop, flexing at the hips but always avoiding full lumbar flexion to minimize posterior ligamentous involvement. (For a more comprehensive discussion, see McGill and Kippers, 1994; McGill and Norman, 1987; Potvin, Norman, and McGill, 1991).

Yet another spine-sparing lifting posture is the golfer's lift, which reduces spine loads for repeated lifting of light objects (see figure 8.6, a-b) (Kinney, Callaghan, and McGill, 1996). The hips act as a fulcrum in which one leg is cantilevered behind with isometric muscle contraction, forming a counterweight to rotate the upper body back to upright.

What Are the Ways to Reduce the Reaction Moment?

A popular instruction is to hold the load close to the torso when handling material. This is biomechanically wise; a reduced lever arm of the load requires lower internal tissue loads necessary to support the reaction moment. But phrasing the principle in terms of holding the load close restricts the notion to lifting tasks. The real biomechanical principle is to reduce the reaction moment. When phrased this way, the principle is now applicable to any task involving the exertion of external force.

Directing the Pushing Force Vector Through the Spine

When pushing a cart handle, directing the pushing force vector through the lumbar spine reduces the reaction moment and therefore the tissue loads and spine load (see figure 8.7). In contrast, a pushing force directed through the shoulder would not be directed through the low back (see figures 8.8, a-b); this forms a **transmissible vector**. The large perpendicular distance from this force to the spine causes a high reaction torque that is balanced by muscular force; this imposes a corresponding compressive

Figure 8.7 Pushing through the hands but directing the force vector to pass through the low back minimizes the low back moment (and minimizes the muscular loads).

penalty on the spine. Redirecting the transmissible vector through the low back reduces this moment arm (and the moment), the muscle forces, and the spine compressive load. Other examples from everyday activities include the technique used to open a door; directing the pulling force through the low back is sparing (see figure 8.9, a-c). Another example is vacuuming, which is often reported by our patients to exacerbate their symptoms. Holding the handle to the side creates a large moment arm for the pushing/pulling forces; this heavily loads the back (see figure 8.10a). Directing the push/pull forces through the low back

Figure 8.8 Pushing or pulling forces that pass by the spine with a large moment arm ensure a high moment and corresponding high muscle forces and spine loads *(a)*. Pushing forces directed through the low back minimize the moment and spine load *(b)*.

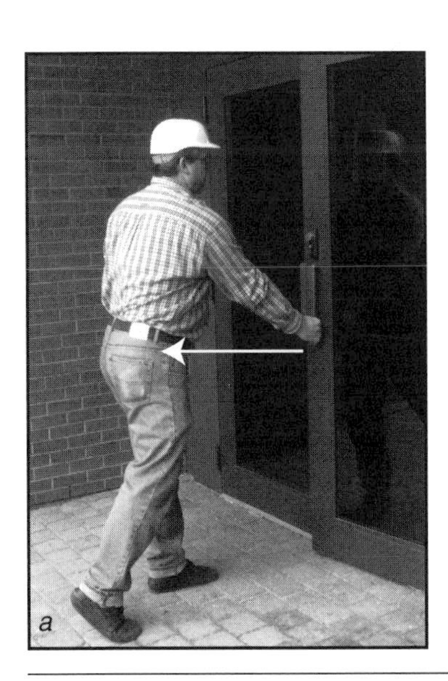

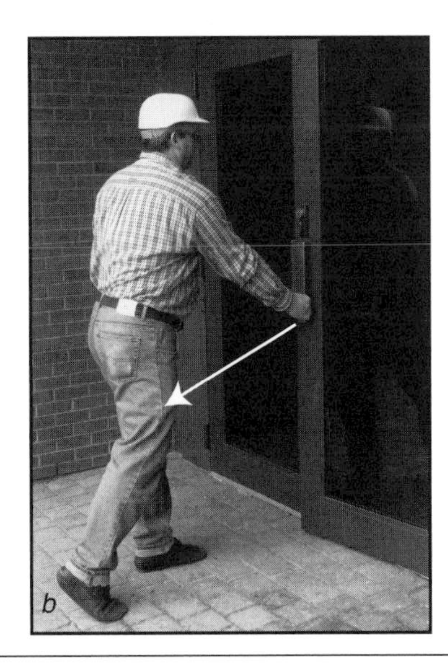

Figure 8.9 Opening a door by directing hand forces through the low back spares the spine *(a)*. Many people are not taught how to optimize this principle during the performance of daily tasks. Instead, they sacrifice their spine by opening the door so that the force is lateral to the lumbar spine creating a twisting torque *(b)* or produce a pulling force that passes over the low back *(c)*.

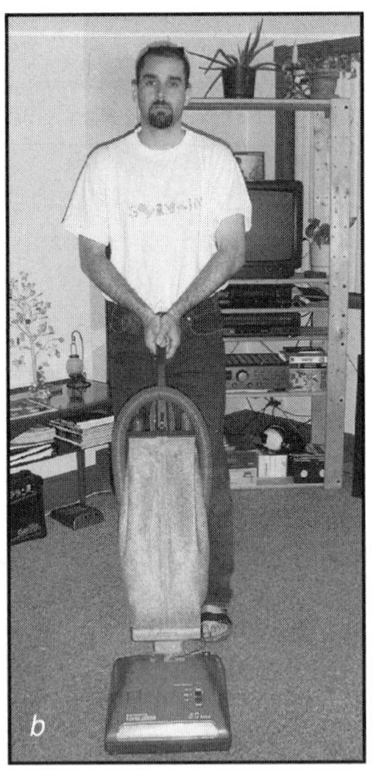

Figure 8.10 *(a)* Vacuuming is often reported as problematic for bad backs. The transmissible vector in this example has a large moment arm with respect to the back and requires large twisting torque forces (causing pain). *(b)* The transmissible vector is directed through the low back, minimizing the load and the pain and enabling vacuuming.

minimizes the moment arm (see figure 8.10b) and removes the loads that ensure that patients remain patients! These examples demonstrate how this powerful, but rarely practiced technique of skilled control of the transmissible vector, spares the back.

Diverting the Force Around the Lumbar Spine

Still other very effective ways exist to reduce the reaction moment. The next examples demonstrate how workers skillfully spare the low back by diverting force around the lumbar region when lifting. The workers lifting the shaft in a paper factory in figure 8.11 are minimizing the reaction moment by lifting the shaft with their thigh (by plantar flexing the foot). In this way, the weight of the shaft is directed down the thigh to the floor, bypassing the upper body linkage and spine. Prior to redesigning the job, the workers had to lift the shaft with arm forces, causing low back problems and motivating our involvement. Figure 8.12, a through c illustrates

Figure 8.11 This task requires a worker to lift a heavy shaft while another worker slides on a core from the end. A high incidence of back troubles motivated our consulting recommendation to install a bar for the foot, which is plantar flexed to raise the shaft now supported on the thigh. The forces now transmit directly down the leg, completely bypassing the arm and spine linkage.

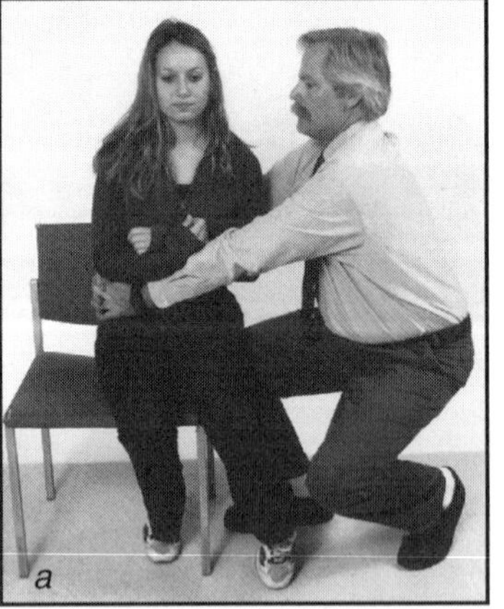

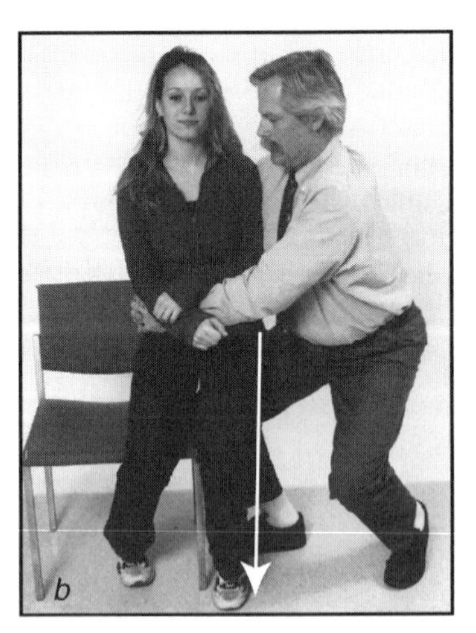

Figure 8.12 Lifting the patient can be performed with minimal loading down the spine provided the height of the seat is close to the height of the lifter's thigh when in a squat position *(a)*. The patient's pelvis is pulled and slid onto the lifter's thigh *(b)*, and then the lifter stands up, hugging the patient's pelvis and minimizing the forces up the arms and down the spine. This spine-sparing technique is reserved for patients who are able to stand.

Figure 8.13 *(a)* Shoveling snow with a large moment arm for the load on the shovel loads the back. *(b)* Resting the arm on the thigh directs the forces to the ground, bypassing the arm and spine linkage.

how a worker can perform the same maneuver to lift a patient. Observe that the lifter is pulling the patient's pelvis onto his thigh and standing along with the patient. (Note that patient lifts cannot be performed with universal technique since every patient is different and can offer variable assistance.) Minimal forces are transmitted down the spine. Shoveling snow by resting the forearm on the thigh involves the same spine-sparing principle (see figure 8.13, a-b).

Reducing the Load

Finally, workers use skill to reduce the actual load and hand forces when lifting large objects. Some tasks can be performed by lifting only half of the object at a time. For example, when loading long logs onto the back of a truck, a worker could lift just one end (effectively handling only half of the full weight of the log) and place it on the bed. The worker could then walk around to the other end of the log and lift while sliding the log into the bed (see figure 8.14). Each lift is half of the total load. A sequence of unloading a refrigerator demonstrates the same technique; the full weight of the object is never lifted (see figure 8.15, a-f)! The worker lifting the mini-refrigerator (shown in figure 8.16, a-b) tilts it up onto an edge, raising its center of gravity together with the initial lifting height. Lifting from this higher starting position reduces the necessary moment. The concept of minimizing the reaction moment is so much more robust than simply telling people to hold the load close to the body.

Figure 8.14 This girl loads the log into the truck by lifting only half its weight at a time. First she lifts one end onto the truck. Then she lifts the other end and slides the log onto the bed.

Figure 8.15 This worker unloads the refrigerator from the trailer without having to lift its full weight. He "walks" the refrigerator on its corners to the edge of the trailer where he balances it over the lip of the bed. *(a)* Next, he slides the refrigerator down the trailer's tailgate *(b)*. He walks the refrigerator clear of the trailer *(c)*, and leaves it standing upright while he retrieves the dolly. He pushes one edge of the refrigerator up just enough to slide the dolly under it *(d)* and *(e)*. Finally, he and a co-worker wheel the appliance away *(f)*.

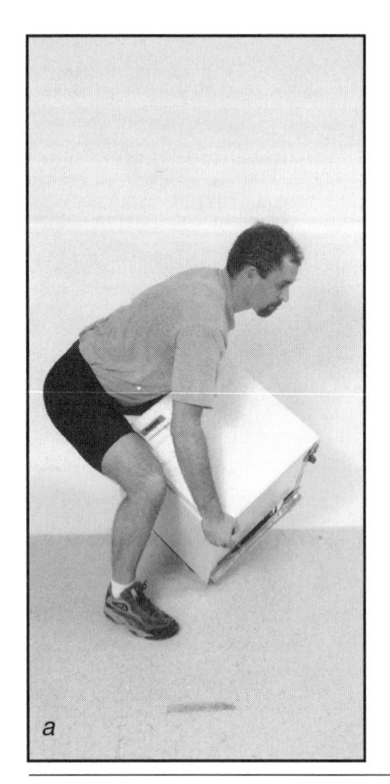

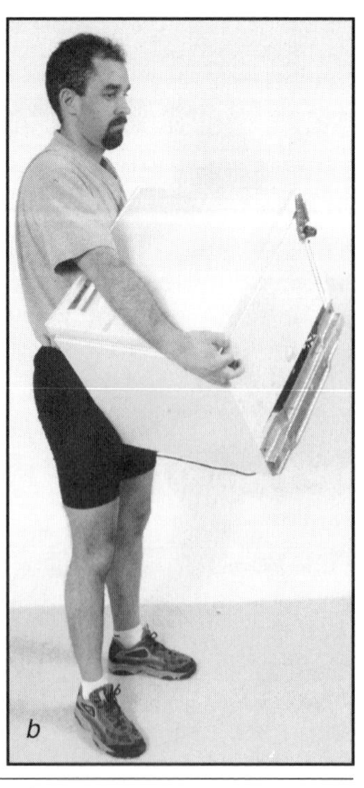

Figure 8.16 This man lifts the mini-refrigerator by tilting it up onto an edge *(a)*, raising its center of gravity together with the initial lifting height. Lifting from this higher starting position reduces the necessary moment so the man can reach the standing position without damage to his back *(b)*.

Should One Avoid Exertion Immediately After Prolonged Flexion?

Recall that prolonged flexion causes both ligamentous creep and a redistribution of the nucleus within the annulus (see chapter 4). In this way the spine tissues have a loading memory. Further, recent evidence from Jackson and colleagues (2001) suggests that prolonged flexion modifies the extensor neurological response and causes muscle spasming, at least until the ligamentous creep has been restored. In this case it was seven hours! Spine stability would be compromised during this period, and the risk of annulus damage remains temporarily high. If possible, after prolonged stooping or sitting activities, people should spend time standing upright before attempting more strenuous exertions. For example, it would be very unwise for the gardener

Figure 8.17 This gardener would be unwise to stand and immediately lift a heavy object after spending a long period of time in a stooped posture. Rather, standing for a brief period and even extending will prepare the disc and posterior passive tissues to reduce the risk of injury.

Figure 8.18 This shipper/receiver's work is characterized by periods of sitting followed immediately by lifting when a truck pulls up to the loading dock. Again, standing and waiting for a few minutes prior to the exertions will reduce the risk of injury.

in figure 8.17 to stand and lift a bag of peat moss or for the shipper/receiver in figure 8.18 to rise and immediately begin loading pallets. How long should one wait before performing an exertion? Data from Twomey and Taylor (1982) derived from cadaveric spines suggest that age delays recovery of the spinal tissues. McGill and Brown (1992) noted that residual laxity remained in the passive tissues even after a half hour of standing following prolonged flexion—although the flexion was extreme in an occupational context. At least 50% of the joint stiffness returned after two minutes of standing following the session of prolonged flexion.

As noted in chapter 4, because the mechanics of the joints are modulated by previous loading history, one should never move immediately to a lifting task from a stooped posture or after prolonged sitting. Rather, one should first stand, or even consciously extend the spine, for a short period. Obviously, people in certain occupations such as emergency ambulance personnel cannot follow this guideline. They arrive at an accident scene without the luxury of time to warm up their backs and may have to perform nasty lifts—such as a 150-kg heart attack victim out of a bathtub! For these individuals, the only strategy is to avoid a fully flexed spine posture while driving so that the spine remains best prepared to withstand the imposed loads. This may be done with accentuated lumbar pads in the ambulance seats and with worker training.

Should Intra-Abdominal Pressure (IAP) Be Increased While Lifting?

Generally the answer is no: at least IAP should not be increased consciously. Recall the discussion concluding that IAP does not reduce spine loading but does act to stiffen it against buckling. By successfully completing the rehabilitation training advocated in the final section of this book, people can train their breathing and IAP to be independent of exertion. In this way any specific instructions regarding breathing and exertion become moot points. In most cases IAP will rise naturally, and no further conscious effort is required.

A final caveat is required here. Very strenuous lifts, if they must be performed, will require the buildup of IAP to increase torso stiffness and ensure stability (Cholewicki, Juluru, and McGill, 1999). On the other hand, we know that a substantially increased introthoracic pressure (as occurs with lifting) will compromise venous return (Mantysaari, Antila, and Peltonen, 1984). Further, breath holding during exertion raises both systolic and diastolic blood pressure (Haslam et al., 1988), which can be a concern for some. Blackout is not uncommon in strenuous lifting even though it is not clear which mechanism is responsible (MacDougall et al., 1985). Reitan (1941) proposed that blackout may be due to elevated central nervous system fluid pressure (IAP also raises CNS fluid pressure in the spine and up to the head), whereas Hamilton, Woodbury, and Harper (1944) proposed that an increase in cerebrospinal fluid pressure might actually serve a protective function (i.e., the consequent decrease in transmural pressure across the cerebral vessels could actually decrease the risk of vascular damage). At this point, given these issues and a lack of full understanding, moderate IAP may be warranted with the understanding of the negative side effects. Extreme lifting efforts involving conscious increases in IAP should not be performed at work.

Are Twisting and Twisting Lifts Particularly Dangerous?

While several researchers have identified twisting of the trunk as a factor in the incidence of occupational low back pain (Frymoyer et al., 1983; Troup et al., 1981), the mechanisms of risk require some explanation. The kinematic act of twisting has been confused in the literature with the kinetic variable of generating twisting torque. Twisting torque in the torso can be accomplished whether the spine is twisted or not.

Generally, twisting to moderate degrees without high twisting torque is not dangerous. Some have hypothesized based on an inertia argument that twisting quickly will impose dangerous axial torques upon braking the axial rotation of the trunk at the end range of motion. Farfan and colleagues (1970) proposed that twisting of the disc is the only way to damage the collagenous fibers in the annulus leading to failure. They reported that distortions of the neural arch permitted such injurious rotations. Shirazi-Adl and colleagues (1986) conducted more detailed analyses of the annulus under twist. They supported Farfan's contention that twisting indeed can damage the annulus at end range but also noted that twisting is not the sole mechanism of annulus failure. In contrast, some research has suggested that twisting in vivo is not dangerous to the disc as the facet in compression forms a mechanical stop to rotation well before the elastic limit of the disc is reached; thus, the facet is the first structure to sustain torsional failure (Adams and Hutton, 1981). In a study of ligament involvement during twisting, Ueno and Liu (1987) concluded that the ligaments were under only negligible strain during a full physiological twist. However, an analysis of the L4-L5 joint by McGill and Hoodless (1990) suggested that posterior ligaments may become involved if the joint is fully flexed prior to twisting.

Generating twisting torque is a different issue (see figure 8.19, a-b). Since no muscle has a primary vector direction designed to create twisting torque, all muscles are

Figure 8.19 *(a)* Twisting the torso is occurring at the same time that twisting torque is required, which is a dangerous combination. *(b)* Generating the twisting torque but restricting the torso twist is a spine-sparing strategy (and can also enhance performance).

activated in a state of great cocontraction. This results in a dramatic increase in compressive load on the spine when compared to an equivalent torque about another axis. For example, data from a combination of our previous studies indicate that supporting 50 Nm in the extension axis imposes about 800 N of spinal compression. The same 50 Nm in the lateral bend axis results in about 1400 N of lumbar compression, but 50 Nm in the axial twist axis would impose over 3000 N (McGill, 1997). It appears that the joint pays dearly to support even small axial torques when extending during the lifting of a load.

To conclude, the generation of axial twisting torque when the spine is untwisted does not appear to be of particular concern. Nor is the act of twisting over a moderate range without accompanying twisting torque. But generating high torque while the spine is twisted appears to create a problematic combination and a high risk. This is of particular concern in several sports and will be addressed in that context in chapter 9.

Is Lifting Smoothly and Not Jerking the Load Always Best?

The answer to this question is no. We have all heard that a load should be lifted smoothly and not jerked. This recommendation was most likely made on the basis that accelerating a load upward increases its effective mass by virtue of an additional inertial force acting downward together with the gravitational vector. However, this may not always be the case. It is possible to lift a load by transferring momentum from an already moving segment. Autier and colleagues (1996) showed that when compared to novice lifters, expert materials handlers sometimes choose techniques that make more efficient use of momentum transfer. They do not always lift slowly and smoothly.

Troup and Chapman (1969) referred to the concept of momentum transfer during lifting, as has Grieve (1975), who coined the term *kinetic lift*. Later, McGill and Norman (1985) documented that smaller low back moments were possible in certain cases using a skillful transfer of momentum. For example, if a load is awkwardly placed, perhaps on a work table 75 cm (30 in.) from the worker, a slow, smooth lift would necessitate the generation of a large lumbar extensor torque for a lengthy duration of time—a situation that is most strenuous on the back. However, the worker could lift this load with a very low lumbar extensor moment or quite possibly no moment at all. If the worker leaned forward and placed his hands on the load, with bent elbows, the elbow extensors and shoulder musculature could thrust upward, initiating upward motion of the trunk to create both linear and angular momentum in the upper body (note that the load has not yet moved). As the worker straightens his arms, coupling takes place between the load and the large trunk mass (as the hands then start to apply upward force on the load), transferring some, or all, of the body momentum to the load and causing it to be lifted with a jerk. This mechanical solution was proven to be effective in a very early experiment in my research career. The person shown in figure 8.20, a and b is

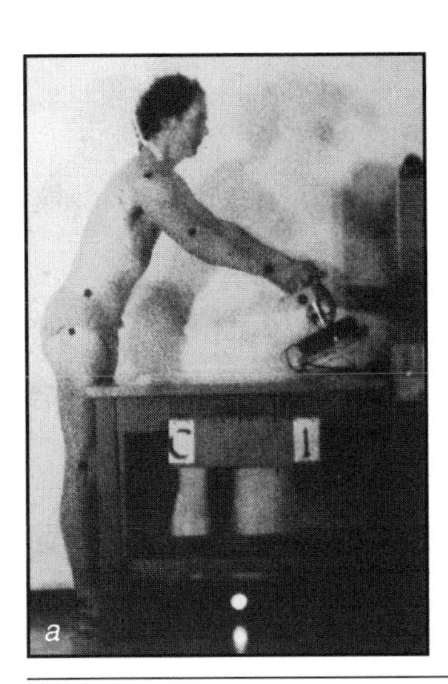

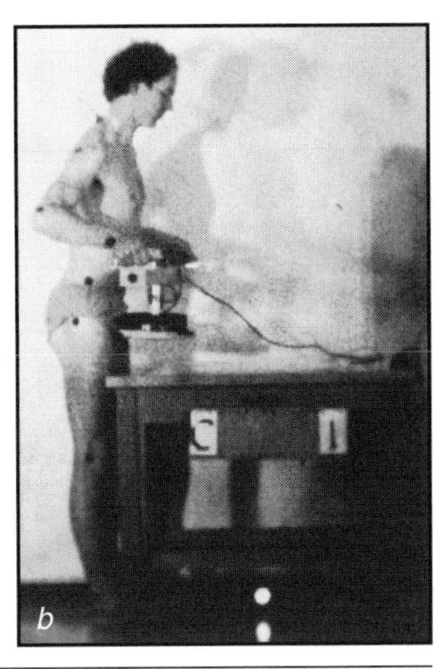

Figure 8.20 Lifting an awkwardly placed load slowly and smoothly as common wisdom suggests compromises the back. Rather, accelerating the torso first and then transferring the momentum to the load as the arms straighten can reduce the spine load. Accelerating the torso and the load at the same time is poor technique that violates this principle and causes higher spine loading. These photographs, *(a)* and *(b)*, are from an early experiment in the author's research career.

demonstrating a task in which the load can be lifted slowly, which would load the low back unduly, or with the kinetic lift technique, which, if correctly performed, will spare the back. (Obviously the markers on the model's body were to assist the measurement of body segment movement.) This highly skilled "inertial" technique is observed quite frequently throughout industry and in some athletic events such as competitive weightlifting, but it must be stressed that such lifts are conducted by highly practiced and skilled individuals. In most cases, acceleration of loads to decrease low back stress in the manner described is not suitable for the average individual conducting the lifting chores of daily living. The momentum-transfer technique is a skilled movement that requires practice; it is feasible only for awkwardly placed lighter loads and cannot be justified for heavy lifts.

Another mechanical variable should be integrated into the analysis of a dynamic technique. The tissue property of viscoelasticity enables tissues to sustain higher loads when loaded quickly (Burstein and Frankel, 1968). Troup (1977) suggested that viscoelasticity may be used to increase the margin of safety for spine injury during a higher strain rate but cautioned that incorporating this principle into lifting technique should depend on the rate of increase in spinal stress, the magnitude of peak stress, and its duration. Tissue viscoelasticity means that under faster loading the tissues do not have time to deform, even when the magnitude of the force is high. In this way the critical levels of tissue deformation required to cause damage are not reached. But given variability in response to load rate among the tissues, and among individuals, no specific guidance pertaining to actual load rate can be offered here.

The instruction to always lift a load smoothly may not invariably result in the least risk of injury. Indeed, it is possible to skillfully transfer momentum to an awkwardly placed object to position the load close to the body quickly and minimize the extensor torque required to support the load. In addition, tissue viscoelasticity can be protective during higher load rates. Clearly, reducing the extensor moment required to support the hand load is paramount in reducing the risk of injury; this is best accomplished by keeping the load as close to the body as possible.

Is There Any Way to Make Seated Work Less Demanding on the Back?

Prolonged sitting is problematic for the back. Unfortunately, this fact seems to be rather unknown in the occupational world. Those recovering from back injuries who return to modified work are often given "light duties" that involve prolonged sitting. While such duties are perceived as being easy on the back, they can be far from it. Even though the returning worker states that he cannot tolerate sitting, that in fact he would be more comfortable walking and even lifting, he is accused of malingering. This is the result of a misunderstanding of sitting mechanics.

Sitting Studies

Epidemiological evidence presented by Videman, Nurminen, and Troup (1990) documented the increased risk of disc herniation in those who perform sedentary jobs characterized by sitting. Known mechanical changes associated with the seated posture include the following:

- Increase in intradiscal pressure when compared to standing postures (Nachemson, 1966)
- Increases in posterior annulus strain (Pope et al., 1977)
- Creep in posterior passive tissues (McGill and Brown, 1992), which decreases anterior/posterior stiffness and increases shearing movement (Schultz et al., 1979)

- Posterior migration of the mechanical fulcrum (Wilder et al., 1988), which reduces the mechanical advantage of the extensor musculature (resulting in increased compressive loading)

These changes caused from prolonged sitting have motivated occupational biomechanists attempting to reduce the risk of injury to consider the duration of sitting as a risk factor when designing seated work. A recently proposed guideline suggested a sitting limit of 50 minutes without a break, although this proposal will be tested and evaluated in the future.

Strategies to Reduce Back Troubles During Prolonged Sitting

We have developed a three-point approach for reducing back troubles associated with prolonged sitting:

1. Use an ergonomic chair, but use it properly (very few actually do). Many people think that they should adjust their chair to create the ideal sitting posture. Typically, they adjust the chair so that the hips and knees are bent to 90° and the torso is upright (see figure 8.21). In fact, this is often shown as the ideal posture in many ergonomic texts. This may be the ideal sitting posture but for no longer than 10 minutes! Tissue loads must be migrated from tissue to tissue to minimize the risk of any single tissue accumulating microtrauma. This is accomplished by changing posture. Thus, an ergonomic chair is one that facilitates easy posture changes over a variety of joint angles (see figure 8.22). Callaghan and McGill (2001a) documented the range of spine postures that people typically adopt to avoid fatigue. Some have three or four preferred angles. The primary recommendation is to continually change the settings on the chair. Many workers continue to believe that there is a single best posture for sitting and are reluctant to try others. This is, of course, unfortunate, as the ideal sitting posture is a variable one. Many employees need to be educated as to how to change their chairs and to the variety of postures that are possible.

Figure 8.21 The "ideal sitting posture" (90° angles at the hips, knees, and elbows) described in most ergonomic guides. This is erroneous; the ideal sitting posture is one that involves variable postures.

2. Get out of the chair. There simply is no substitute for getting out of the chair. Some guidelines suggest performing exercise breaks while seated and some even go as far as to suggest flexing the torso in a stretch. This is both nonsense and disastrous! A rest break must consist of the opposite activity to reduce the imposed stressors. Extension relieves posterior annulus stress, but more flexion while seated increases it. The recommended break that we have developed involves standing from the chair and maintaining a relaxed standing posture for 10 to 20 seconds. At this stage, some may choose to perform neck rolls and arm windmills to relieve neck and shoulder discomfort from their desk work. The main objective is to buy some time to allow redistribution of the nucleus and reduce annular stresses. The person then raises the arms over the head (see figure 8.23, a-c) and then pushes the hands upward to the ceiling. By inhaling deeply, one will find that the low back is fully extended. In this way, the person has taken his back through gentle and progressive lumbar extension without having been taught lumbar position awareness or even understanding the concept. Some will argue that in their jobs they cannot stand and take a break; they must continue their seated work. These people generally will need to be shown the opportunities

Figure 8.22 Good posture for prolonged sitting is a variable one that migrates the internal loads among the various tissues. Possible sitting posture options are shown.

for standing. For example, they could choose to stand when the phone rings and speak standing. With these simple examples, they will soon see the opportunities to practice this part of good spine health.

3. Perform an exercise routine at some time in the workday. Midday would be ideal, but first thing in the morning is unwise (see the previous guideline). A good general back routine is presented in the last section of this book.

Some Short-Answer Questions

The following questions and their answers provide further guidance to reduce the risk of occupational LBD.

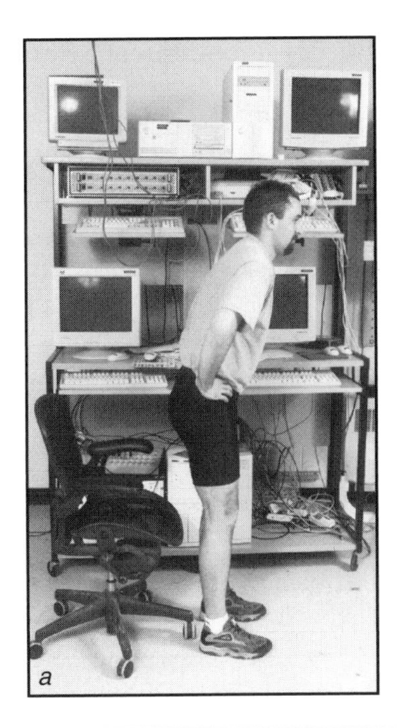

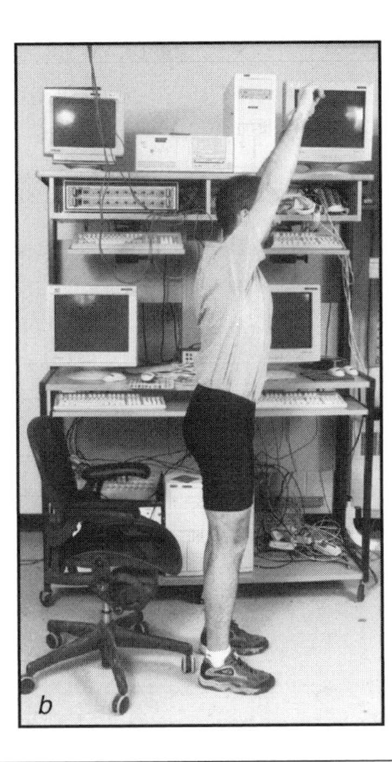

Figure 8.23 A strategy to thwart the accumulation of disc stresses from prolonged sitting is to stand (note the forward antalgic posture often observed after sitting) *(a)*, reach for the ceiling *(b)*, stretch pushing the hands upward, and then inhale deeply *(c)*. This sequence when performed slowly causes a gentle and progressive extension of the lumbar region and dispels the stresses of sitting.

- **Is it advisable to maintain a reasonable level of fitness?** As much as we would all like to believe that higher levels of fitness are protective for low back troubles, it is argued by some that the literature is not strongly supportive. This is for several reasons. Many clinical trials in which the intervention was designed to enhance fitness actually chose exercises that unknowingly increased the risk of spine damage. For example, many assumed that enhancing abdominal strength was a good idea and addressed this goal by prescribing sit-up exercises. Sit-ups will damage the backs of most people; they will not increase back health. Perhaps this has acted as an artifact biasing the literature. Interestingly enough, the most recent studies that have used biomechanical evidence to develop exercises—particularly stabilization exercises—have shown them to be efficacious. In fact, a recent review of preventive interventions by Linton and van Tulder (2001) suggests that well-chosen exercise is the most powerful strategy for preventing occupational back troubles. A stable spine maintained with healthy and wise motor patterns, along with higher muscle endurance, is protective.
- **Should one lift or perform extreme torso bending shortly after rising from bed?** The answer is no. Recall the biomechanics of daily changes in the spine as discussed in chapter 4: the discs are hydrophilic and imbibe fluid overnight. This is why people are taller in the morning than when they retire at night. This is also why it is much more difficult to bend in the morning to put one's socks on compared to taking them off at night; the bending stresses are much higher together with the risk of disc damage. This diurnal variation in spine length and the ability to flex forward have been well documented. As previously noted, Snook and colleagues (1998) found the strategy to restrict forward spine flexion in the morning to be very effective in reducing symptoms in a group of back-troubled patients.

• **Should workers adopt a lifting strategy to recruit the lumbodorsal fascia or involve the hydraulic amplifier mechanism?** As noted on page 80 in chapter 4, these mechanisms have been shown to be untenable proposals for sparing the back. While some still attempt to train workers to invoke these strategies, they have little scientific support; in many cases such strategies will be detrimental.

• **Should the trunk musculature be cocontracted to stabilize the spine?** As noted in chapter 6, the answer is generally yes. On page 71 in chapter 4, we noted that although such coactivation imposes some penalty on the spine, it is best for the spine to pay this price to enhance stiffness and resistance to buckling and to reduce the risk of other unstable behavior. How much cocontraction is necessary? As noted in chapter 6, in most tasks, sufficient stability is ensured with very modest amounts of cocontraction—somewhere in the magnitude of 5% to 10% of MVC for the abdominal wall and other antagonists. Achieving added stiffness in the spine through cocontraction will help prevent injuries that can otherwise result even during the lifting of a very light object, sneezing, or tying one's shoe. Of course, where there is the potential for surprise loading or handling nonrigid material or heavier loads, the coactivation magnitudes must increase. Damaged joint tissues also require higher levels of cocontraction to avoid unstable joint behavior. The necessary levels of coactivation would depend specifically on the task and the history of the individual. Thus, lightly cocontracting the stabilizing musculature upon exertion is a reasonable guideline; this should become automatic following stability training for those who do not naturally stiffen.

LBD Prevention for Employers

Not all prevention strategies can be implemented solely by workers. Employers, too, play a role, which is outlined here.

• **Provide protective clothing to facilitate the least stressful postures.** Workers sometimes handle material that is too noxious or too hazardous to hold close to their bodies. For example, I have been involved in cases in which workers held dirty material away from the body to keep their shirts clean, unnecessarily loading their backs. The solution was to provide protective coveralls to spare workers' backs. Leather aprons are helpful if the material is sharp to foster holding the load against the abdomen, as for sheet metal workers, for example. Knee pads are good for prolonged work on the ground. Once the employer has provided the necessary protective clothing, workers will figure out the variety of working postures that can spare their backs.

• **Should abdominal belts be prescribed to manual materials handlers?** Chapter 8 contains a thorough discussion of this topic.

• **Optimize containers and packaging of raw material to spare workers' backs.** Often in the design of the industrial process, sparing the joints of the workers is not considered. When considering the industrial process, see if handled materials can be bulk containerized. Can raw material be handled in smaller bundles or in bundles of different dimensions? Sometimes it is as simple as contacting the suppliers to find alternate ways of packaging material that foster handling in a way that conserves the body joints. The purchasing department plays a role in reducing injury! Bins and containers with folding sides help if parts must be picked out of the bin.

• **Encourage workers to practice lifting and work/task kinematic patterns.** Some individuals simply do not move, bend, and work in ways that spare their spines. In a recent study (McGill et al., in press), we noted that workers who had a previous history of back troubles were more likely to adopt motion patterns that resulted in

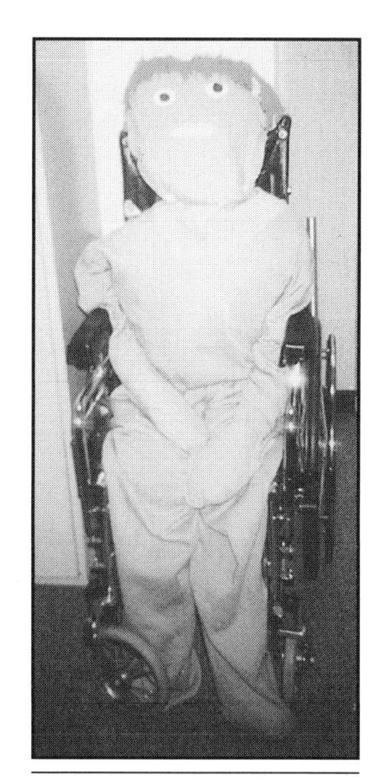

Figure 8.24 Nurses at a patient care facility fabricated this dummy to practice their patient lifts—an example of worker professionalism.

higher spine loads! Kinematic patterns need to be practiced and grooved into movement repertoires. (Remember the expression quoted earlier: "Practice does not make perfect, but practice makes permanent.") Some people must practice the spine-conserving motions every day—especially before attempting a particularly strenuous task. Even high-performance athletes must continually regroove motion patterns daily. Some worker groups have attempted to fabricate their own job-specific training and practice equipment. An example of this type of worker professionalism is shown in figure 8.24, which shows a dummy sitting in a wheelchair. Nurses built the dummy from plumbing pipes and use it to practice one- and two-person patient lifts. This is part of the worker professionalism ethic noted in the following section.

• **Optimize worker rest breaks.** A properly designed rest break consists of the opposite activity (and consequently the opposite loading) from that required by the job. For example, the sedentary secretary will be best served by a dynamic rest break. The welder, on the other hand, will be better served with rest and perhaps a stretch. The following example illustrates a violation of this principle that caused grief. Back in the 1960s, operators in a power plant monitored the process from a chair and had to respond to a vigilance buzzer on their desk that went off every 10 minutes (see figure 8.25, a-b). At each buzzer interval they would stand and walk around the control panel making adjustments. There was no history of back troubles. The room recently became obsolete and was replaced with an updated control room. The design team for the new control room believed that rising from the chair every 10 minutes was too strenuous. Consequently, the job was redesigned so that the operators were able to stay seated to perform all operations. These operators worked 12-hour shifts. Having the workers sit for this period of time (and removing the need to get out of the chair regularly) resulted in an increase in low back problems. The power plant then hired a consulting group who recommended rest breaks that consisted of having the workers sit on an exercise bike. The workers' backs failed to improve because the consultants failed to understand that the rest break must not exacerbate or replicate the forces of the job. In this case, sitting on a bike was not a break from sitting on the job.

Figure 8.25 The first control room *(a)* was built in the 1960s and required the operator to stand every 10 minutes to respond to the vigilance buzzer and make an adjustment to the analog instrumentation. There were no reported back troubles. This room was replaced with a new layout *(b)* based on the designers' assumption that standing every 10 minutes was too strenuous; the operators were able to sit for hours. A huge low back problem emerged.

- **Involve workers in the ergonomic process.** Design teams often neglect to consult with the expert on a particular job—the worker who has done it for years. Quite often the worker knows of a good solution, and it is simply a matter of listening and facilitating the intervention. An added benefit, psychologists claim, is that workers are more likely to comply with the intervention if they perceive themselves to be a major part of the solution.
- **Design work to be variable.** As several previous examples have documented, accumulation of tissue stress is thwarted by a change in posture, loading, or activity. Human beings were not made to perform repetitive work that emphasizes only a few tissues. Nor were humans designed not to be stressed. Research has established that tissues adapt and remodel in response to load (e.g., bone: Carter, 1985; ligament: Woo, Gomez, and Akeson, 1985; disc: Porter, 1992; vertebrae: Brinckmann, Biggemann. and Hilweg, 1989). Too little activity can be as problematic as too much. Krismer and colleagues' recent study (2001) strongly reinforces the idea so frequently stated in this text that the object of good work design is not to make every job easier; in fact, some jobs should be made more demanding for optimal health. In the Krismer study, students who reported low back pain were distinguished from those who did not by several risk factors. Two of those factors included
    - watching TV or playing computer games more than two hours per day (sedentary) and
    - regularly going beyond personal limits in sports activities (too much loading).

Good occupational health from a musculoskeletal perspective is achieved by performing a variety of tasks with well-designed rest activities, along with all of the traditional components such as proper nutrition, sleep, avoidance of smoking, and so on. Design work to be variable!

As illustrated in the previous examples and guidelines, management plays a role in reducing back troubles in workers. It is a mistake to think that management does not need to understand the science to justify specific injury prevention approaches. As we saw previously, injury prevention involves a thorough understanding of the industrial process, the way in which goods and materials are purchased, provision of protective equipment, appropriate training, the consideration of the cost/benefit of intervention, and so forth. Training of both workers and management ensures the best results.

Injury Prevention Primer

To use biomechanics to its full potential in injury prevention, workers and employers must have a reasonable understanding of the concepts as we understand them today. Workers must be educated in the biomechanically justifiable principles described earlier using examples with which they are familiar. Armed with the general principles, they can tackle any job and devise the best joint-conserving strategies. The intention is to enhance the industrial process and enable workers to retire in good health. The highlights of this chapter are summarized here:

- **First and foremost, design work/tasks that facilitate variety.** Perhaps the single most important guideline should be this: Don't do too much of any one thing. Both too much and too little loading are detrimental.
    - Too much of any single activity leads to trouble. Relief of cumulative tissue strains is accomplished with posture changes or, better yet, other tasks that have different musculoskeletal demands.

- While the tasks of many jobs cannot be changed, the working routines and arrangement of tasks within a job can be designed scientifically to incorporate this principle. Sometimes it is as easy as continually changing the sitting posture.

- **During all loading tasks, avoid a fully flexed or bent spine and rotate the trunk using the hips (preserving a neutral curve in the spine).** Doing this has the following benefits:
  - Disc herniation cannot occur.
  - Ligaments cannot be damaged, as they are slack.
  - The anterior shearing effect from ligament involvement is minimized, and the posterior supporting shear of the musculature is maximized.
  - Compressive testing of lumbar motion units has shown increases in tolerance with partial flexion but decreased ability to withstand compressive load at full flexion.
- **During lifting, choose a posture to minimize the reaction torque on the low back (either stoop or squat or somewhere in between), but keep the external load close to the body.**
  - A neutral spine is still maintained, but sometimes the load can be brought closer to the spine with bent knees (squat lift) or relatively straight knees (stoop lift). The key is to reduce the torque that has been shown to be a dominant risk factor.
  - When exerting force with the hands or shoulders, try to direct the line of the force through the low back. This will reduce the reaction torque and the spine load from muscle contraction.
- **Consider the transmissible vector.** Attempt to direct external forces through the low back, minimizing the moment arm, which causes high torques and crushing muscle forces. This principle should be applied when using pulling forces when opening a door, vacuuming, and performing other household chores.
- **Use techniques that minimize the actual weight of the load being handled.** The log-lifting example given in this chapter demonstrates how an entire log can be lifted into the back of a truck by lifting no more than half of its weight at any point.
- **Allow time for the disc nucleus to "equilibrate," ligaments to regain stiffness, and the stress on the annulus to equalize after prolonged flexion (e.g., sitting or stooping), and do not immediately perform strenuous exertions.**
  - After prolonged sitting or stooping, spend time standing.
  - This principle can be adapted to many special jobs, but some workers do not have the luxury of being able to take the time to allow the disc nucleus to equilibrate. For example, ambulance drivers are often called on to lift heavy loads immediately after significant periods of driving. A strategy for them is to use a lumbar support in their seat driving to the incident so that their spines are not flexed. Thus it can be prepared for the load with minimal disc equilibration (part of the process of warm up).
- **Avoid lifting or spine bending shortly after rising from bed.**
  - Forward-bending stresses on the disc and ligaments are higher after rising from bed compared with later in the day (at least one or two hours after rising), causing discs to become injured at lower levels of load and degree of bending.

- This principle is problematic for some occupations such as firefighting in which workers are often aroused from sleep to attend a fire. Such workers should not sit in a slouched posture with the spine flexed when traveling to the scene but rather sit upright. In this way the spine will be best prepared for strenuous work without a warm up.

- **Prestress and stabilize the spine even during light tasks.**
  - Lightly cocontract the stabilizing musculature to remove the slack from the system and stiffen the spine even during light tasks such as picking up a pencil. The exercises shown in chapter 13 were chosen to groove these motor patterns.
  - Mild cocontraction and the corresponding increase in stability increase the margin of safety of material failure of the column under axial load.
- **Avoid twisting and the simultaneous generation of high twisting torques.**
  - Twisting reduces the intrinsic strength of the disc annulus by disabling some of its supporting fibers while increasing the stress in the remaining fibers under load.
  - Since there is no muscle designed to produce only axial torque, the collective ability of the muscles to resist axial torque is limited, and they may not be able to protect the spine in certain postures. The additional compressive burden on the spine is substantial for even a low amount of axial torque production.
  - Generating twisting torque with the spine untwisted may not be as problematic nor is twisting lightly without substantial torque.
- **Use momentum when exerting force to reduce the spine load (rather than always lifting slowly and smoothly, which is an ill-founded recommendation for many skilled workers).**
  - This is a skill that sometimes need to be developed.
  - This strategy is dangerous for heavy loads and should not be used for lifting them.
  - It is possible that a transfer of momentum from the upper trunk to the load can start moving an awkwardly placed load without undue low back involvement.
- **Avoid prolonged sitting.**
  - Prolonged sitting is associated with disc herniation.
  - When required to sit for long periods, adjust posture often, stand up at least every 50 minutes, extend the spine, and/or, if possible, walk for a few minutes.
  - Organize work to break up bouts of prolonged sitting into shorter periods that are better tolerated by the spine.
- **Consider the best rest break strategies.** Customize this principle for different job classifications and demands.
  - Workers engaged in sedentary work would be best served by frequent, dynamic breaks to reduce tissue stress accumulation.
  - Workers engaged in dynamic work may be better served with longer and more restful breaks.
- **Provide protective clothing to foster joint-conserving postures**. Provide coveralls for dirty material handling, heavy aprons for sharp metals, knee pads for those who work at ground level, and so on.
- **Practice joint-conserving kinematic movement patterns.** Some workers need to constantly regroove motion patterns such as locking the lumbar spine when lifting and rotating about the hips.

- **Maintain a reasonable level of fitness.**
- **These guidelines may be combined for special situations.** For example, some people have difficulty rolling over in bed when their backs are painful. Nearly all can be taught to manage their pain and still accomplish this task by combining a momentum transfer with the minimal twisting guidelines (see figure 8.26, a-d).

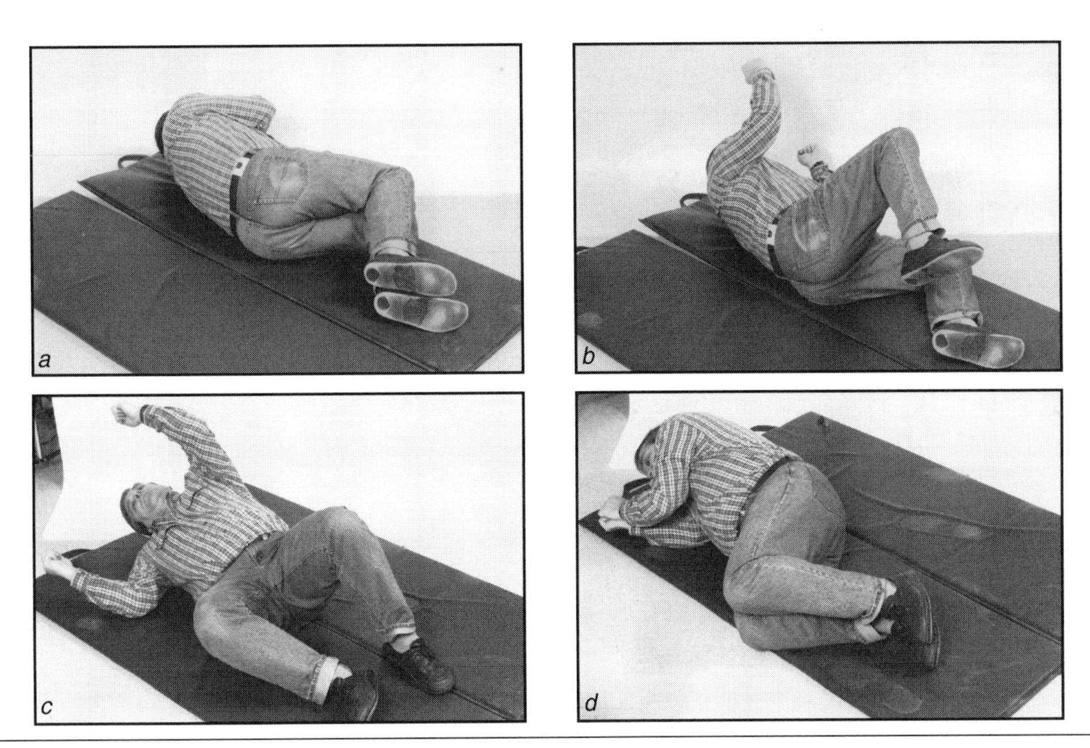

Figure 8.26 Rolling over in bed can be taught to those who maintain that it is too painful an activity. The figure illustrates rolling from the left side to the right. While lying on one side, the patient braces the torso *(a)* so that the spine does not twist in the steps that follow. Then the upper arm and leg are raised together with the lower arm and leg prying off the floor *(b)*. This is performed quickly enough to generate momentum that will carry the patient through the roll *(c)*. The patient should now be resting comfortably on his other side *(d)*.

A Note for Consultants

Acting as a consultant, I have made many mistakes, some of which motivated the following tips.

- **Don't fall into the trap of thinking that you are the expert and that you know what is best for the workers (unless you have done the job for years yourself).** Always consult the worker. Successful job incumbents have developed personal strategies for working that assist them in avoiding fatigue and injury. Their insights are the result of thousands of hours of performing the task, and they can be very perceptive. Try to accommodate them.
- **Do not take the instructions for a specific worker verbatim from the preceding Injury Prevention Primer.** Instead, explain the relevant biomechanical principle in language and terminology that are familiar to the worker.
- **Do not focus only on the most demanding tasks**. Given the links among different tasks from a tissue load perspective, you can often obtain better solutions by considering the full complement of exposures. In a similar vein, some consultants tend to focus on a single metric of risk (low back compression, for example) or rely on

only a few simple solutions. The average ergonomist probably does not have the specific training necessary to achieve the best solutions for low back problems. Perhaps I am biased since in recent years I am asked to consult only when consultants' poorly conceived ergonomic approaches have failed. I am requested to become involved when the company faces lawsuits or other issues that have raised the stakes. Remember that many solutions are neither simple nor unidimensional, regardless of your training. Use the Injury Prevention Primer as a checklist to evaluate whether potential exists for better and more comprehensive solutions.

• **Do not focus exclusively on the musculoskeletal issue.** Rather, look for the opportunities that lie in enhancing the industrial process. Any management board will recognize the worth of a consultant who makes the process more efficient, produces a higher-quality product, and/or reduces their injury compensation costs.

• **Finally, never make a recommendation that is not feasible to implement**, whether for monetary or any other reasons.

References

Aaras, A. (1994) The impact of ergonomic intervention on individual health and corporate prosperity in a telecommunications environment. *Ergonomics,* 37 (10): 1679-1696.

Adams, M.A., and Hutton, W.C. (1981) The relevance of torsion to the mechanical derangement of the lumbar spine. *Spine,* 6: 241-248.

Autier M., Lortie, M., and Gagnon M. (1996) Manual handling techniques: Comparing novices and experts. *International Journal of Industrial Ergonomics,* 17: 419-429.

Baldwin, M.L., Johnson, W.G., and Butter, R.J. (1996) The error of using returns-to-work to measure the outcomes of health care. *American Journal of Industrial Medicine,* 29 (6): 632-641.

Biering-Sorensen, F. (1984) Physical measurements as risk indicators for low-back trouble over a one-year period. *Spine,* 9: 106-119.

Brinkmann, P., Biggemann, M., and Hilweg, D. (1989) Prediction of the compressive strength of human lumbar vertebrae. *Clinical Biomechanics,* 4 (Suppl. 2): S1-S27.

Burnett, A.F., Khangure, M., Elliot, B.C., Foster, D.H., Marshall, R.N., and Hardcastle, P. (1996) Thoracolumbar disc degeneration in young fast bowlers in cricket. A follow-up study. *Clinical Biomechanics,* 11: 305-310.

Burstein, A.H., and Frankel, W.H. (1968) The viscoelastic properties of some biological material. *Annals of New York Academy of Science,* 146: 158-165.

Cady, L.D., Bischoff, D.P., O'Connell, E.R., Thomas, P.C., and Allan, J.H. (1979) Strength and fitness and subsequent back injuries of firefighters. *Journal of Occupational Medicine,* 21: 269.

Callaghan, J.P., and McGill, S.M. (2001a) Low back joint loading and kinematics during standing and unsupported sitting. *Ergonomics,* 44 (3): 280-294.

Callaghan, J., and McGill, S.M. (2001b) Intervertebral disc herniation. Studies on a porcine model exposed to highly repetitive flexion/extension motion with compressive force. *Clinical Biomechanics,* 16 (1): 28-37.

Carter, D.R. (1985) Biomechanics of bone. In: Nahum, H.M., and Melvin, J. (Eds.), *Biomechanics of trauma.* Norwalk, CT: Appleton Century Crofts.

Cholewicki, J., Juluru, K., and McGill, S.M. (1999) The intraabdominal pressure mechanisms for stabilizing the lumbar spine. *Journal of Biomechanics,* 32: 13-17.

Farfan, H.F., Cossette, J.W., Robertson, G.H., Wells, R.V., and Kraus, H. (1970) The effects of torsion on the lumbar intervertebral joints: The role of torsion in the production of disc degeneration. *Journal of Bone and Joint Surgery,* 52A (3): 469-497.

Farmer, M.E., Locke, B.Z., Moscicki, E.K., Dannenburg, A.L., Larson, D.B., and Radloff, L.S. (1988) Physical activity and depressive symptoms. The NHANES I epidemiologic follow-up study. *American Journal of Epidemiology,* 128: 1340-1351.

Frank, J.W., Brooker, A.S., DeMaio, S.E., et al. (1996) Disability resulting from occupational LBP: Part II. What do we know about secondary prevention? A review of the scientific evidence on prevention after disability begins. *Spine,* 21: 2918-2917.

Frymoyer, J.W., Pope, M.H., Clements, J.H., Wilder, D.G., MacPherson, B., and Ashikaga, T. (1983) Risk factors in low back pain. *Journal of Bone and Joint Surgery,* 65A: 213-218.

Garg, A., and Herrin, G. (1979) Stoop or squat: A biomechanical and metabolic evaluation. *American Institute of Industrial Engineers Transactions,* 11: 293-302.

Grieve, D.W. (1975) Dynamic characteristics of man during crouch and stoop lifting. *Biomechanics iv,* (eds. Nelson, R.C., and Morehouse, C.A.) (pp 19-29) Baltimore: University Park Press.

Hamilton, W.F., Woodbury, R.A., and Harper, H.T. (1944) Arterial, cerebrospinal and venous pressures in man during cough and strain. *American Journal of Physiology,* 141: 42-50.

Haslam, D., McCartney, N., McKelvie, R., and MacDougall, D. (1988) Direct measurements of arterial blood pressure during formal weight lifting in cardiac patients. *Journal of Cardiopulmonary Rehabilitation,* 8: 213-225.

Hughes, J.R. (1984) Psychological effects of habitual aerobic exercise: A critical review. *Preventive Medicine,* 13: 66-78.

Jackson, M., Solomonow, M., Zhou, B., Baratta, R.V., and Harris, M. (2001) Multifidus EMG and tension-relaxation recovery after prolonged static lumbar flexion. *Spine,* 26 (7): 715-723.

Kinney, S.E., Callaghan, J., and McGill, S.M. (1996) Lumbar spine movement and muscle activity using the golfer's lifting technique. In: *Evidence based Ergonomics,* 28th Annual Conference of the Human Factors Association of Canada, Kitchener, pp. 73-78.

Krause, N., Dasinger, L.K., and Neubrauser, F. (1998) Modified work and return to work: A review of the literature. *Journal of Occupational Rehabilitation,* 8 (2): 113-139.

Krismer, M., Trobos, S., Hanna, R., Sollner, W., Schonthaler, C., Auckenthaler, T., and Watzdorf, M. (2001) Prevalence and risk factors of low back pain in school children: A cross sectional study. In: *Abstracts,* International Society for Study of the Lumbar Spine, Edinburgh, Scotland.

Linton, S.J., and van Tulder, M.W. (2001) Preventative interventions for neck and back pain problems. *Spine,* 26 (7): 778-787.

Loisel, P., Abenhaim, L., Durand, P., Esdaile, J.M., Suissa, S., Gosselin, L., Simard, R., Turcotte, J., and Lemaire, J. (1997) A population-based, randomized clinical trial on back pain management. *Spine,* 22 (24): 2911-2918.

Luoto, S., Helioraara, M., Hurri, H., and Alavanta, M. (1995) Static back endurance and the risk of low back pain. *Clinical Biomechanics,* 10: 323-324.

MacDougall, D., Tuxen, D., Sale, D., Moroz, J., and Sutton, J.R. (1985) Arterial blood pressure response in heavy resistance exercise. *Journal of Applied Physiology,* 58 (3): 785-790.

Mantysaari, M., Antila, K., and Peltonen, T. (1984) Relationship between the changes in heart rate and cardiac output during the valsalva manoeuver. *Acta Physiologica Scand*inavica, (Suppl.), 537: 45-49.

McGill, S.M. (1997) Biomechanics of low back injury: Implications on current practice and the clinic. *Journal of Biomechanics,* 30 (5): 465-475.

McGill, S.M., and Brown, S. (1992) Creep response of the lumbar spine to prolonged lumber flexion. *Clinical Biomechanics,* 7: 43-46.

McGill, S.M., Grenier, S., Preuss, R., and Brown, S. (in press) Psychophysical, personal and biomechanical sequellae to low back pain.

McGill, S.M., and Hoodless, K. (1990) Measured and modelled static and dynamic axial trunk torsion during twisting in males and females. *Journal of Biomedical Engineering,* 12: 403-409.

McGill, S.M., and Kippers, V. (1994) Transfer of loads between lumbar tissues during the flexion relaxation phenomenon. *Spine,* 19 (19): 2190-2196.

McGill, S.M., and Norman, R.W. (1985) Dynamically and statically determined low back moments during lifting. *Journal of Biomechanics,* 18 (12): 877-885.

McGill, S.M., and Norman, R.W. (1987) Effects of an anatomically detailed erector spinae model on L4/L5 disc compression and shear. *Journal of Biomechanics,* 20 (6): 591-600.

Nachemson, A.L. (1966) The load on lumbar discs in different positions of the body. *Clinical Orthopaedics and Related Research,* 45: 107-122.

Pope, M.H., Hanley, E.N., Matteri, R.E., Wilder, D.G., and Frymoyer, J.W. (1977) Measurement of intervertebral disc space height. *Spine,* 2: 282-286.

Porter, R.W. (1992) Is hard work good for the back? The relationship between hard work and low back pain-related disorders. *International Journal of Industrial Ergonomics,* 9: 157-160.

Potvin, J., Norman, R.W., and McGill, S. (1991) Reduction in anterior shear forces on the L4/L5 disc by the lumbar musculature. *Clinical Biomechanics,* 6: 88-96.

Reitan, M. (1941) On the movements of fluid inside the cerebrospinal space. *Acta. Radiol. Scand.*, 22: 762-779.

Ross, C.E., and Hayes, D. (1988) Exercise and psychologic well-being in the community. *American Journal of Epidemiology,* 127: 762-771.

Scannell, J., and McGill, S.M. (in press) Elastic equilibrium of the lumbar spine.

Schultz, A.B., Warwick, D.N., Berkson, M.H., and Nachemson, A. (1979) Mechanical properties of the human lumbar spine motion segments. Part 1: Responses to flexion, extension, lateral bending and torsion. *Journal of Biomechanical Engineering,* 101: 46-52.

Shirazi-Adl, A., Ahmed, A.M., and Shrivastava, S.C. (1986) Mechanical response of a lumbar motion segment in axial torque alone and combined with compression. *Spine,* 11 (9): 914-927.

Snook, S.H., Webster, B.S., McGarry, R.W., Fogleman, M.T., and McCann, K.B. (1998) The reduction of chronic nonspecific low back pain through the control of early morning lumbar flexion: A randomized controlled trial. *Spine,* 23 (23): 2601-2607.

Suni, J.H., Oja, P., Miilunpalo, S.I., Pasanen, M.E., Vuori, I.M., Bos, K. (1998) Health-related fitness test battery for adults: Association with perceived health, mobility, and back function and symptoms. *Archives of Physical Medicine and Rehabilitation,* 79(5): 559-569.

Troup, J.D.G., and Chapman, A.E. (1969) The strength of the flexor and extensor muscles of the trunk. *Journal of Biomechanics,* 2: 49-62.

Troup, J.D.G. (1977) Dynamic factors in the analysis of stoop and crouch lifting methods: A methodological approach to the development of safe materials handling standards. *Orthop. Clin. N.Am.*, 8 (1): 201-209.

Troup, J.D.G., Martin, J.W., and Lloyd, D.C. (1981) Back pain in industry: A prospective survey. *Spine,* 6: 61-69.

Twomey, L., and Taylor, J. (1982) Flexion creep deformation and hysteresis in the lumbar vertebral column. *Spine,* 7: 116-122.

Ueno, K., and Liu, Y.K. (1987) A three-dimensional nonlinear finite element model of lumbar intervertebral joint in torsion. *Journal of Biomechanical Engineering,* 109: 200-209.

Videman, T., Nurminen, M., and Troup, J.D. (1990) Lumbar spinal pathology in cadaveric material in relation to history of back pain, occupation and physical loading. *Spine,* 15: 728-740.

Wilder, D.G., Pope, M.H., and Frymoyer, J.W. (1988) The biomechanics of lumbar disc herniation and the effect of overload and instability. *Journal of Spinal Disorder,* 1: 16-32.

Winkel, J., and Westgaard, R.H. (1996) Editorial: A model for solving work related musculoskeletal problems in a profitable way. *Applied Ergonomics,* 27: 71-78.

Woo, S.L.-Y., Gomez, M.A., and Akeson, W.H. (1985) Mechanical behaviors of soft tissues: Measurements, modifications, injuries, and treatment. In: Nahum, H.M., and Melvin, J. (Eds.), *Biomechanics of trauma* (pp. 109-133). Norwalk, CT: Appleton Century Crofts.

Young, R.J. (1979) The effect of regular exercise on cognitive functioning and personality. *British Journal of Sports Medicine,* September 13 (3): 110-117.

Reducing the Risk in Athletes: Guidelines

This chapter applies the same scientifically justifiable guidelines outlined in chapter 8 to reduce the risk of back troubles, although this chapter addresses athletic situations. Athletes and teams from a variety of sporting activities—from world-class professionals to amateurs—have sought my advice as a low back injury consultant. In many cases their bad backs were ending their careers. But as we have seen in preceding chapters, success in dealing with bad backs requires efforts to address both the cause of the troubles and the most appropriate rehabilitative therapy. In many cases, addressing the cause meant that athletes had to change their technique. Without exception, they had to change the way they trained. Sometimes this resulted in small decrements in their performances, but these technique changes enabled them to remain competitive and financially rewarded. The athletic activities of the many athletes I have worked with have varied from team sports, such as basketball, hockey, baseball, and football, to individual sports such as golf.

Upon completion of this chapter, you will be able to formulate scientifically based guidelines that reduce the risk of low back troubles in athletes.

Reducing the Risk in Athletes

The examples contained in the following sections illustrate how biomechanical principles can be applied effectively to virtually any sport. The following questions are often asked about how to minimize athletic back troubles. The answers should be used as a checklist to determine the most effective approach to reduce the irritants of a specific athlete's back troubles.

- **Should athletes avoid end range of spine motion during exertion?** Generally, the answer is yes—if the sporting objective allows for this strategy. In previous chapters I documented the many complications that arise from taking the lumbar spine to the end range of motion in any activity. Athletic tasks often have the additional stressor caused by high rotational velocities in the spine, which often force the passive tissues to experience impulsive loading as they act to create the mechanical

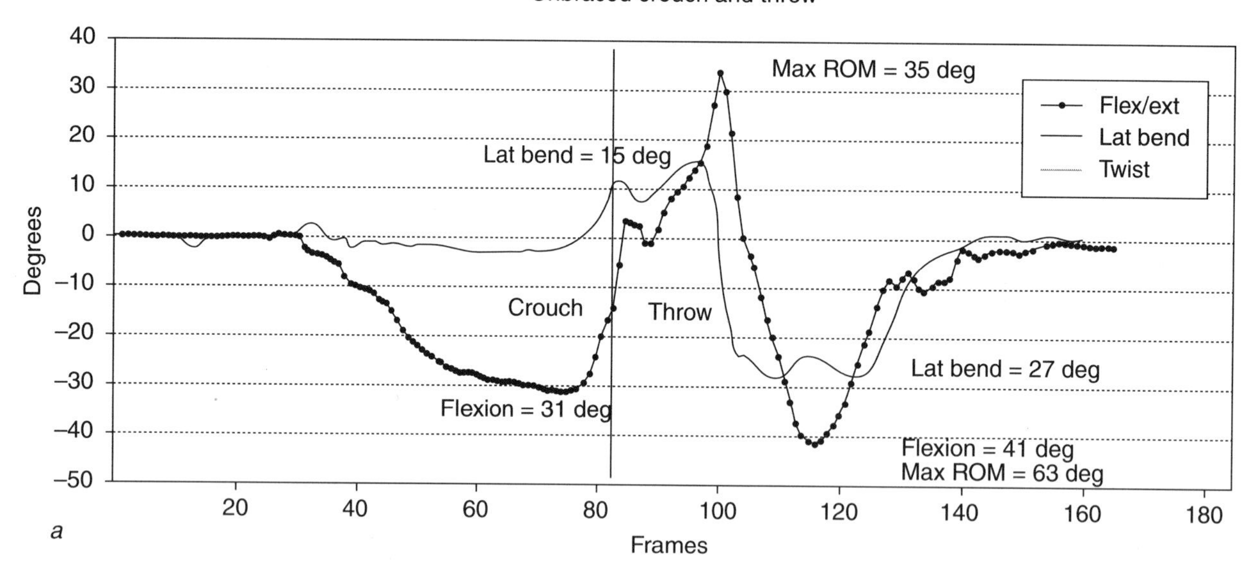

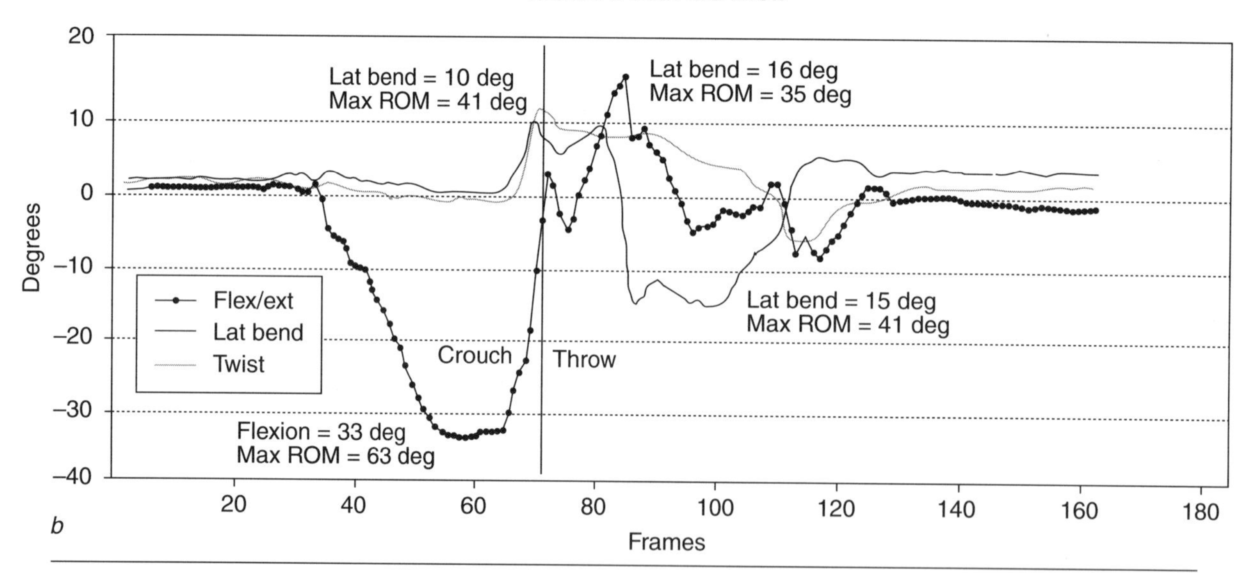

Figure 9.1 This baseball catcher's low back was troubled, in particular, by rising out of the crouch and throwing the ball to second base to deal with a runner stealing second base. Documentation of the lumbar kinematics revealed large excursions in flexion extension axis *(a)*. In fact, the lumbar spine was extended to its full range of motion (35°) and flexed to 41°. Teaching abdominal bracing together with spine position and motion awareness reduced the range of lumbar motion to 16° in extension and 33° in flexion *(b)*. As a result, the symptoms lessened.

stop to motion. Following are examples in which technique change has salvaged some athletes' spines:

- A baseball catcher was experiencing debilitating spine troubles when rising from the crouch and throwing the ball to address runners stealing second base. A combination of abdominal bracing and learning a spine kinematic motion profile that reduced the spine range of motion helped greatly (see figure 9.1, a-b). The athlete's throwing speed was compromised to a small degree, but the athlete remained competitive.
- Golfers typically take their spines through full ranges of motion in the lateral bend and axial twist axes. In an example from a professional golfer, the spine reached

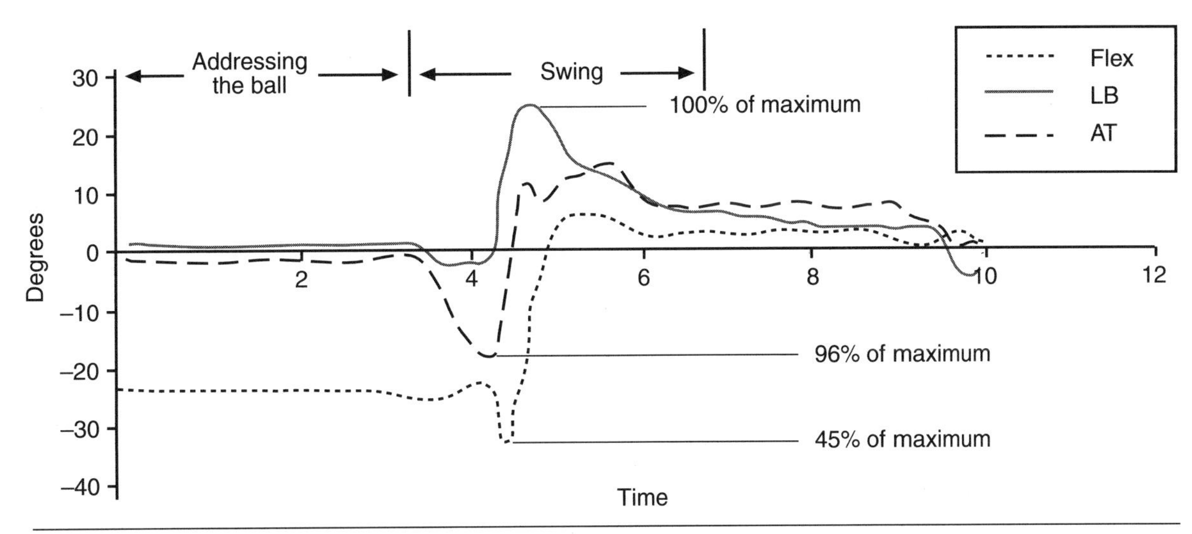

Figure 9.2 This professional golfer suffered chronic debilitating low back troubles and sought our expertise. The lumbar kinematic analysis of the swing revealed that the spine was taken to 100% of the lateral bending capability and 96% of the twist capability with each shot. The only way to help this back was to remove the cause—that is, reduce this motion. Retraining the abdominals to brace and reducing the backswing reduced the symptoms.

100% of the lateral bend capability and 96% of the twisting capability. This spine brutality was performed with each swing of the golf club—at least 200-300 times per day! There was little hope for recovery without reducing this loading. The spine-sparing approach was to reduce the backswing and groove abdominal bracing patterns that locked the rib cage to the pelvis on the follow-through. Spine range of motion was diminished (see figure 9.2). This golfer lost a few yards from the ball distance during a maximum drive effort, but as professional golfers play a game of accurate ball placement, this reduction paled in comparison to the possibility of losing the capability to play at a high level of competition.

• **How do you use technique to reduce the reaction moment?** The reaction moment about the low back results from the applied force vector and the perpendicular distance from the force to the lumbar spine. A smaller reaction moment results in lower internal tissue loads necessary to support the moment; this spares the spine. Generally, however, in sporting activities a reduced force is undesirable. The better option is to reduce the lever arm, or the perpendicular distance of the external force to the fulcrum in the low back (see p. 166). Martial artists take advantage of this technique, as do football linemen. Directing the external force vector through the lumbar spine minimizes the moment and the resulting compressive forces (see figure 9.3). The external force is still high for maximum performance, but the loads on the spine are minimized.

Figure 9.3 This football player in attempting to tackle the ball carrier is developing high spine loads, as the hand forces are not directed through the low back. Generally, better technique is to direct the forces through the low back while applying unstabilizing forces to the opponent.

• **Is there a problem with prolonged sitting on the bench?** Recall that prolonged flexion exacerbates discogenic back troubles together with ligament-based syndromes. The position of most athletes when sitting on the bench results in prolonged full lumbar flexion (see figure 9.4a). The spine postures adopted by some of the tall basketball players are tragically comical. Sporting tradition dictates this posture even though it is so detrimental to many troublesome back conditions. Two issues to consider about bench-sitting are as follows:

- Simply having athletes with back troubles, in particular discogenic back disorders, not sit on the bench in this way reduces their symptoms. Some find relief from standing and pacing while others find relief from sitting in chairs with an elevated seat pan that is angulated forward to minimize spine flexion (figure 9.4b).
- For years physiologists have shown the muscular, vascular, and ventilatory changes from warming up. Warm-ups affect spine mechanics, too. A recent study documented the increased spine compliance in a cohort of varsity volleyball players (Green, Grenier, and McGill, in press) following a warm-up. After sitting on the bench for 20 minutes, however, this compliance was lost. Twenty minutes is a typical sitting time for volleyball players (and for athletes in many other sports as well). The lasting effect of the warm-up is quite short-lived if activity is interrupted with rest—particularly seated rest. Players should reprepare their backs following a prolonged period of seated rest. Again, this includes sitting in higher chairs and standing and sometimes pacing prior to resuming play.

• **Should athletes train shortly after rising from bed?** The answer is no—if a large amount of lumbar motion is required. Researchers have documented the increased annulus stresses after a bout of bed rest. In particular, full bending exercises are contraindicated. Yet many athletes and laypeople alike get up in the morning and perform spine stretches, sit-ups, and so on. This is the most dangerous time of the day to undertake such activities. It is standard athletic lore for rowers, for example, to rise very early and train on the water since the water is calm this time of day. Rowing requires large amounts of spine flexion during the catch phase. (We have had to train rowers with disc troubles to adopt a technique strategy that avoids full lumbar flexion.) Rowers pay dearly for their early-morning flat water; their backs would be much better served by training later in the day.

Figure 9.4 Sitting on the bench with a flexed lumbar spine *(a)* is problematic for two reasons: this posture creates and/or exacerbates a posterior disc bulge, and it causes loss of the compliance obtained from warming up the back. Sitting in taller chairs with an angulated seat pan *(b)* reduces lumbar flexion and the associated concerns.

- **Should athletes breathe during a particular phase in the exertion?** Many recommend that athletes exhale when raising a weight—or the opposite. We hear, for example, that when performing the bench press exercise one should exhale upon the exertion. What evidence is this based on? This pattern will not transfer to the athletic situation in a way that will ensure sufficient spine stability. As I emphasized in chapter 6 on stability, the grooving of transferable stabilizing motor patterns requires that lung ventilation become independent of exertion. The spine must be stabilized regardless of whether the individual is inhaling or exhaling. We recommend that the athlete train to breathe independently of the exertion. The rare exception may be for the one-time maximum effort that requires supreme spine stiffness for a brief period (for example, weightlifting). Hopefully, this would be a rare occurrence for most athletes in training.
- **Should athletes wear abdominal belts during training?** Evidence suggests that people change their motor patterns when using a belt. In a study examining baggage handlers (workers who load and unload airplanes), Reddell and colleagues (1992) noted increased injury rates in those who were former belt wearers but had then stopped. Granata, Marras, and Davis (1997) noted gross technique changes when people wore belts. Many adopt belts in training for one of three reasons:
  - They have observed others wearing them and have assumed that it will be a good idea for them to do so.
  - Their backs are becoming sore and they believe that a back belt will help.
  - They want to lift a few more pounds.

None of these reasons are consistent with either the literature or the objective of good health. If one must lift a few more pounds, wear a belt. If one wants to groove motor patterns to train for other athletic tasks that demand a stable torso, it is probably better not to wear one. Chapter 10 offers a thorough discussion of the research that supports this recommendation.

- **Should the trunk musculature be cocontracted to stabilize the spine?** The answer is generally yes. Many athletic techniques require core stabilization to more effectively transmit forces throughout the body segment chain. Stiffening the torso is taught for many tasks. This approach also reduces the risk of low back injury. The optimal amount of stiffening depends on the task. Larger amounts of stiffness can be needed in the following situations:
  - When the athlete might experience unexpected loading, as in contact sports
  - When combinations of simultaneous moments and applied forces develop, as in a basketball player setting a post
  - After speed and acceleration of body segments are required, as in a golf swing or delivering a blow in martial arts
- **Is specificity of practice regimens important for low backs?** Spine-sparing kinematic patterns and stabilizing motor patterns need to be practiced and grooved into movement repertoires. Since practice makes permanent, it is crucial that routines done in practice replicate as closely as possible the movements and skills demanded by the sport. You don't want the "permanent" you achieve to be incorrect! Ask, for example, will an exercise with free weights or an exercise machine that isolates a body part better replicate the conditions of the sports task? Is strength really the primary requirement for a particular task? Is spine range of motion a requirement, or does the athletic task at hand require a stiff spine to transmit forces from the upper body through to the legs and to the ground? Every athlete and coach must examine the requirements and objectives of their performance tasks. Time and again I have

seen training that included objectives not required by the athlete's particular event that greatly compromise spine health. For example, recently in England, Professor Fred Yeadon simulated some gymnastics moves and then increased various components of fitness to see if performance was enhanced. For some tumbling and somersault tasks, increasing strength did not help; rather, takeoff speed was the component that enabled more rotations in the air. Unfortunately, the principles of bodybuilding have penetrated training for so many athletes, leading some to train to look good rather than to focus on regimens directed solely at enhancing performance. The no pain–no gain approach is the misdirected nemesis of many a bad back. Remember, a strong spine in isolation may not necessarily be a stable spine. A stable spine, maintained with healthy and wise motor patterns and higher muscle endurance, protects against back troubles and generally enhances performance.

- **Can imagery help protect low backs?** The steps in using imagery for developing back-sparing patterns for both performance and health are listed in chapter 11. While these are for spine position awareness and grooving motor patterns, they can be very helpful to athletes.

What Coaches Need to Know

Many athletes' backs are ruined in training—not in competition. This sometimes results from the circumstances that are traditional to the sport, circumstances that must be changed. A good example comes from Australia where cricket bowlers typically developed pars fractures. After scientific research clarified the mechanism of injury, legislation was developed that now dictates the maximum number of bowls a cricket bowler can perform in a certain period.

Athletic preparation by its very nature requires overload, which is counter to training for health objectives. As such, it exposes athletes to more risk. When coaches add bodybuilding regimens to their athletes' programs, they can further increase the risk to their athletes. Coaches are well advised to define clearly the training outcome objectives and design a training program to attain them in a way that spares the back and other joints. The process of developing a stable core described in chapters 13 and 14 is designed to spare the spine. A back-injured athlete has no chance for success.

Coaches who ask me to provide training programs for their athletes and teams are often amazed to find that I cannot do so without a penetrating analysis of the individual(s) and the task-specific objectives. Optimal preparation of an athlete is not achieved by following a cookbook recipe; every athlete and event or sport is different. Even the experts have to continually improve their skill and practice to optimally prepare different athletes.

References

Granata, K.P., Marras, W.S., and Davis, K.G. (1997) Biomechanical assessment of lifting dynamics, muscle activity and spinal loads while using three different style lifting belts. *Clinical Biomechanics,* 12 (2): 107-115.

Green, J., Grenier, S., and McGill, S.M. (in press) Low back stiffness is altered with warmup and bench rest: Implications for athletes.

Reddell, C.R., Congleton, J.J., Huchinson R.D., and Montgomery J.F. (1992) An evaluation of a weightlifting belt and back injury prevention training class for airline baggage handlers. *Applied Ergonomics,* 23 (5): 319-329.

The Question of Back Belts

Chapter 9 addressed the use of belts for athletic endeavors. This chapter will focus on occupational belt use (see figure 10.1, a-c for the typical types of belts used and tested). After reading this chapter, you will be able to make decisions on who should wear a belt and to justify the guidelines for their prescription and use.

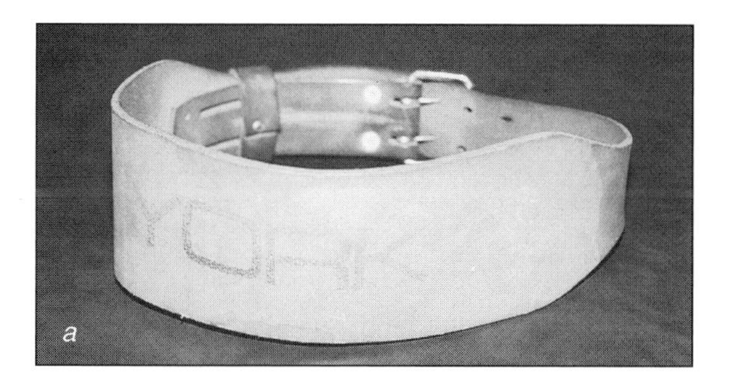

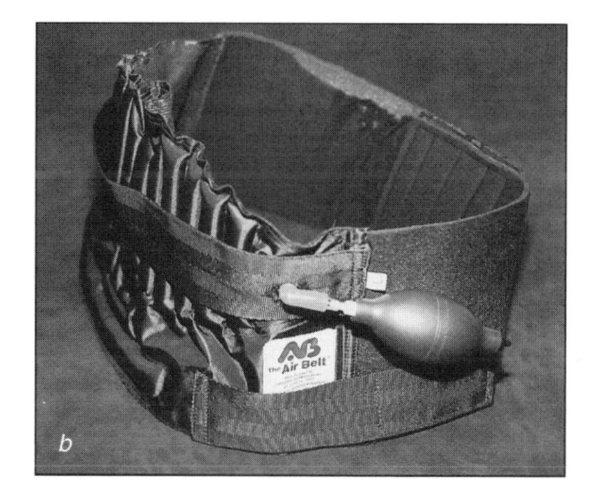

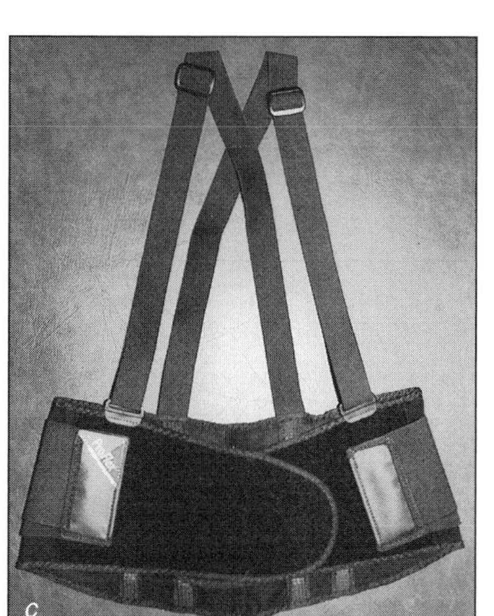

Figure 10.1 Several types of belts are worn and have been tested: *(a)* the leather belt, *(b)* the inflatable cell belt, and *(c)* the stretch belt with suspenders are a few examples.

Issues of the Back Belt Question

The average patient must be confused when he observes both Olympic athletes and back-injured people wearing abdominal belts. Several years ago I conducted a review of the effects of belt wearing (McGill, 1993) and summarized my findings as follows:

- Those who have never had a previous back injury appear to have no additional protective benefit from wearing a belt.
- It would appear that those who were injured while wearing a belt risk a more severe injury.
- Belts appear to give people the perception they can lift more and may in fact enable them to lift more.
- Belts appear to increase intra-abdominal pressure and blood pressure.
- Belts appear to change the lifting styles of some people to either decrease the loads on the spine or increase the loads on the spine.

In summary, given the assets and liabilities of belt wearing, I do not recommend them for healthy individuals either in routine work or exercise participation. However, the temporary prescription of belts may help some individual workers return to work.

Manufacturers of abdominal belts and lumbar supports continue to sell them to industry in the absence of a regulatory requirement to conduct controlled clinical trials similar to those required of drugs and other medical devices. Many claims have been made as to how abdominal belts could reduce injury. For example, some have suggested that belts perform the following functions:

- Remind people to lift properly
- Support shear loading on the spine that results from the effect of gravity acting on the handheld load and mass of the upper body when the trunk is flexed
- Reduce compressive loading of the lumbar spine through the hydraulic action of increased intra-abdominal pressure associated with belt wearing
- Act as a splint, reducing the range of motion and thereby decreasing the risk of injury
- Provide warmth to the lumbar region
- Enhance proprioception via pressure to increase the perception of stability
- Reduce muscular fatigue

These ideas, among others, will be addressed in this chapter. The following section addresses the science regarding the occupational use of belts and concludes with evidence-based guidelines.

The 1994 NIOSH report, "Workplace Use of Back Belts," contained critical reviews of a substantial number of scientific reports evaluating back belts. The report concluded that back belts do not prevent injuries among uninjured workers nor are they protective for those who have not been injured. While this is generally consistent with our position stated in 1993, my personal position on belt prescription is somewhat more moderate.

Scientific Studies

In the following sections I have subdivided the scientific studies into clinical trials and those that examined biomechanical, psychophysical, and physiological changes from belt wearing. Finally, based on the evidence, I recommend guidelines for the prescription and use of belts in industry.

Clinical Trials

Many clinical trials reported in the literature were fraught with methodological problems and suffered from the absence of a matched control group, no posttrial follow-up, limited trial duration, and insufficient sample size. As a result, I will review only a few clinical trials in this chapter, while acknowledging the extreme difficulty in executing such trials.

• **Walsh and Schwartz (1990)** divided 81 male warehouse workers into three groups:

- A control group (*n* = 27)
- A group that received a half-hour training session on lifting mechanics (*n* = 27)
- A group that received a one-hour training session and wore low back orthoses while at work for the subsequent six months (*n* = 27)

Instead of using more common types of abdominal belts, this research group used orthoses with hard plates that were heat molded to the low back region of each individual. Given the concern that belt wearing was hypothesized to cause the abdominals to weaken, the researchers measured the abdominal flexion strength of the workers both before and after the clinical trial. The control group and the training-only group showed no changes in abdominal flexor strength, nor any change in lost time from work. The third group, which received training and wore the belts, showed no changes in abdominal flexor strength or accident rate but did show a decrease in lost time. However, the increased benefit appeared only to accrue to those workers who had a previous low back injury. Van Poppel and colleagues (1998) reached a similar conclusion in a study of 312 airline baggage handlers.

• **Reddell and colleagues (1992)** studied 642 baggage handlers who worked for a major airline. They divided the baggage handlers into four treatment groups:

- A control group (*n* = 248)
- A group that received only a belt (*n* = 57)
- A group that received only a one-hour back education session (*n* = 122)
- A group that received both a belt and a one-hour education session (*n* = 57)

The trial lasted eight months, and the belt used was a fabric weightlifting belt 15 cm (6 in.) wide posteriorly and approximately 10 cm (4 in.) wide anteriorly. The researchers noted no significant differences among treatment groups for total lumbar injury incident rate, lost workdays, or workers' compensation rates. Although the lack of compliance by a significant number of subjects in the experimental group was cause for consideration, those who began wearing belts but discontinued their use had a higher lost-day case injury incident rate. In fact, 58% of workers belonging to the belt-wearing groups discontinued wearing belts before the end of the eight-month trial. Further, an increase in the number and severity of lumbar injuries occurred following the trial in those who previously wore belts.

• **Mitchell and colleagues (1994)** conducted a retrospective study administered to 1316 workers who performed lifting activities in the military. While this study relied on self-reported physical exposure and injury data over six years prior to the study, the authors did note that the costs of back injuries that occurred while wearing a belt were substantially higher than the costs of injuries sustained while not wearing a belt.

- **Kraus and colleagues (1996)**, in a widely reported study, surveyed the low back injury rates of nearly 36,000 employees of the Home Depot stores in California from 1989 to 1994. During this study period the company implemented a mandatory back belt use policy. Injury rates were recorded. Even though the authors claim that belt wearing reduced the incidence of low back injury, analysis of the data and methodology suggests that a much more cautious interpretation may be warranted. The data show that while belt wearing reduced the risk in younger males and those older than 55 years, belt wearing appeared to *increase* the risk of low back injury for men working longer than four years by 27% (although the large confidence interval required an even larger increase for statistical significance) and in men working less than one year. However, of greatest concern is the lack of scientific control to ferret out the true belt-wearing effect: there was no comparable non-belt-wearing group, which is critical given that the belt-wearing policy was not the sole intervention at Home Depot. For example, over the period of the study, the company increased the use of pallets and forklifts, installed mats for cashiers, implemented postaccident drug testing, and enhanced worker training. In fact, a conscious attempt was made to enhance safety in the corporate culture. This was a large study, and the authors deserve credit for the massive data reduction and logistics. However, despite the title and claims that back belts reduce low back injury, this uncontrolled study cannot answer the question about the effectiveness of belts.
- **Wassell and colleagues (2000)**, in response to the huge promotion of the Kraus et al. (1996) paper by some interest groups and the study's methodological concerns, replicated Kraus and coworkers' work under the sponsorship of NIOSH. These researchers, however, used better scientific control in order to evaluate only the effect of the belts. Surveying over 13,873 employees at newly opened stores of a major retailer, in which 89 stores required employees to wear belts and 71 stores had only voluntary use, they discovered that the belts failed to reduce the incidence of back injury claims. This study has more power than the Kraus study.

In summary, difficulties in executing a clinical trial are acknowledged. The Hawthorne effect is a concern, as it is difficult to present a true double-blind paradigm to workers since those who receive belts certainly know so. In addition, logistical constraints on duration, diversity in occupations, and sample size are problematic. However, the data reported in the better-executed clinical trials cannot support the notion of universal prescription of belts to all workers involved in manual handling of materials to reduce the risk of low-back injury. Weak evidence suggests that those already injured may benefit from belts (or molded orthoses) with a reduced risk of injury recurrence. However, evidence does not appear to support uninjured workers wearing belts to reduce the risk of injury; in fact, the risk of injury seems to increase during the period following a trial of belt wearing. Finally, some evidence suggests that the cost of a back injury may be higher in workers who wear belts than in workers who do not.

Biomechanical Studies

Researchers who have studied the biomechanical issues of belt wearing have focused on spinal forces, intra-abdominal pressure (IAP), load, and range of motion. The most informative studies are reviewed in this section.

IAP and Low Back Compressive Load

Biomechanical studies have examined changes in low back kinematics and posture in addition to issues of specific tissue loading. Two studies in particular (Harman et al., 1989, and Lander, Hundley, and Simonton, 1992) suggested that wearing an abdominal belt can increase the margin of safety during repetitive lifting. Both of these papers reported ground reaction force and measured intra-abdominal pressure while subjects repeatedly lifted barbells. Both reports observed an increase in intra-abdominal pressure in subjects who wore abdominal belts. These researchers assumed that intra-abdominal pressure is a good indicator of spinal forces, which is highly contentious. Nonetheless, they assumed the higher recordings of intra-abdominal pressure indicated an increase in low back support that in their view justified the use of belts. Neither study measured or calculated spinal loads.

Several studies have questioned the hypothesized link between elevated intra-abdominal pressure and reduction in low back load. For example, using an analytical model and data collected from three subjects lifting various magnitudes of loads, McGill and Norman (1987) noted that a buildup of intra-abdominal pressure required additional activation of the musculature in the abdominal wall, resulting in a net increase in low back compressive load and not a net reduction of load, as researchers had previously thought. In addition, Nachemson and colleagues (1986) published some experimental results that directly measured intradiscal pressure during the performance of Valsalva maneuvers, documenting that an increase in intra-abdominal pressure increased, not decreased, the low back compressive load. Therefore, the conclusion that an increase in intra-abdominal pressure due to belt wearing reduces compressive load on the spine seems erroneous. In fact, such an increase may have no effect or may even increase the load on the spine.

IAP and Low Back Muscles

Several studies have put to rest the belief that IAP affects low back extensor activity. McGill and colleagues (1990) examined intra-abdominal pressure and myoelectric activity in the trunk musculature while six male subjects performed various types of lifts either wearing or not wearing an abdominal belt (a stretch belt with lumbar support stays, Velcro tabs for cinching, and suspenders for when subjects were not lifting). Wearing the belt increased intra-abdominal pressure by approximately 20%. Further, the authors hypothesized that if belts were able to help support some of the low back extensor moment, one would expect to measure a reduction in extensor muscle activity. There was no change in activation levels of the low back extensors nor in any of the abdominal muscles (rectus abdominis or obliques).

In a study that examined the effect of belts on muscle function, Reyna and colleagues (1995) examined 22 subjects for isometric low back extensor strength and found belts provided no enhancement of function (although this study was only a four-day trial and did not examine the effects over a longer duration).

Ciriello and Snook (1995) examined 13 men over a four-week period lifting 29 metric tonnes in four hours twice a week both with and without a belt. Median frequencies of the low back electromyographic signal (which is sensitive to local muscle fatigue) were not modified by the presence or absence of a back belt, strengthening the notion that belts do not significantly alleviate the loading of back extensor muscles. Once again, this trial was not conducted over a very long period of time.

Belts and Range-of-Motion Restriction

In 1986 Lantz and Schultz observed the kinematic range of lumbar motions in subjects wearing low back orthoses. While they studied corsets and braces rather than abdominal belts, they did report restrictions in the range of motion, although the restricted motion was minimal in the flexion plane.

In another study McGill, Seguin, and Bennett (1994) tested flexibility and stiffness of the lumbar torsos of 20 male and 15 female adult subjects, both while they wore and did not wear a 10-cm (4-in.) leather abdominal belt. The stiffness of the torso was significantly increased about the lateral bend and axial twist axes with wearing belts but not when subjects were rotated into full flexion. Thus, these studies seem to indicate that abdominal belts help restrict the range of motion about the lateral bend and axial twist axes but do not have the same effect when the torso is forced in flexion, as in an industrial lifting situation.

Posture of the lumbar spine is an important issue in injury prevention for several reasons. In particular, Adams and Hutton (1988) showed that the compressive strength of the lumbar spine decreases when people approach the end range of motion in flexion. Therefore, if belts restrict the end range of motion, one would expect the risk of injury to be correspondingly decreased. While the splinting and stiffening action of belts occurs about the lateral bend and axial twist axes, stiffening about the flexion/extension axis appears to be less.

A more recent data set presented by Granata, Marras, and Davis (1997) supports the notion that some belt styles are better in stiffening the torso in the manner described previously—namely, the taller elastic belts that span the pelvis to the rib cage. Furthermore, these authors also documented that a rigid orthopedic belt generally increased the lifting moment while the elastic belt generally reduced spinal load. Nevertheless, the authors noted a wide variety in subject response. (Some subjects experienced increased spinal loading with the elastic belt.) Even in well-controlled studies, belts appear to modulate lifting mechanics in some positive ways in some people and in negative ways in others.

Studies of Belts, Heart Rate, and Blood Pressure

Hunter and colleagues (1989) monitored the blood pressure and heart rate of five males and one female performing dead lifts and one-arm bench presses and riding bicycles while wearing and not wearing a 10-cm (4-in.) weight belt. Subjects were required to hold in a lifting posture a load of 40% of their maximum weight in the dead lift for two minutes. The subjects were required to breathe throughout the duration so that no Valsalva effect occurred. During the lifting exercise blood pressure (up to 15 mmHg) and heart rate were both significantly higher in subjects wearing belts. Given the relationship between elevated systolic blood pressure and an increased risk of stroke, Hunter and colleagues (1989) concluded that individuals who may have cardiovascular system compromise are probably at greater risk when undertaking exercise while wearing back supports than when not wearing them.

Subsequent work conducted in our own laboratory (Rafacz and McGill, 1996) investigated the blood pressure of 20 young men performing sedentary and very mild activities both with and without a belt. (The belt was the elastic type with suspenders and Velcro tabs for cinching at the front.) Wearing this type of industrial back belt significantly increased diastolic blood pressure during quiet sitting and standing both with and without a handheld weight, during a trunk rotation task, and during a squat lifting task. Evidence increasingly suggests that belts increase blood pressure!

Over the past decade I have been asked to deliver lectures and participate in academic debate on the back belt issue. On several occasions occupational medicine personnel have approached me after hearing the effects of belts on blood pressure and intra-abdominal pressure and have expressed suspicions that long-term belt wearing at their particular workplace may possibly be linked with higher incidents of varicose veins in the testicles, hemorrhoids, and hernias. As of this writing, there has been no scientific and systematic investigation of the validity of these suggestions. Rather than wait for strong scientific data to either lend support to these ideas or dismiss them, it may be prudent to simply state concern. This will motivate studies in the future to track the incidence and prevalence of these pressure-related disorders to assess whether they are indeed linked to belt wearing.

Psychophysical Studies

Some have expressed concern that wearing belts fosters an increased sense of security that may or may not be warranted. Studies based on the psychophysical paradigm allow workers to select weights that they can lift repeatedly using, their own subjective perceptions of physical exertion. McCoy and colleagues (1988) examined 12 male college students while they repetitively lifted loads from floor to knuckle height at the rate of three lifts per minute for a duration of 45 minutes. They repeated this lifting bout three times, once without a belt and once each with two different types of abdominal belts: a belt with a pump and air bladder posteriorly and the elastic stretch belt previously described in the McGill, Norman, and Sharratt (1990) study. After examining the various magnitudes of loads that subjects had selected to lift in the three conditions, the researchers noted that wearing belts increased the load that subjects were willing to lift by approximately 19%. This evidence may lend some support to the theory that belts give people a false sense of security.

Summary of Prescription Guidelines

My earliest report on back belts (McGill, 1993) presented data and evidence that neither completely supported nor condemned the wearing of abdominal belts for industrial workers. After more laboratory studies and field trials, my position (which has been implemented by several governments and corporations) has not changed.

Given the available literature, it would appear the universal prescription of belts (i.e., providing belts to all workers in an industrial operation) is not in the best interest of globally reducing both the risk of injury and compensation costs. Uninjured workers do not appear to enjoy any additional benefit from belt wearing and, in fact, may be exposing themselves to the risk of a more severe injury if they were to become injured. Moreover, they may have to confront the problem of weaning themselves from the belt. However, if some individual workers perceive a benefit from belt wearing, they should be allowed to wear a belt conditionally but only on trial. The mandatory conditions for prescription (for which there should be *no* exception) are as follows:

1. Given the concerns regarding increased blood pressure and heart rate and issues of liability, all candidates for belt wearing should be screened for cardiovascular risk by medical personnel.
2. Given the concern that belt wearing may provide a false sense of security, belt wearers must receive education on lifting mechanics (back school). All too often, belts are being promoted to industry as a quick fix to the injury problem. Promotion of belts, conducted in this way, is detrimental to the goal of reducing injury as it redirects the focus from the cause of the injury. Education programs should include information on how tissues become injured, techniques to minimize musculoskeletal loading, and what to do about feelings of discomfort to avoid disabling injury.
3. Consultants and clinicians should not prescribe belts until they have conducted a full ergonomic assessment of the individual's job. The ergonomic approach should examine and attempt to correct the cause of the musculoskeletal overload and provide solutions to reduce the excessive loads. In this way, belts should only be

used as a supplement for a few individuals, while a greater plantwide emphasis should be on the development of a comprehensive ergonomics program.

4. Belts should not be considered for long-term use. The objective of any small-scale belt program should be to wean workers from the belts by insisting on mandatory participation in comprehensive fitness programs and education on lifting mechanics, combined with ergonomic assessment. Furthermore, consultants would be wise to continue vigilance in monitoring former belt wearers for a period of time following belt wearing, given that this period appears to be characterized by an elevated risk of injury.

References

Adams, M.A., and Hutton, W.C. (1988) Mechanics of the intervertebral disc. In: Ghosh, P. (Ed.), *The biology of the intervertebral disc*. Boca Raton, FL: CRC Press.

Ciriello, V.M., and Snook, S.H. (1995) The effect of back belts on lumbar muscle fatigue. *Spine,* 20 (11):1271-1278.

Granata, K.P., Marras, W.S., and Davis, K.G. (1997) Biomechanical assessment of lifting dynamics, muscle activity and spinal loads while using three different style lifting belts. *Clinical Biomechanics,* 12 (2): 107-115.

Harman, E.A., Rosenstein, R.M., Frykman, P.N., and Nigro, G.A. (1989) Effects of a belt on intra-abdominal pressure during weight lifting. *Medicine and Science in Sports and Exercise,* 2 (12): 186-190.

Hunter, G.R., McGuirk, J., Mitrano, N., Pearman, P., Thomas, B., and Arrington, R. (1989) The effects of a weight training belt on blood pressure during exercise. *Journal of Applied Sport Science Research,* 3 (1): 13-18.

Kraus, J.F., Brown, K.A., McArthur, D.L., Peek-Asa, C., Samaniego, L., and Kraus, C. (1996) Reduction of acute low back injuries by use of back supports. *International Journal of Occupational and Environmental Health,* 2: 264-273.

Lander, J.E., Hundley, J.R., and Simonton, R.L. (1992) The effectiveness of weight belts during multiple repetitions of the squat exercise. *Medicine and Science in Sports and Exercise,* 24 (5): 603-609.

Lantz, S.A., and Schultz, A.B. (1986) Lumbar spine orthosis wearing. I. Restriction of gross body motion. *Spine,* 11 (8): 834-837.

McCoy, M.A., Congleton, J.J., Johnston, W.L., and Jiang, B.C. (1988) The role of lifting belts in manual lifting. *International Journal of Industrial Ergonomics,* 2: 259-266.

McGill, S.M. (1993) Abdominal belts in industry: A position paper on their assets, liabilities and use. *American Industrial Hygiene Association Journal,* 54 (12): 752-754.

McGill, S.M., and Norman, R.W. (1987) Reassessment of the role of intra-abdominal pressure in spinal compression. *Ergonomics,* 30 (11): 1565-1588.

McGill, S., Norman, R.W., and Sharratt, M.T. (1990) The effect of an abdominal belt on trunk muscle activity and intra-abdominal pressure during squat lifts. *Ergonomics,* 33 (2): 147-160.

McGill, S.M., Seguin, J.P., and Bennett, G. (1994) Passive stiffness of the lumbar torso in flexion, extension, lateral bend and axial twist: The effect of belt wearing and breath holding. *Spine,* 19 (6): 696-704.

Mitchell, L.V., Lawler, F.H., Bowen, D., Mote, W., Asundi, P., and Purswell, J. (1994) Effectiveness and cost-effectiveness of employer-issued back belts in areas of high risk for back injury. *Journal of Occupational Medicine,* 36 (1): 90-94.

Nachemson, A.L., Andersson, G.B.J., and Schultz, A.B. (1986) Valsalva maneuver biomechanics. Effects on lumbar trunk loads of elevated intra-abdominal pressures. *Spine,* 11 (5): 476-479.

National Institute for Occupational Safety and Health (NIOSH) (1994, July) Workplace use of back belts. U.S. Department of Health and Human Services, Centers for Disease Control and Prevention.

Rafacz, W., and McGill, S.M. (1996) Abdominal belts increase diastolic blood pressure. *Journal of Occupational and Environmental Medicine,* 38 (9): 925-927.

Reddell, C.R., Congleton, J.J., Huchinson, R.D., and Montgomery, J.F. (1992) An evaluation of a weightlifting belt and back injury prevention training class for airline baggage handlers. *Applied Ergonomics,* 23 (5): 319-329.

Reyna, J.R., Leggett, S.H., Kenney, K., Holmes, B., and Mooney, V. (1995) The effect of lumbar belts on isolated lumbar muscle. *Spine,* 20 (1): 68-73.

Van Poppel, M.N.M., Koes, B.W., van der Ploeg, T., et al. (1998) Lumbar supports and education for the prevention of low back pain in industry: A randomized controlled trial. *Journal of the American Medical Association,* 279: 1789-1794.

Walsh, N.E., and Schwartz, R.K. (1990) The influence of prophylactic orthoses on abdominal strength and low back injury in the work place. *American Journal of Physical Medicine and Rehabilitation,* 69 (5): 245-250.

Wassell, J.T., Gardner, L.I., Landsittel, D.P., Johnston, J.J., and Johnston, J.M. (2000) A prospective study of back belts for prevention of back pain and injury. *Journal of the American Medical Association,* 284 (21): 2727-2734.

Low Back Rehabilitation

The evidence developed in previous chapters on low back function is used to justify better rehabilitation approaches.

Building Better Rehabilitation Programs for Low Back Injuries

The evidence presented in this book supports the establishment of spine stability first, followed (sometimes) by spine mobility in some back-injured patients. People who are not helped by this spine stabilization approach are generally those considered to be "failed" backs. Most patients, from those looking for functional enhancement and pain relief to athletes seeking performance enhancement, benefit from the spine stabilization approach. Evidence presented throughout this book is unanimous: a spine does not behave like a knee or shoulder, and approaches that work with these joints are often not effective for back therapy. Loading throughout the range of motion—which works well for joints in the extremities—is the nemesis of many backs, at least in the early stages of rehabilitation. During this period, training for strength is more often counterproductive. Unfortunately, the principles used in bodybuilding pervade rehabilitation "clinical wisdom." This approach hypertrophies muscle at the expense of developing functional motor/motion patterns needed for optimal health. In this chapter several general recommendations to maximize the chance for successful rehabilitation are discussed, followed by considerations of the stages of patient progression and guidelines for developing the best exercise regimen for each patient.

Upon completion of this chapter you will understand how to get bad backs to respond and how to develop better programs.

Finding the Best Approach

Given the wide variety of low back patients, one cannot expect to be successful in low back rehabilitation by treating everyone with the same cookbook program. The following concepts will help guide clinical decisions to individualize—and thus optimize—the rehabilitation program.

• **Training for health versus performance.** The notion that athletes are healthy is generally a myth, at least from a musculoskeletal point of view. Training for superior athletic performance demands substantial overload of the muscles and tissues of the joints. An elevated risk of injury is associated with athletic training and performance. Unfortunately, many patients observe the routines used by athletes to enhance performance and wrongly conclude that copying them will help their own backs. Training for health requires quite a different philosophy; it emphasizes muscle endurance, motor control perfection, and the maintenance of sufficient spine stability in all expected tasks. While strength is not a targeted goal, strength gains do result.

• **Symbiotic integration of prevention and rehabilitation approaches.** The best therapy rigorously followed will not produce results if the cause of the back troubles is not addressed. Part II provided guidelines for reducing the risk of back troubles: removing the source that exacerbates tissue overload cannot be overstressed. Linton and van Tulder (2001) demonstrated the efficacy of exercise for prevention; exercise satisfies the objective for both better prevention and rehabilitation outcome.

• **Establishing a slow, continuous improvement in function and pain reduction.** The return of function and reduction of pain, particularly for the chronic bad back, is a slow process. The typical pattern of recovery is akin to the stock market pricing history. Daily, and even weekly, price fluctuations eventually result in higher prices (see figure 11.1). Patients have good days and bad days. Many times lawyers have hired private investigators to make clandestine videos of back troubled people performing tasks that appear inconsistent. I am hired to provide comment. Some of these people are true malingerers and get caught. Others are simply having a good day on the day they are videotaped. I see all sorts of movement pathology consistent with their chronic history, and they are exonerated.

• **Keeping a journal of daily activities.** Documenting daily back pain and stiffness is essential in identifying the link with mechanical scenarios that exacerbate the troubles. Two critical components should be recorded in a daily journal: how the back feels and what tasks and activities were performed. When patients encounter repeated

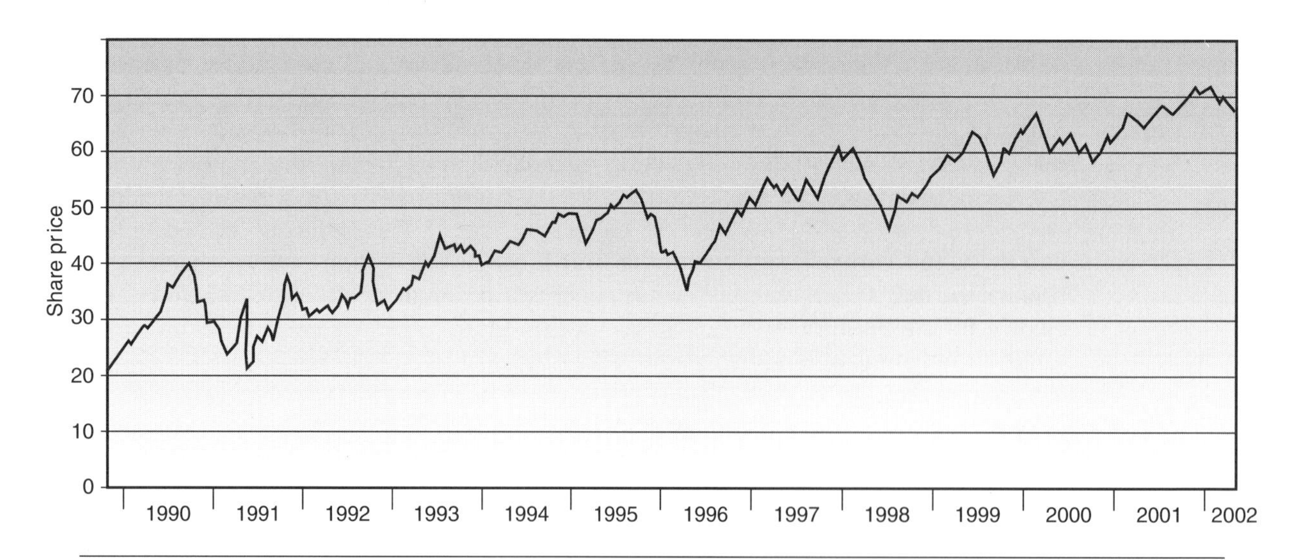

Figure 11.1 The positive slope. The establishment of improvement in back symptoms is often similar to the movement of high-performing stocks on the stock market. While stocks form a noisy signal with their short-term fluctuations, mimicking the day-to-day variations in back symptoms, the average progress over the longer duration has a positive slope. Rehabilitation challenges should never be increased unless a "positive slope" has been established for symptoms and function.

setbacks, they should try to identify a common task or activity that preceded the trouble. Likewise, even when progress is slow, patients will be encouraged to see some progress nonetheless. Without referring to the diary, patients sometimes do not realize they are improving.

- **Ensuring the "positive slope" in progress.** Over the next two chapters I will introduce the "big three" exercises in different forms. We designed these exercises to spare the spine of large loads and to groove stabilizing motor patterns. Use the three to establish a positive slope in patient improvement. Once the slope is established, you may choose to add new exercises one at a time. The patient may tolerate some exercises well and others not so well. If the improvement slope is lost after adding a new activity, remove it, go back to the big three, and reestablish the positive slope. If the patient requires advanced objectives for athletic performance, perhaps spine mobility, you may add exercises to achieve such objectives after establishing the positive slope. How long should each stage be? There is no single answer for all backs. Some will progress quickly, while others will require great patience. This is the job of the clinician—to determine the initial challenge, to gauge progress and enhance the challenge accordingly, and to keep the patient motivated, even during periods of no apparent progress. The great clinicians blend keen clinical skills and experience with scientifically founded guidelines and knowledge.

- **Is the patient willing to make a change?** Obviously, the patient must change the current patterns that caused him to become a back patient. This will require motivation, which is not always easy to establish. Others have listed the importance of and steps in developing a change in motivation and attitude (e.g., Ranney, 1997). Briefly, such a program begins with the setting of goals—for example, returning to a specific job or partaking in a leisure activity. The employer's role in enhancing motivation is to ensure that modified work is available together with the opportunity for graduated return to duty. Employers can also play a role in motivation by fostering a culture in which worker success equates to company success, which in turn helps the worker. The second step in a motivation program is to formulate a realistic plan for reaching the goal established in the first step. It is beyond the mandate of this book to develop the components of maintaining and enhancing motivational opportunities at each stage of recovery.

Stages of Patient Progression

Before we can undertake to remove the activities that exacerbate low back troubles, we must determine what they are. This is a crucial part of the rehabilitation process. Uncovering the activities that cause back troubles begins with a patient interview. Some clinicians also perform provocative testing at this time as well. Chapter 12 thoroughly discusses how to interview and test the patient. Once this has been done, the rehabilitation can proceed. The clinician may choose to overlap the stages involved in this process or conduct them in parallel. At all times, however, the objective is to establish and maintain the positive slope of continual improvement.

Stage 1. Awareness of Spine Position and Muscle Contraction (Grooving Motion/Motor Patterns)

Some people are very "body aware" and are able to adopt a neutral spine or a flexed spine on command. Others can be frustratingly clueless.

Distinguishing Hip Flexion From Lumbar Flexion

The initial objective of this stage is to have the patient consciously separate hip rotation from lumbar motion when flexing the torso. For the more difficult cases we typically begin by demonstrating on ourselves lumbar flexion versus rotation about the hips. Other techniques that we have found particularly helpful are as follows:

- Have your patient place one hand on the tummy while placing the other over the lumbar surface. This way she can feel whether the spine is locked and motion is occurring about the hips (see figure 11.2).
- Sometimes patients are receptive to being coached while using a practice load. The dummy constructed by nurses to help them rehearse proper patient lifting and shown in figure 8.24 on page 179 is an example of such a practice load.
- Other patients respond well to photos of people correctly doing tasks that they will also be called on to do in the course of their jobs or their everyday activities. Figure 11.3, a and b shows examples of such photos.

Figure 11.2 For people who are not "body aware" and are unable to adopt a neutral or a flexed spine on command, we suggest *(a)* rehearsing the spine-neutral position and hip (not lumbar) flexion while doing knee bends before *(b)* exertion.

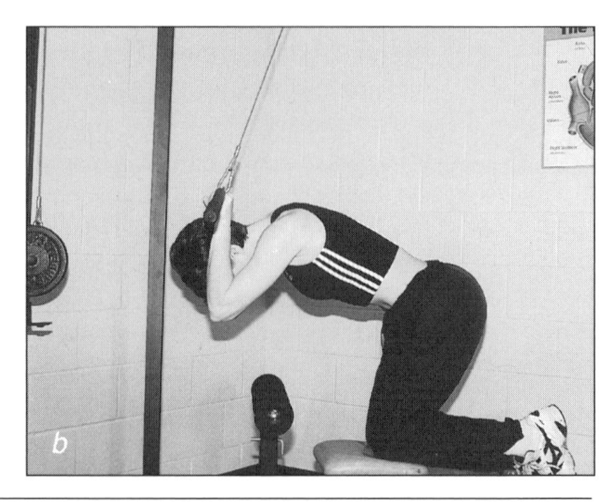

Figure 11.3 Sometimes patients are receptive to task-specific illustrations showing spine and hip postures. A rescue worker is applying a pulling force to the victim with the spine flexed *(a)*. Discussion with the patient of this flexed spine posture together with one that would better spare the spine is very helpful. Also discussed are exercise postures such as the neutral lumbar posture adopted by this patient performing cable pulls *(b)*.

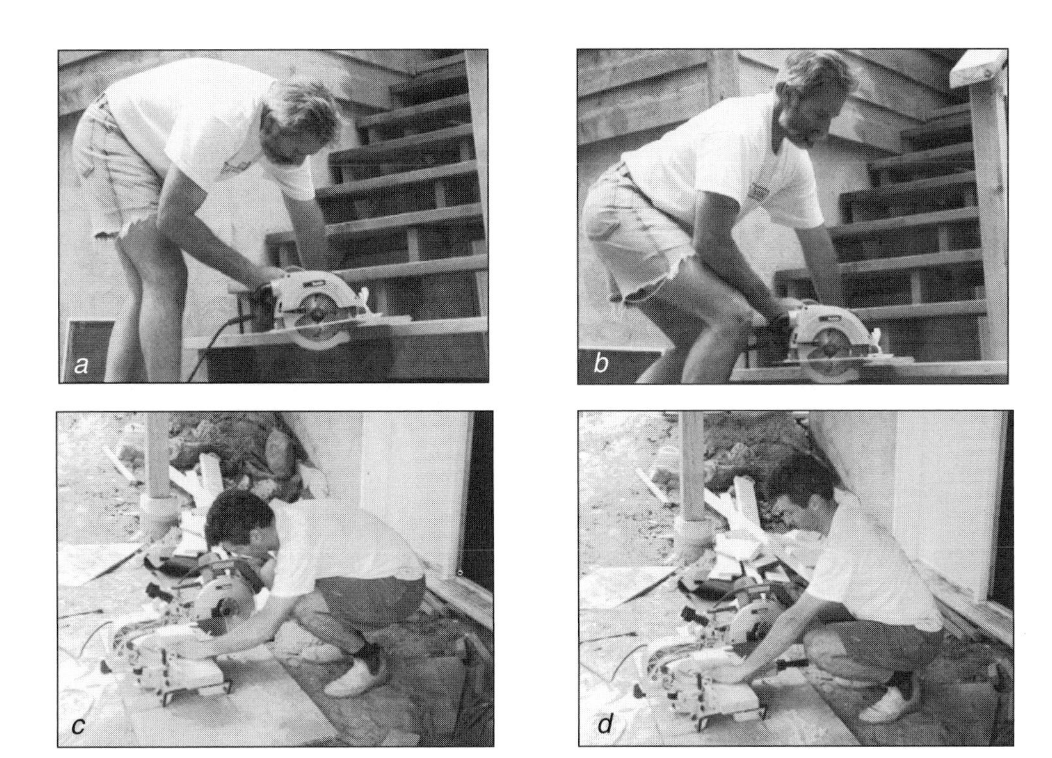

Figure 11.4 Workers relate to tasks with which they are familiar. These photos of construction tasks are helpful for construction workers, especially since they can readily see the difference between the incorrect *(a)* and *(c)* and correct *(b)* and *(d)* spine postures.

- Before (incorrect) and after (correct) photos, such as those in figure 11.4, a through d can be especially helpful.
- Yet another technique is to place a stick along the spine with the instruction to flex the torso forward using the hips but maintain contact with the stick over the length of the spine (see figure 11.5, a-c). While this begins in the clinic, other, and more complex, bending tasks can also help to groove these patterns into the general motion patterns used in work and other daily activities.
- When all of these attempts fail, we resort to the final technique: having the patient perform the "midnight movement" (rolling the pelvis); this is lumbar motion. (No figure here!) Interestingly, some patients who found sex painful never associated pelvis tilting with lumbar flexion. Pointing this out to them often facilitates their making the next leap in spine position awareness and in being able to avoid these painful motions and postures.

Once you begin to see that any or several of the techniques just listed are helping these "difficult" patients learn how to achieve and maintain neutral spine position during their daily activities, you may have them attempt some tasks to see if the concept of the neutral spine is becoming ingrained. One such task is to take the spine stick and ask the patient to knock down an imaginary spider web overhead in the corner of the room. If he loses the lumbar neutral posture, point it out to him so he can correct it and obviously continue spine position awareness training.

Some people have a very difficult time remembering the protective neutral spine pattern. We tell these types of patients to

1. stop before an exertion (perhaps prior to lifting a household item),
2. place the hands on the tummy and lumbar region,

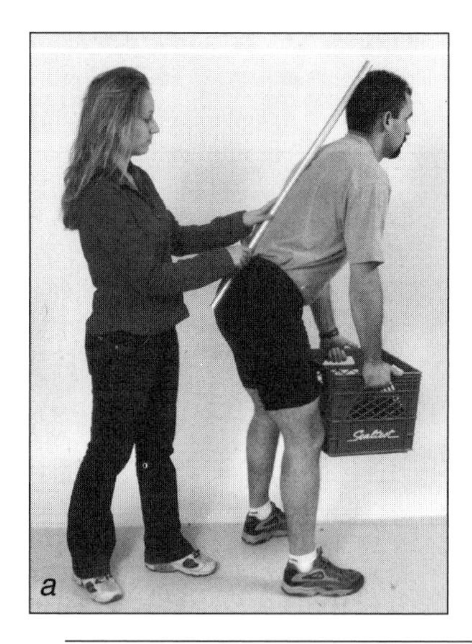

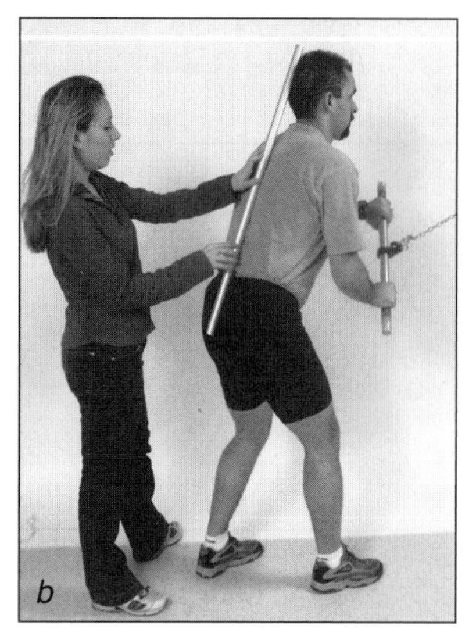

Figure 11.5 Placing a stick along the spine with the instruction to flex forward but maintain contact with the stick over the length of the spine will help patients separate hip rotation from lumbar motion when flexing the torso. Examples of motions that can be used are sagittal plane symmetric lifts *(a)*, various three-dimensional cable pull exercises *(b)*, and specific work tasks that are familiar to the patient *(c)*.

3. practice a few knee bends with the motion about the hips and not the lumbar spine, and
4. then perform the lift (see figure 11.2, a-b).

This practice is effective for many people.

Maintaining Mild Abdominal Contraction

Maintaining a mild contraction of the abdominal wall can also help ensure sufficient spine stability. This maneuver is called abdominal hollowing in many back care circles, but we prefer to avoid that terminology, as it suggests to most people either pulling in or puffing out the abdominal wall. When the contraction is performed correctly, no geometric change at all occurs in the abdominal wall. In other words, rather than "hollowing in" the abdominal wall, the patient simply activates the muscles to make them stiff. We call this contraction "abdominal bracing," whereas when we speak of abdominal hollowing, we are referring to the pulling in of the abdominal wall. See below.

A Note on Abdominal Hollowing and Bracing

Some confusion exists in the interpretation of the literature regarding the issue of abdominal hollowing and abdominal bracing. Richardson's group observed that the hollowing (see figure 11.6a) of the abdominal wall recruits the transverse abdominis. On the other hand, an isometric abdominal brace (see figure 11.6b) coactivates the transverse abdominis with the external and internal obliques to ensure stability in virtually all modes of possible instability (Juker et al., 1998). Note that in bracing the wall is neither hollowed in nor pushed out. In this way, abdominal bracing is superior to abdominal hollowing in ensuring stability. With this background, Richardson and colleagues (1999) noted that the recruitment of transverse abdominis is impaired following injury. The group developed a therapy program designed to reeducate the motor system to activate transverse abdominis in a normal way in

LBD patients. Thus, while hollowing may act as a motor reeducation exercise, the act of hollowing does not ensure stability. Some clinical practitioners may have misinterpreted Richardson's work to suggest that abdominal hollowing should be recommended to patients who require enhanced stability in order to perform daily activities. This is misguided. Abdominal bracing, which activates the three layers of the abdominal wall (external oblique, internal oblique, transverse abdominis) with no "drawing in," is much more effective than abdominal hollowing at enhancing spine stability (McGill, 2001; Grenier and McGill, in press).

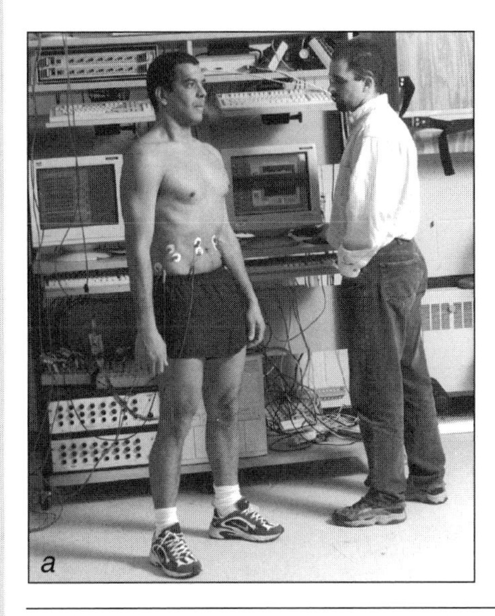

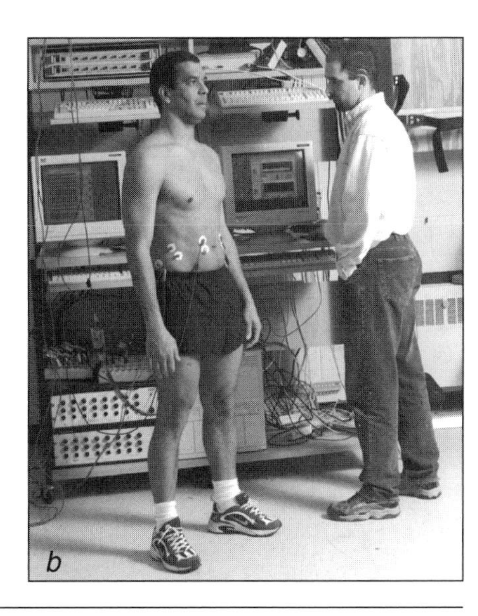

Figure 11.6 Hollowing involves the sucking in of the abdomen to activate transverse abdominis *(a)*. On the other hand, the abdominal brace activates the three layers of the abdominal wall (external oblique, internal oblique, transverse abdominis), with no "drawing in" *(b)*.

Two mechanisms can shed some light on this issue of hollowing versus bracing. First, the supporting "guy wires" are more effective when they have a wider base (see figure 11.7, a-c), that is, when the abdomen is not hollowed. Second, the obliques must be active to provide stiffness with crisscrossing struts, which measurably enhances stability. The lumbar torso must prepare to withstand all manner of possible loads, including steady-state loading (which may be a complex combination of flexion/extension, lateral bend, and axial twisting moments) and sudden, unexpected complex loads together with loads that develop from planned ballistic motion. Stiffness is required in every rotation and translation axis to eliminate the possibility of unstable behavior. The abdominal brace ensures sufficient stability using the oblique cross-bracing, although high levels of cocontraction are rarely required—probably about 5% MVC cocontraction of the abdominal wall during daily activities and up to 10% MVC during rigorous activity.

A quantitative comparison of the hollow and the brace is clearly seen in an individual standing upright with loads in the hands. Simply hollowing can cause the stability index to drop to low levels or even negative levels, which indicates the possibility of instability (see figure 11.8). Bracing increases the positive stability index value. The subject in these figures is typical in that even though the hollow was taught to target the transverse abdominis, all abdominal muscles were activated when measured. Thus stability was created while attempting "hollowing" although true bracing is superior to create lumbar stability. If a true "hollow" is accomplished with just the

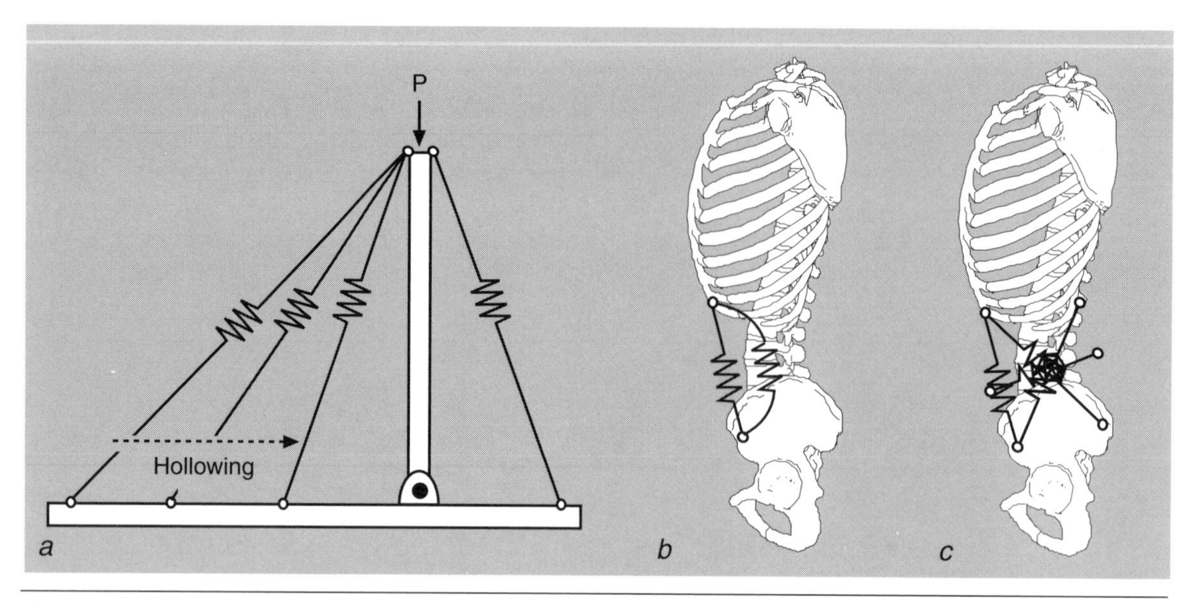

Figure 11.7 Hollowing the muscles reduces the size of the base of the guy wires, together with the incidence angle where they attach to the spine *(a)*. This inherently reduces their contribution to spine stiffness in various modes, which compromises spine stability *(b)*. Bracing assists in keeping a wide base to the guy wires and recruits the oblique muscle to supply cross-bracing struts for stability in all axes *(c)*.

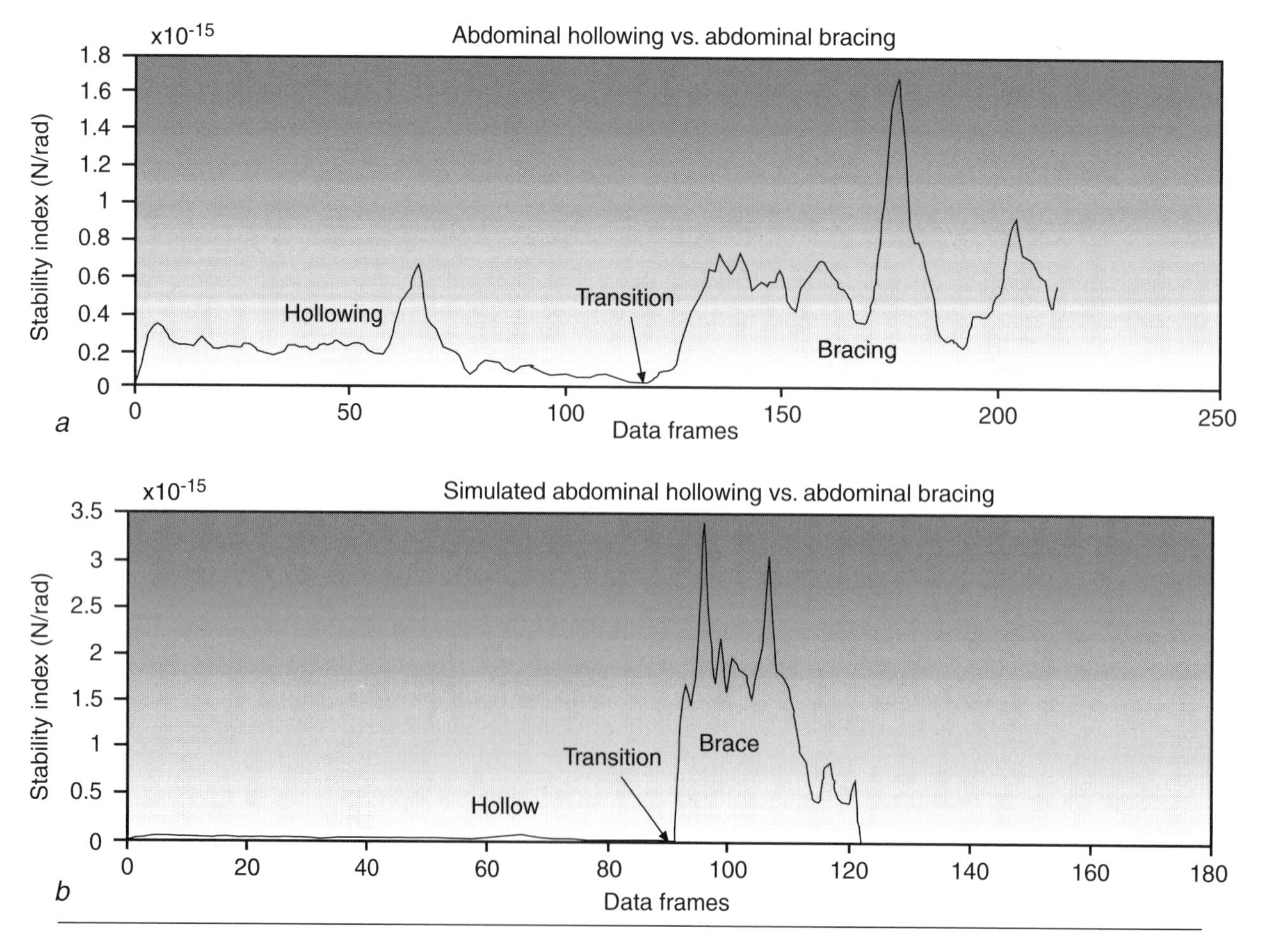

Figure 11.8 An example comparing the hollow (*a*) with the brace (higher stability) in an individual standing upright, arms at the sides, with loads placed in the hands. A "perfect" hollow was shown to be inferior to the brace (*b*) in this analysis of simulated muscle activation.

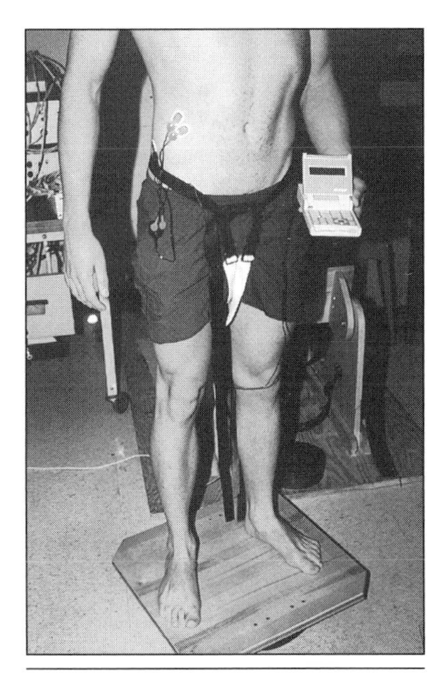

Figure 11.9 EMG biofeedback devices are an economical way to provide feedback to the patient regarding the level of abdominal activation during any type of functional task, from standing on a wobble board to getting into a car.

transverse abdominis, as simulated in figure 11.8b, stability is low compared with a brace where the three layers of the abdominal wall are activated. Simply hollowing causes the stability index to drop to negative levels when the load is placed in the hands compressing the spine. A negative value indicates instability is possible. In contrast, bracing maintains a positive stability index value, eliminating the possibility for buckling.

Teaching Abdominal Bracing

Generally, to demonstrate abdominal bracing to the patient, we stiffen one of our own joints, such as an elbow, by simultaneously activating the flexors and extensors. The patient then palpates the joint both before and after we stiffen it. Then we ask the patient to attempt to stiffen her own joint through simultaneous activation of flexors and extensors. Once she can successfully stiffen various peripheral joints, we demonstrate (again on ourselves, with patient palpation) the same technique in the torso, achieving abdominal bracing. Finally, we again ask her to replicate the technique in her own torso. Occasionally, we use a portable EMG monitor so the patient can learn through biofeedback what, for example, 5%, 10%, or 80% of maximum contraction feels like (see figure 11.9). We use similar devices to teach patients how to maintain the contraction while on a wobble board and in functional situations such as when picking up a child, getting on and off the toilet, and getting in and out of cars.

Mental Imagery

The use of mental imagery is useful for both spine position and muscle activation awareness. Following is a general protocol that we have adapted from the imagery literature for use with spine training.

Steps of Mental Imagery

1. Focus on feeling the surface under the feet or buttocks. Whatever body part is touching a surface, be aware of the sensation.
2. Practice simple motions such as tightening and then relaxing specific muscles in different areas of the body. Then graduate to performing the abdominal brace.
3. Palpate, and have the patient self-palpate, the muscle involved while he is attempting to tighten and relax it. Sometimes a full-body mirror is helpful. The focus for the patient is on the specific muscle(s) involved.
4. Perform motions slowly, chunking them into segments and sequences, then visualize the total motion. For example, beginning with a simple task such as a forward reach, visualize the neutral spine, then activating the extensors and the bracing abdominals, and finally the motion about the hips.
5. Practice the imagery independent of physical action. Of course, the patient will have already been successful in learning spine position awareness, proper muscle control, and desirable motion patterns.

Source: Kathryn McGill, sports psychology consultant

Lumbar Spine Proprioceptive Training

Given our research on the importance of spine position awareness to spare the spine and our experience in teaching positioning, we became interested in proprioceptive training for the back. The fact that very little evidence was available to validate the use of proprioceptive rehabilitation for the lumbar spine motivated our recent work on spine proprioception (Preuss and McGill, in press). The purpose of this work was to quantify the effects of a six-week rehabilitation program designed to improve lumbar spine position sense and sitting balance. Twelve subjects with a previous history of LBP were evenly split into a training group and a control group. Subjects in the control group received no intervention, while the subjects in the training group received a 20-minute rehabilitation session three times per week emphasizing spine stabilization exercises with a neutral spine. Lumbar spine repositioning error in four-point kneeling and sitting balance for the training group showed significant improvement over the study. This small but initial study demonstrated that proprioception and position awareness in the lumbar spine can improve through active rehabilitation.

Stage 2. Stabilization Exercises to Groove Stabilizing Motor Patterns and Build Muscle Endurance

Generally the next stage should begin with stabilization exercise. The trick is to find an appropriate starting level. For the chronic "basket cases" who arrive at our university clinic having failed traditional therapy, we generally undershoot what many would consider an appropriate loading level. The typical progressive improvement philosophy of "work hardening" with weekly improvement goals will not work for these people. They will have good and bad days, but the general positive slope in improvement must be established—however gradual. The difficult cases will not tolerate the weekly improvement goals of traditional work hardening programs; they require more patience. Once we can document a positive slope in improvement, then we can increase the rehabilitation challenge. Specific exercises are illustrated in chapter 13.

Stage 3. Ensuring Stabilizing Motion Patterns and Muscle Activation Patterns During All Activities

The clinician must clarify the range of activities (daily living, occupational, athletic, etc.) for which the individual patient must be prepared. This is obtained through interview. While there is no standard form, we begin by documenting the patients' daily routine, which includes their occupational demands and those of daily living. Previous chapters offered many examples in which the spine is spared with an appropriate posture and muscle activation pattern. The guidelines to selecting exercises discussed in the next section will help you determine what exercises to prescribe for patients in this stage. In addition, we spend time rehearsing daily activities to be sure the patient is learning and utilizing spine-sparing motion and motor (muscle activation) patterns.

An Important Reminder

Remember, since half the battle is to remove the irritants, make sure to follow the recommendations of the previous chapters.

Summary: Stages of Patient Progression

1. Identify and remove the exacerbating activities.
2. Record in a journal daily how the back feels as well as the tasks and activities performed.
3. Develop spine position awareness (hips vs. lumbar motion) and the ability to maintain abdominal bracing. Groove motion/motor patterns.
4. Begin the appropriate spine exercise and appropriate stabilization/mobilization tasks.
5. Develop muscular endurance.
6. Transfer to daily activities.

Note that training athletes in our program involves the same stages with the addition of two stages—training strength and power (see chapter 14).

A Note on Soft Tissue Treatments

Patients often present with local muscle spasms and "odd-feeling" local muscle texture as perceived by a good manual medicine clinician. Further, these spasms and local neurocompartment disorders are associated with larger dysfunctions of the agonist and synergist muscles involved in a movement. In many cases these dysfunctions delay recovery or prevent complete recovery. Clinicians use a variety of soft tissue treatments to remove spasm and release the tissues that can impede attaining more normal muscular and joint function. Documenting them is beyond the scope of this book. The reader is simply alerted to their potential significance and role in rehabilitation.

Guidelines for Developing the Best Exercise Regimen

The reported effectiveness of various training and rehabilitation programs for the low back is quite variable, with some claiming great success while others report no success or even negative results (Faas, 1996; Koes et al., 1991). The discrepancy regarding the effectiveness and safety of exercise programs is probably due to clinicians prescribing inappropriate exercises because they do not understand the tissue loading that results during various tasks. Resist the urge to enhance mobility just to adhere with the disability rating system. That system is for legislative convenience only. Rather, judge your success by how well you are able to reduce patients' pain and restore their ability to complete tasks.

Developing a Sound Basis for Exercise Prescription

I have selected and evaluated the exercises in the following chapters based on tissue loading evidence and the knowledge of how injury occurs to specific tissues (described in previous chapters). Choosing exercises, however, still involves the best educated guess that is developed through clinical experience. The following example illustrates the need for quantitative analysis in order to evaluate the safety of certain exercises.

We have all been told to perform sit-ups and other flexion exercises with the knees flexed—but on what evidence? Several hypotheses have suggested that this disables the psoas and/or changes the line of action of the psoas. MRI-based data (Santaguida and McGill, 1995) demonstrated that the psoas line of action does not change due to lumbar or hip posture (except at L5/S1) as the psoas laminae attach to each vertebra and follow the changing orientation of the spine. However, there is no doubt that the psoas is shortened with the flexed hip, modulating force production. But the question remains, Is there a reduction in spine load with the legs bent? In 1995 McGill examined 12 young men using the laboratory technique described previously and observed no major difference in lumbar load as the result of bending the knees (average moment of 65 Nm in both straight and bent knees; compression of 3230 N with straight legs and 3410 N with bent knees; shear of 260 N with straight legs and 300 N with bent knees). Compressive loads in excess of 3000 N certainly raise questions of safety. This type of quantitative analysis is necessary to demonstrate that whether to perform sit-ups using bent knees or straight legs is probably not as important as whether to prescribe sit-ups at all! There are better ways to challenge the abdominals.

Basic Issues in Low Back Exercise Prescription

Several exercises are required to train all the muscles of the lumbar torso, but which exercises are best for a given individual? Making this determination will depend on a number of variables, such as the individual's fitness level, training goals, history of previous spinal injury, and other factors. However, depending on the purpose of the exercise program, several principles apply. For example, an individual beginning a postinjury program would be advised to avoid loading the spine throughout the range of motion, while a trained athlete may indeed achieve higher performance levels by doing so. Another general rule of thumb is to preserve the normal low back curve (similar to that of upright standing) or some variation of this posture that minimizes pain. While in the past many clinicians have recommended performing a pelvic tilt when exercising, this is not justified; we now know that the pelvic tilt increases spine tissue loading, as the spine is no longer in static–elastic equilibrium. Thus, the pelvic tilt appears to be contraindicated when challenging the spine. Basic issues you should consider when prescribing exercises for low back rehab are discussed in the following sections.

Flexibility

Whether to train for optimization of spine flexibility depends on the person's injury history and exercise goal. Generally, for the injured back, spine flexibility should not be emphasized until the spine has stabilized and has undergone endurance and strength conditioning—and some may never reach this stage! Despite the notion held by some, there is little quantitative data to support the idea that a major emphasis on trunk flexibility will improve back health and lessen the risk of injury. In fact, some exercise programs that have included loading of the torso throughout the range of motion (in flexion/extension, lateral bend, or axial twist) have had negative results (e.g., Nachemson, 1992), and greater spine mobility has been associated with low back trouble in some cases (e.g., Biering-Sorensen, 1984). Further, research has shown that spine flexibility has little predictive value for future low back trouble (e.g., Sullivan, Shaof, and Riddle, 2000). The most successful programs appear to emphasize trunk stabilization through exercise with a neutral spine (e.g., Hides, Jull, and Richardson, 2001; Saal and Saal, 1989) while stressing mobility at the hips and knees (Bridger, Orkin, and Henneberg, 1992, demonstrate advantages for sitting and standing, while McGill and Norman, 1992, outline advantages for lifting). Finally, removing lumbar flexion from morning activities substantially improves patients, on average (Snook et al., 1998). Despite this evidence, many patients are still instructed to "pull their knees" to their chest in the morning and perform toe touches (see figure 11.10, a-c). The destabilizing consequences of full flexion were described in chapters 4 and 5.

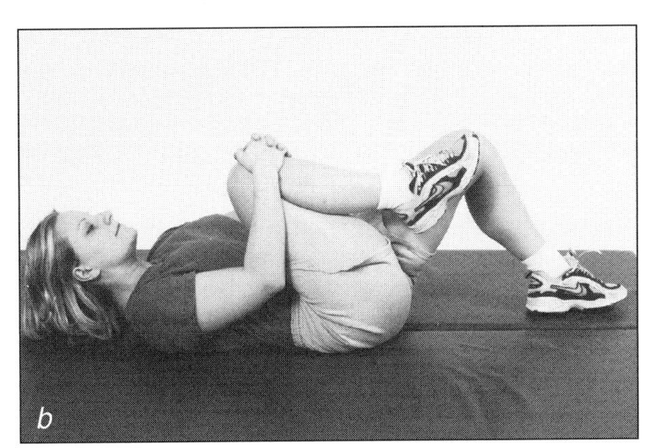

Figure 11.10 "Silly stretches," such as *(a)* and *(b)* pulling the knees to the chest and *(c)* toe touches, are often prescribed to patients to do in the morning. These can cause instability!

For these reasons, torso flexibility exercises should be limited to unloaded flexion and extension for those concerned with safety. Those interested in specific athletic activities may sometimes be an exception to this rule. (Of course, spine flexibility may be more desirable in athletes who have never suffered back injury.)

Strength

In general, strength seems to have little to do with back health even though increasing torso muscle strength is a popular objective of low back rehabilitation protocols. Back muscle strength in particular has not been found to be a significant predictor of first-time injury. (In the context of cause and effect, predicting first-time injury offers special insight.) Only Troup, Martin, and Lloyd (1981) found, while testing torso muscles, that reduced dynamic strength was a predictor of recurring back pain. However, in a prospective study, Leino (1987) found that neither isometric nor dynamic trunk strength predicted the development of low back troubles over a 10-year follow-up period. The Biering-Sorensen (1984) study, previously noted, found that isometric back strength did not predict the appearance of low back trouble in previously healthy subjects over a one-year follow-up. Holmstrom and Moritz (1992) recorded reduced isometric trunk extensor endurance times in male workers with low back disorders compared to those without but found no differences in isometric flexion or extension strengths. Strength appears to have little, or a very weak, relationship with low back health. Strength is for athletic performance objectives. In contrast, muscle endurance, when separated from strength, appears to be linked with better back health.

Most recent data have suggested that while having a history of low back troubles is not related to reduced strength, it is related to a perturbed flexion-to-extension strength ratio (McGill et al., in press). This difference in the ratio appeared to be mainly influenced by greater extensor strength relative to flexor strength in those with troubles.

Endurance

Two cross-sectional studies, those of Nicolaisen and Jorgensen (1985) and Alaranta et al. (1994), found reduced extensor endurance in workers who reported low back troubles. Both Biering-Sorensen (1984) and Luoto and colleagues (1995) suggested that while isometric strength was not associated with the onset of back troubles, poor static back endurance scores are. Some have expressed concern that patients with poor muscle endurance scores have poor scores from a lack of effort (a psychological variable) rather than any physiological limitation. A recent study by Mannion and colleagues (2001) suggested that of the total variance measured in the endurance scores of their back patients, 40% was explained by physiological fatigue (quantified as declines in the EMG power spectrum), while only 10% was explained by psychological variables (quantified as motivation and fear of pain variables, from a questionnaire). A recent study (McGill et al., in press) has suggested that having a history of low back troubles appears to be associated with a different flexion-to-extension endurance ratio, with the extensors having less endurance and the flexors having more endurance. This imbalance in endurance also appeared between the right and left side lateral musculature as evidenced by the asymmetry in right and left endurance holding times (e.g., RSB/LSB ratio of .93 for those with an LBD history vs. 1.05 for those without). The next issue addressed the question of whether strength and endurance are related. Interestingly, the flexor strength (Nm) to endurance (sec) ratio was between 3 and 3.5 for the flexors in both "normal" backs and in those with a history of troubles, and for the extensors of the normals. The ratio was much larger for the extensors (5.3, p = .033) in those with a history of LBD.

In summary, several studies have suggested that diminished trunk extensor endurance and not strength is linked to low back troubles. Recent data confirm this notion and enhance it further by suggesting that the balance of endurance between flexor and extensor muscles and the balance between right and left sides of the torso appear to be linked to a history of back troubles severe enough to result in work loss. Note that these were losses that linger long after the disabling episode. In those with a history of disabling LBD, the average length of time since the last work loss episode was 261 weeks (standard deviation = 275), while the average length of time lost from work in that episode was seven days (standard deviation = 10). The lesson for exercise prescription is that graduated, progressive exercise programs (i.e., of longer duration and lower effort), which emphasize endurance, thus seem preferable over strengthening exercises.

Aerobic Exercise

Mounting evidence supporting the role of aerobic exercise in both reducing the incidence of low back injury (Cady et al., 1979) and treating low back patients (Juker et al., 1998) is compelling. Recent investigation into loads sustained by the low back tissues during walking (Nutter, 1988) confirms very low levels of supporting passive tissue load coupled with mild, but prolonged, activation of the supporting musculature. Callaghan, Patla, and McGill (1999) documented that fast walking with the arms swinging results in lower oscillating spine loads. When tolerable, aerobic exercise, particularly fast walking, appears to enhance the effects of back-specific exercise.

Order of Exercises Within a Session

Because the spine has a loading memory, a prior activity can modulate the biomechanics of the spine in a subsequent activity. For example, if a person sat in a slouched posture for a period of time sufficient to cause ligamentous and disc creep, she would have residual ligament laxity for a period of time. (We have measured laxity of over a half hour in some cases [McGill and Brown, 1992].) The nucleus volume appears to redistribute upon adopting a standing posture (Krag et al., 1987). This redistribution takes time. If the spine is flexed in one maneuver, then it probably should return to neutral or extension for the next.

Viscosity is another property of biological tissues—in this case a frictional resistance to motion within the spine and torso tissues. This is why motion exercises are usually performed first as part of a warm-up; once the viscous friction has been reduced, subsequent motion can be accomplished with less stress.

A final consideration is the need to continually groove healthy, joint-conserving, and stabilizing motor patterns. Depending on the exercise objectives, we often begin an exercise/training session with some spine-stabilization exercises to groove the patterns that will continue over to other exercises in the program. In summary, understanding spine biomechanics can optimize the ordering of tasks in a training session.

Establishing Grooved Patterns

We have found that patients are best served when we establish the grooved patterns for spine stability at the beginning of the session. On the other hand, in a performance-oriented training program, participants then move on to torso-stabilization exercises at the end of the session.

Breathing

Debate continues regarding the entrainment of breathing during exertion. Should one exhale or inhale during a particular phase of movement or exertion?

In the rare cases of very heavy lifting or maximal exertions (which would not be part of a rehabilitation program), high levels of intra-abdominal pressure are produced by breath holding using the Valsalva maneuver. This elevated IAP, when combined with high levels of abdominal wall cocontraction (bracing), ensures spine stiffness and stability during these extraordinary demands.

Another motivation given for striving to achieve higher IAP is the need to reduce the transmural gradient in the cranium to lessen the risk of blackout or stroke (McGill, Sharratt, and Seguin, 1995). The explanation for this risk reduction is as follows:

- Building IAP is associated with a rise in the central nervous system (CNS) fluid pressure in the spine, which forms an open vessel to the CNS and brain.
- Upon exertion, enormous elevation in blood pressure occurs (documented in weightlifters to be well over 400 mmHg).
- This pressure in the cranial vessels creates a large transmural pressure gradient that is reduced if the CNS fluid pressure is likewise elevated, reducing the load on the vascular vessels.

Although this explanation is valid, the mechanism should be considered only for extreme weightlifting challenges—not for rehabilitation exercise.

When designing rehabilitation exercise, a major objective is to establish spine-stabilization patterns. An important feature of stable and functional backs is the ability to cocontract the abdominal wall (abdominal brace) independently of any lung ventilation patterns. Good spine stabilizers maintain the critical symmetrical muscle stiffness during any combination of torque demands and breathing patterns (such as when playing a basketball game, for example). Poor stabilizers allow abdominal contraction levels to cycle with breathing at critical moments where stability is needed. Grooving muscular activation patterns so that a particular direction in lung air flow is entrained to a particular part of an exertion is not helpful. This would be of little carryover value to other activities; in fact, it would be counterproductive.

Breathe Freely

Train to breathe freely while maintaining the stabilizing isometric abdominal wall contractions.

Time of Day for Exercise

As pointed out in part II, the intervertebral discs are highly hydrated upon rising from bed, the annulus is subjected to much higher stresses during bending under these conditions, and the end plates fail at lower compressive loads as well. Thus, performing spine-bending maneuvers at this time of day is unwise. Yet many manual medicine physicians continue to suggest to patients to perform their therapeutic routines first thing in the morning. This appears to be due to convenience and ignorance. Because the discs generally lose 90% of the fluid that they will lose over the course of a day within the first hour after rising from bed, we suggest simply avoiding this period for exercise (that is, bending exercise) either for rehabilitation or performance training. While there hasn't been a study on the enhancements obtained during exercise routines as a function of time of day, Snook and colleagues (1998) did prove that the conscious avoidance of forward spine flexion in the morning improved their patients' back troubles.

No Early-Morning Bending

Don't perform bending exercises in the first hour or two after rising.

Notes for Rehabilitation Exercise Prescription

Exercise professionals face the challenge of designing exercise programs that consider a wide variety of objectives. Consider these guidelines:

1. While some experts believe that exercise sessions should be performed at least three times per week, low back exercises appear to be most beneficial when performed daily (e.g., Mayer et al., 1985).
2. The no pain–no gain axiom does not apply when exercising the low back, particularly when applied to weight training. Scientific and clinical wisdom would suggest the opposite is true.
3. Research has shown that general exercise programs that combine cardiovascular components (such as walking) are more effective in both rehabilitation and injury prevention (e.g., Nutter, 1988).

4. Diurnal variation in the fluid level of the intervertebral discs (discs are more hydrated early in the morning after rising from bed) changes the stresses on the disc throughout the day. People should not perform full-range spine motion under load shortly after rising from bed (e.g., Adams and Dolan, 1995).
5. Low back exercises performed for health maintenance need not emphasize strength with high-load, low-repetition tasks. Rather, more repetitions of less demanding exercises will enhance endurance and strength. There is no doubt that back injury can occur during seemingly low-level demands (such as picking up a pencil) and that injury from motor control error can occur. While the chance of motor control errors that result in inappropriate muscle forces appears to increase with fatigue, evidence also indicates that passive tissue loading changes with fatiguing lifting (e.g., Potvin and Norman, 1992). Given that endurance has more protective value than strength (Luoto et al., 1995), strength gains should not be overemphasized at the expense of endurance.
6. No set of exercises is ideal for all individuals. An appropriate exercise regimen should consider an individual's training objectives, be they rehabilitation, reducing the risk of injury, optimizing general health and fitness, or maximizing athletic performance. While science cannot evaluate the optimal exercises for each situation, the combination of science and clinical experiential experience will result in enhanced low back health.
7. Both patients and clinicians should be patient and stick with the program. Increased function and reduction in pain may not occur for three months (e.g., Manniche et al., 1988).

References

Adams, M.A., and Dolan, P. (1995) Recent advances in lumbar spine mechanics and their clinical significance. *Clinical Biomechanics*, 10: 3-19.

Alaranta, H., Hurri, H., Heliovaara, M., Soukka, A., and Harju, R. (1994) Non dynamometric trunk performance tests: Reliability and normative data. *Scandinavian Journal of Rehabilitation Medicine,* 26: 211-215.

Biering-Sorensen, F. (1984) Physical measurements as risk indicators for low back trouble over a one year period. *Spine*, 9: 106-109.

Bridger, R.S., Orkin, D., and Henneberg, M. (1992) A quantitative investigation of lumbar and pelvic postures in standing and sitting: Interrelationships with body position and hip muscle length. *International Journal of Industrial Ergonomics,* 9: 235-244.

Cady, L.D., Bischoff, D.P., O'Connell, E.R., et al. (1979) Strength and fitness and subsequent back injuries in firefighters. *Journal of Occupational Medicine*, 21 (4): 269-272.

Callaghan, J.P., Patla, A.E., McGill, S.M. (1999) Low back three-dimensional joint forces, kinematics and kinetics during walking. *Clinical Biomechanics,* 14: 203-216.

Faas, A. (1996) Exercises: Which ones are worth trying, for which patients, and when? *Spine*, 12 (24): 2874-2879.

Grenier, S., and McGill, S.M. (in press) Bracing, not hollowing, the abdominals enhances spine stability.

Hides, J.A., Jull, G.A., and Richardson, C.A. (2001) Long-term effects of specific stabilizing exercises for first-episode low back pain. *Spine*, 26: E243-248.

Holmstrom, E., and Moritz, U. (1992) Trunk muscle strength and back muscle endurance in construction workers with and without back disorders. *Scandinavian Journal of Rehabilitation Medicine,* 24: 3-10.

Juker, D., McGill, S.M., Kropf, P., and Steffen, T. (1998) Quantitative intramuscular myoelectric activity of lumbar portions of psoas and the abdominal wall during a wide variety of tasks. *Medicine and Science in Sports and Exercise,* 30 (2): 301-310.

Koes, B.W., Bouter, L.M., Beckerman, H., et al. (1991) Physiotherapy exercises and back pain: A blinded review. *British Medical Journal*, 302: 1572-1576.

Krag, M.H., Seroussi, R.E., Wilder, D.G., and Pope M.H. (1987) Internal displacement distribution from in vitro loading of human thoracic and lumbar spinal motion segments: Experimental results and theoretical predictions. *Spine*, 12 (10): 1001.

Leino, P., Aro, S., and Hasan, J. (1987) Trunk muscle function and low back disorders. *Journal of Chronic Disease,* 40: 289-296.

Linton, S.J., and van Tulder, M.W. (2001) Preventative interventions for neck and back pain problems. *Spine*, 26 (7): 778-787.

Luoto, S., Heliovaara, M., Hurri, H., et al. (1995) Static back endurance and the risk of low back pain. *Clinical Biomechanics*, 10: 323-324.

Manniche, C., Hesselsoe, G., Bentzen, L., et al. (1988) Clinical trial of intensive muscle training for chronic low back pain. *Lancet*, 24: 1473-1476.

Mannion, A.F., Taimela, S., Muntener, M., and Dvorak, J. (2001) Poor back muscle endurance: A psychological or physiological limitation? In: *Abstracts*–International Society for Study of the Lumbar Spine, Edinburgh, Scotland, June 19-23, p. 147.

Mayer, T.G., Gatchel, R.J., Kishino, N., et al. (1985) Objective assessment of spine function following industrial injury: A prospective study with comparison group and one-year follow up. *Spine*, 10: 482-493.

McGill, S.M. (2001) Low back stability: From formal description to issues for performance and rehabilitation. *Exercise and Sports Science Reviews,* 29(1): 26-31.

McGill, S.M. (1995) The mechanics of torso flexion: Situps and standing dynamic flexion manoeuvres. *Clinical Biomechanics*, 10 (4): 184-192.

McGill, S.M., and Brown, S. (1992) Creep response of the lumbar spine to prolonged full flexion. *Clinical Biomechanics*, 7: 43-46.

McGill, S.M., Grenier, S., Preuss, R., and Brown, S. (in press) Asymmetries in torso endurance and strength parameters are associated with a history of low back troubles.

McGill, S.M., and Norman, R.W. (1992) Low back biomechanics in industry—The prevention of injury. In: Grabiner, M.D. (Ed.), *Current issues of biomechanics*. Champaign, IL: Human Kinetics.

McGill, S.M., Sharratt, M.T., and Seguin, J.P. (1995) Loads on spinal tissues during simultaneous lifting and ventilatory challenge. *Ergonomics,* 38: 1772-1792.

Nachemson, A. (1992) Newest knowledge of low back pain: A critical look. *Clinical Orthopaedics*, 279: 8-20.

Nicolaisen, T., and Jorgensen, K. (1985) Trunk strength, back muscle endurance and low back trouble. *Scandinavian Journal of Rehabilitation Medicine,* 17: 121-127.

Nutter, P. (1988) Aerobic exercise in the treatment and prevention of low back pain. *Occupational Medicine,* 3: 137-145.

Potvin, J.R., and Norman, R.W. (1992) Can fatigue compromise lifting safety: Proc. NACOB II. The Second North American Congress on Biomechanics, August 24-28, pp. 513-514.

Preuss, R., and McGill, S. (in press) Improved lumbar spine position sense and sitting balance following a six-week rehabilitation program in individuals with a history of low back pain.

Ranney, D. (1997) Chronic musculoskeletal injuries in the workplace. Philadelphia: W.B. Saunders.

Saal, J.A., and Saal, J.S. (1989) Nonoperative treatment of herniated lumbar intervertebral disc with radiculopathy: An outcome study. *Spine*, 14: 431-437.

Santaguida, L., and McGill, S.M. (1995) The psoas major muscle: A three-dimensional mechanical modelling study with respect to the spine based on MRI measurement. *Journal of Biomechanics*, 28 (3): 339-345.

Snook, S.H., Webster, B.S., McGarry, R.W., Fogleman, M.T., and McCann, K.B. (1998) The reduction of chronic nonspecific low back pain through the control of early morning lumbar flexion: A randomized controlled trial. *Spine,* 23 (23): 2601-2607.

Sullivan, M.S., Shaof, L.D., and Riddle, D.L. (2000) The relationship of lumbar flexion to disability in patients with low back pain. *Physical Therapy*, 80: 241-250.

Troup, J.D.G., Martin, J.W., and Lloyd, D.C.E.F. (1981) Back pain in industry: A prospective survey. *Spine*, 6: 61-69.

Videman, T., Sarna, S., Crites-Battie, M., et al. (1995) The long term effects of physical loading and exercise lifestyles on back-related symptoms, disability, and spinal pathology among men, *Spine*, 20 (b): 669-709.

CHAPTER 12

Evaluating the Patient

Of the several deficits in low back variables identified in earlier chapters, many are the direct result of injury. They include aberrant lumbar motion patterns, perturbed motor patterns of muscle recruitment, and aberrant joint motion with concomitant pain and loss of muscle endurance. Unfortunately, testing for these deficits is not easy. The challenge is to find the tests that can best identify the deficits and that are reasonably safe and do not require expensive or specialized equipment. The tests that come reasonably close to meeting these criteria are described in this chapter.

Virtually any manual medicine textbook describes the typical tests for determining the range of motion (ROM). But of what real importance is this information to the clinician? These numbers are more for the legal determination of disability as defined by the AMA than for aiding in the clinical decision process. Some erroneously think that all patients should have "normal to average" ROM values even though they probably were not "average" prior to becoming a patient. The tests discussed in this chapter offer more useful indicators of pathology to assist you in making treatment decisions.

This chapter provides guidelines for tests designed to quantify patient deficits; the results will form the rehabilitation objectives for the patient, together with clues for designing exercise. It also discusses how patients can help to define their own rehabilitation targets.

First Clinician–Patient Meeting

The items in the following checklist will help determine the type of rehabilitation exercise. You should consider them only once you have screened your patient for all red flag conditions.

1. **Identify the rehabilitation objectives (specific health or performance objectives).** The specific rehabilitation objective determines the acceptable risk-to-benefit ratio. A performance objective carries higher risk. Since the principles of bodybuilding and athletic training are so pervasive, you need to be sure that all patients understand the difference between athletic performance objectives and those for pain reduction and improved daily function.

2. Consider patient age and general condition. Younger patients tend to have more discogenic troubles (from the teens to the fifth decade), while arthritic spines tend to begin developing after 45 years and stenotic conditions after that. Note how patients walk and sit. Are they in noticeably poor condition, either emaciated with little muscle mass or heavy and loose with fat rather than muscle? It is also assumed that patients have been screened and cleared for cardiovascular concerns.

3. Identify occupation and lifestyle details. Generally, you should begin by documenting patients' daily routines: when and how they rise from and retire to bed, meal routines, and exercise and recreational habits. Then direct specific focus toward areas of concern. For example, if the patient reports watching TV for two hours in the evening, ask for details on the type of chair, range of postures used, and so on. After gathering information about the patient's daily routines, inquire about occupational demands. All of this information, when added to the clinical presentation, will help you evaluate common links. Discogenic troubles are linked with prolonged sitting (particularly prolonged driving) and repeated torso flexion. A passive/inactive lifestyle is also associated with disc troubles. Arthritic conditions, facet troubles, and the like are more linked with jobs and activities that involve large ranges of motion and higher loading. Former athletes such as soccer players also fall into this category, although long-distance runners do not since they do not, presumably, take the spine to the end range of motion.

4. Consider the mechanism of injury. Attempts to re-create injury mechanisms are fruitful only when the real mechanisms are understood. These were detailed in chapters 4 and 5. Once identified, the mechanisms can be linked with specific tissue damage (much of which is otherwise not diagnosable). Not only will this assist in designing the therapeutic exercise, but it will also help in teaching patients to avoid loading scenarios that could exacerbate the damage and symptoms. Note that some of these will have acute onset, while others progress slowly. Slow onset may result in some patients being unable to identify the mechanism of injury. Nevertheless, a "culminating event" is usually involved. Careful questioning about events leading up to that event will provide clues as to the mechanisms of injury.

5. Have the patient describe the perceived exacerbators of pain and symptoms. Prompt the patient to describe the tasks, postures, and movements that exacerbate the pain. Examine these reported tasks from a biomechanical perspective to determine which tissues are loaded or irritated. These tissues should be spared in the exercise therapy, and the exacerbating movements minimized.

6. Have the patient describe the type of pain, its location, whether it is radiating, and specific dermatomes and myotomes. Description of the type of pain is sometimes helpful; patients may describe their pain as deep and boring, scratchy, sizzling, at a point, general over the back region, continually changing, and so on. You may need to help some people describe their pain by offering adjectives to choose from. In chapters 4 and 5 I described the link between pain types and specific tissues and syndromes. Keep in mind that changing symptoms over the short time of an examination generally suggests more fibromyalgic syndromes, which can sometimes be resistive to exercise therapies.

7. Take dermatomes and myotomes into account. With radiating symptoms, the dermatomes and myotomes can assist in understanding the involved segmental levels and whether the pain originates from a specific nerve root. For example, direct pressure on the root could indicate a unilateral disc bulge or end-plate fracture that would cause a loss in disc height together with a loss of root outlet foramen size. In this way the spinal level can be linked with the dermatome or myotome but not the actual tissue damage. Further, nerve root pressure can occur at a specific spinal level on the outlet nerve, consistent with a dermatome or myotome, or on the traversing nerve from above if there is pressure on the cauda equina centrally. Thus dermatomes and myotomes are another consideration when forming an opinion from the consistency obtained from several tests that can include medical imaging, provocative tests, and so forth.

8. **Perform provocative tests.** Once you suspect that specific tissues are damaged, you can load them to see if loading produces pain. This is provocative testing. Many patients have more complex presentations, with several tissues involved. Nonetheless, the provocative procedure still indicates which postures, motions, and loads cause pain and thus should be avoided when designing the therapeutic exercise. Generally, patients' descriptions of the activities they find exacerbating of their pain (item 5 of this list) will guide your decision as to which specific tissues to load and stress. For example, lumbar extension with a twist can provoke the facets, while the anterior shear test may be warranted for suspected instability (shown in the next few pages).

Specific Tests

Several specific tests that can help you assess your patients and plan the most appropriate exercise are described in this section. You are the final judge as to which tests are relevant for specific patients.

Testing Muscle Endurance

While Biering-Sorensen (1984) showed that decreased torso extensor endurance predicts those who are at greater risk of future back troubles, recent work has suggested that the balance of endurance among the torso flexors, extensors, and lateral musculature better discriminates those who have had back troubles from those who have not. Because these three muscle groups are involved in spine stability during virtually any task, the endurance should be measured in all three. Simple tests that isolate these groups of muscles are difficult to find; I chose the following tests because each was shown to have high reliability coefficients, at least .98 or higher, when repeated over five consecutive days (McGill, Childs, and Liebenson, 1999).

▶ Lateral Musculature Test

The lateral musculature is tested with the person lying in the full side-bridge position. Legs are extended, and the top foot is placed in front of the lower foot for support. Subjects support themselves on one elbow and on their feet while lifting their hips off the floor to create a straight line over their body length. The uninvolved arm is held across the chest with the hand placed on the opposite shoulder. Failure occurs when the person loses the straight-back posture and the hip returns to the ground.

▶ Flexor Endurance Test

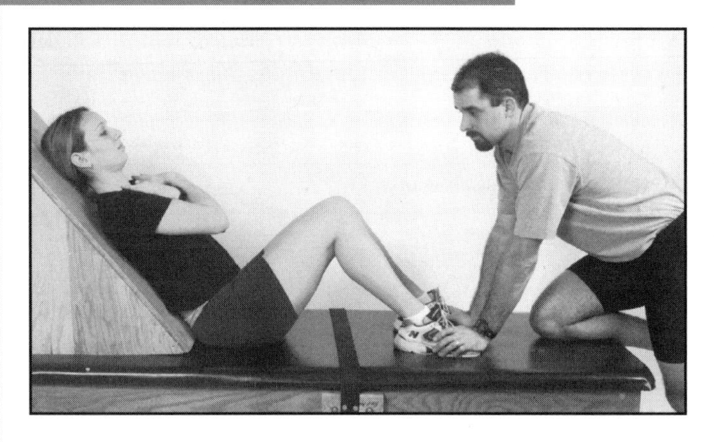

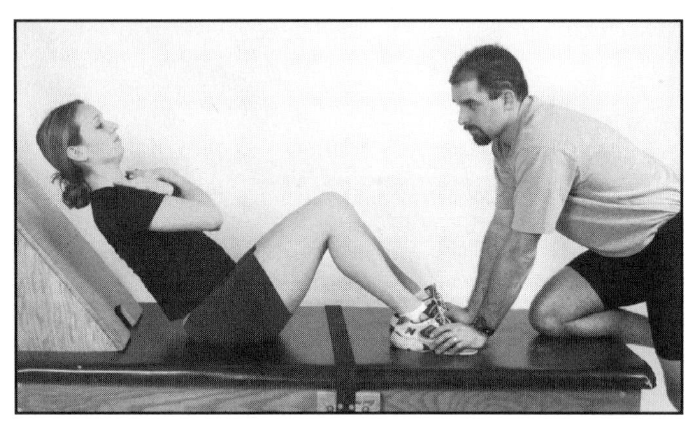

Testing endurance of the flexors (rectus) begins with the person in a sit-up posture with the back resting against a jig angled at 60° from the floor (a). Both knees and hips are flexed 90°, the arms are folded across the chest with the hands placed on the opposite shoulder, and toes are secured under toe straps. To begin, the jig is pulled back 10 cm (4 in.) and the person holds the isometric posture as long as possible (b). Failure is determined to occur when any part of the person's back touches the jig.

▶ Back Extensors Test

The back extensors are tested in the "Biering-Sorensen position" with the upper body cantilevered out over the end of a test bench and with the pelvis, knees, and hips secured. The upper limbs are held across the chest with the hands resting on the opposite shoulders. Failure occurs when the upper body drops from the horizontal position.

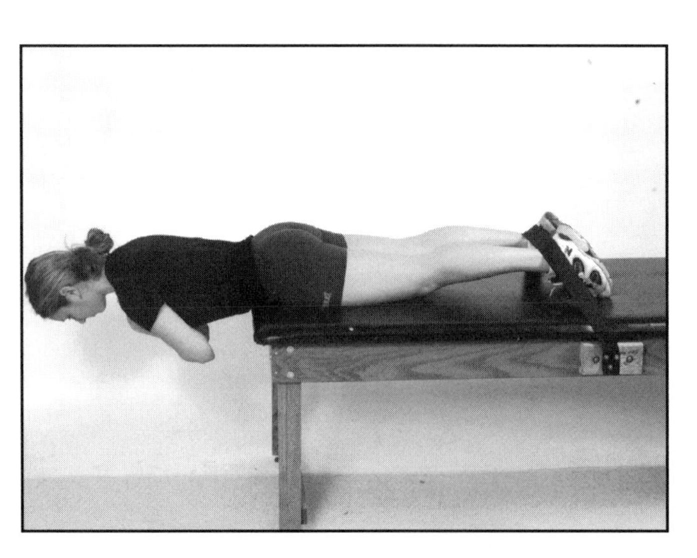

Normative Data

Normative absolute endurance times collected from young healthy individuals are listed in table 12.1. Note that women have greater endurance than men in the extensors. Further, the flexor endurance time (as a proportion of the extensor endurance time) for healthy young men is about .84, while other tests have yielded lower ratios for healthy older men. McGill and colleagues (in press) showed that the relationship of endurance among the anterior, lateral, and posterior musculature is upset once back troubles begin (see table 12.2); that upset in the relationship remains long after the symptoms have resolved. Typically, the extensor endurance is diminished relative to both the flexors and the lateral musculature in those with lingering troubles.

Interpreting Endurance Scores

Interpreting absolute endurance is probably secondary to interpreting the relationship among the three muscle groups (flexors, lateral, and extensors). Normally, the following discrepancies suggest unbalanced endurance (note that the last ratio is a strength-to-endurance ratio).

Right-side bridge/left-side bridge endurance   > 0.05
Flexion/extension endurance   > 1.0
Side bridge (either side)/extension endurance   > 0.75
Extensor strength (Nm)/extensor endurance (sec)   > 4.0

Table 12.1 Mean Endurance Times (sec) and Ratios Normalized to the Extensor Endurance Test Score

Mean age 21 yrs (men: *n* = 92; women: *n* = 137)

	Men			**Women**			**All**		
Task	**Mean**	**SD**	**Ratio**	**Mean**	**SD**	**Ratio**	**Mean**	**SD**	**Ratio**
Extension	161	61	1.0	185	60	1.0	173	62	1.0
Flexion	136	66	0.84	134	81	0.72	134	76	0.77
RSB	95	32	0.59	75	32	0.40	83	33	0.48
LSB	99	37	0.61	78	32	0.42	86	36	0.50
Flexion/extension ratio	0.84			0.72			0.77		
RSB/LSB ratio	0.96			0.96			0.96		
RSB/extension	0.58			0.40			0.48		
LSB/extension	0.61			0.42			0.50		

RSB = right side bridge; LSB = left side bridge

Table 12.2 Mean Endurance Times Comparing Normal Workers With Those Who Have Had Back Disorders

Men, mean age 34 yr (never had back troubles: n = 24; lost work due to LBD: n = 26) from the same workplace. Variables that are significantly different between those with a history and those who have never had troubles have an asterisk. Note that all men were asymptomatic at the time of testing; these are long-lingering deficits subsequent to back troubles.

	No back troubles			History of disabling back troubles		
Task	**Mean**	**SD**	**Ratio**	**Mean**	**SD**	**Ratio**
Extension	103	35	1.0	90	49	1.0
Flexion*	66	23	0.64	84	45	0.93
RSB	54	21	0.52	58	23	0.64
LSB	54	22	0.52	65	27	0.72
Flexion/extension ratio*	0.71	0.26		1.15	0.66	
RSB/LSB ratio*	1.05	0.32		0.93	0.22	
RSB/extension*	0.57	0.29		0.97	1.20	
LSB/extension*	0.58	0.28		1.03	1.16	

Testing for Aberrant Gross Lumbar Motion

The usual practice when testing for aberrant gross lumbar motion is to test for range of motion about the three primary axes of motion by determining the degrees of motion possible from neutral (usually using standing posture as a reference base). For example, a flexion range of motion may be determined to be 45° from neutral. As I mentioned earlier, this single number is of little use as an indicator of pathology or for guiding therapeutic exercise decisions. Other well-known markers are the inability to achieve myoelectric silence in the extensors at full flexion or the inability to achieve full flexion (see figure 12.1). However, pathology may manifest within the moderate range of motion (not at end range) and also may occur in axes other than the primary plane of motion. Generally, this pathology is repeatable and evident in each repeated motion cycle. The following examples illustrate this point.

The patient in figure 12.2 had a normal range of motion and by traditional tests would be classified as normal. Yet during a lateral bend test (from neutral 0° to 23°), every time the spine bent laterally passing 17°, a small flexion "clunk" (instability) of a degree and a half occurred in the flexion axis. This occurred every time the spine passed 17° of lateral bend. The point is that the motion pathology, or instability, was not in the lateral bend axis (or primary motion axis) but in the flexion axis (a secondary axis). Clinicians often miss these very subtle aberrant catches in the "off" axes during examination. The more skilled clinicians sometimes see them or feel them with their hands.

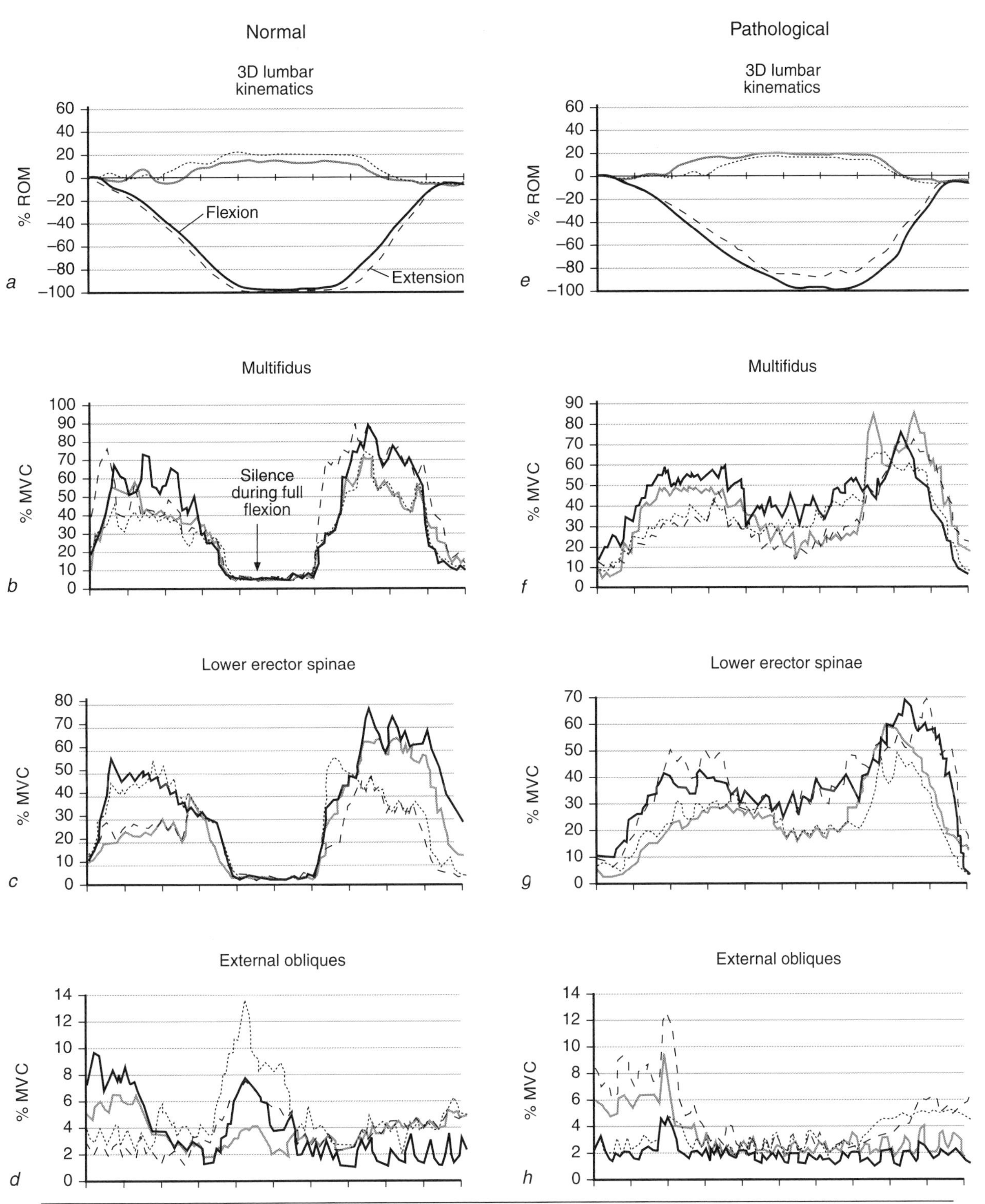

Figure 12.1 Normal 3D motion profiles *(a)* are shown on the left together with EMG patterns. Myoelectric silence is achieved in the lumbar extensors *(b)* and *(c)* during full flexion. This demonstrates the ligaments' and discs' ability to bear the load, suggesting they are in good health. As well, the abdominals are activating *(d)* to do one of two things—either the patient is hollowing in the abdominal wall so as not to let it hang, or he is willing to add to the flexor moment to pull into even more full flexion. This too is a marker of good ligament and posterior disc health. The individual on the right cannot fully flex *(e)* and is unable to achieve lumbar extensor muscle silence *(f) and (g)*. The abdominals are completely unwilling to activate *(h)*, exacerbating the flexion moment. Further testing will reveal troubles in the posterior complex.

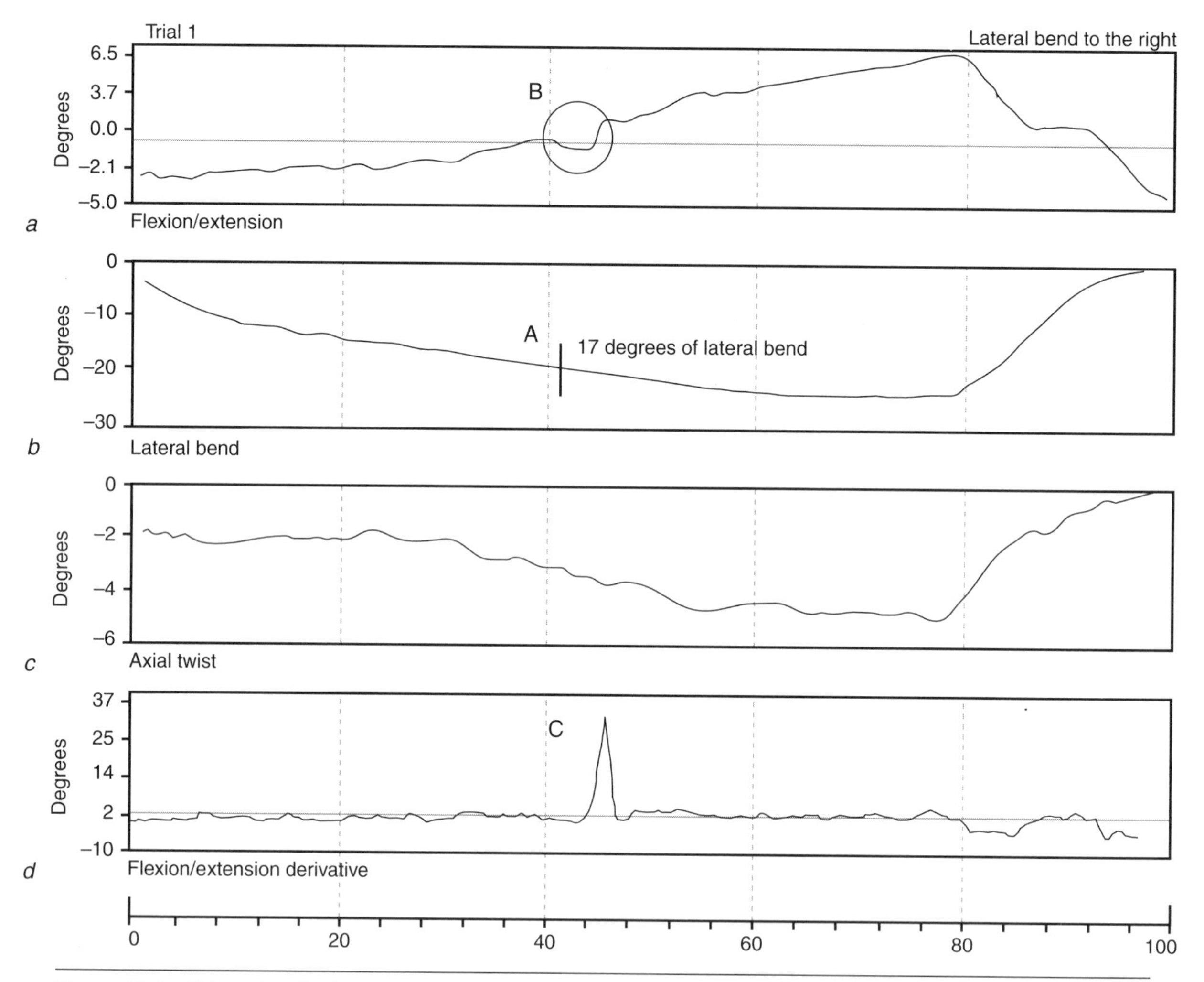

Figure 12.2 This patient had a normal range of motion but demonstrated an "instability catch." During a lateral bend test (from neutral 0° to 23°), every time the spine bent laterally passing 17° (event A) *(b)*, a small flexion hitch of a degree and a half occurred in the flexion axis (event B) *(a)*. This occurred very time the spine passed 17° of lateral bend. There was no deviation in the twist axis *(c)*. The bottom curve represents the first derivative of the flexion curve, which we use to better locate instability. The spike (event C) *(d)* clearly points to the point at which the instability occurred in the flexion axis. In this patient, a muscle activation pattern was identified that removed the clunk or catch.

In figure 12.3, a and b, a "clunk" or catch occurs in the flexion axis as the patient flexes forward and passes 30°; this occurred in every flexion attempt. Sometimes patients will report pain as the catch occurs, and sometimes the clinician notices evidence of muscle activation bursts, but the patient simply moves through the region. We use an electromagnetic instrument that tracks three-dimensional spine motion and graphically displays the results. Sometimes the catches are so subtle that it is far more illuminating to observe the first derivative of the motion axis with the catch. Patients with pain have a much higher incidence of these catches (McGill and Peach, in press) than normals, indicating instability. Normals who demonstrate these catches, although rare, may be at elevated risk for becoming patients according to the aberrant motion model of Richardson and colleagues (1999).

Testing for Lumbar Joint Shear Stability

Chapters 4 and 5 offer a well-developed discussion of the role of the lumbar extensors (longissimus thoracis and iliocostalis lumborum) to support anterior/posterior shear of

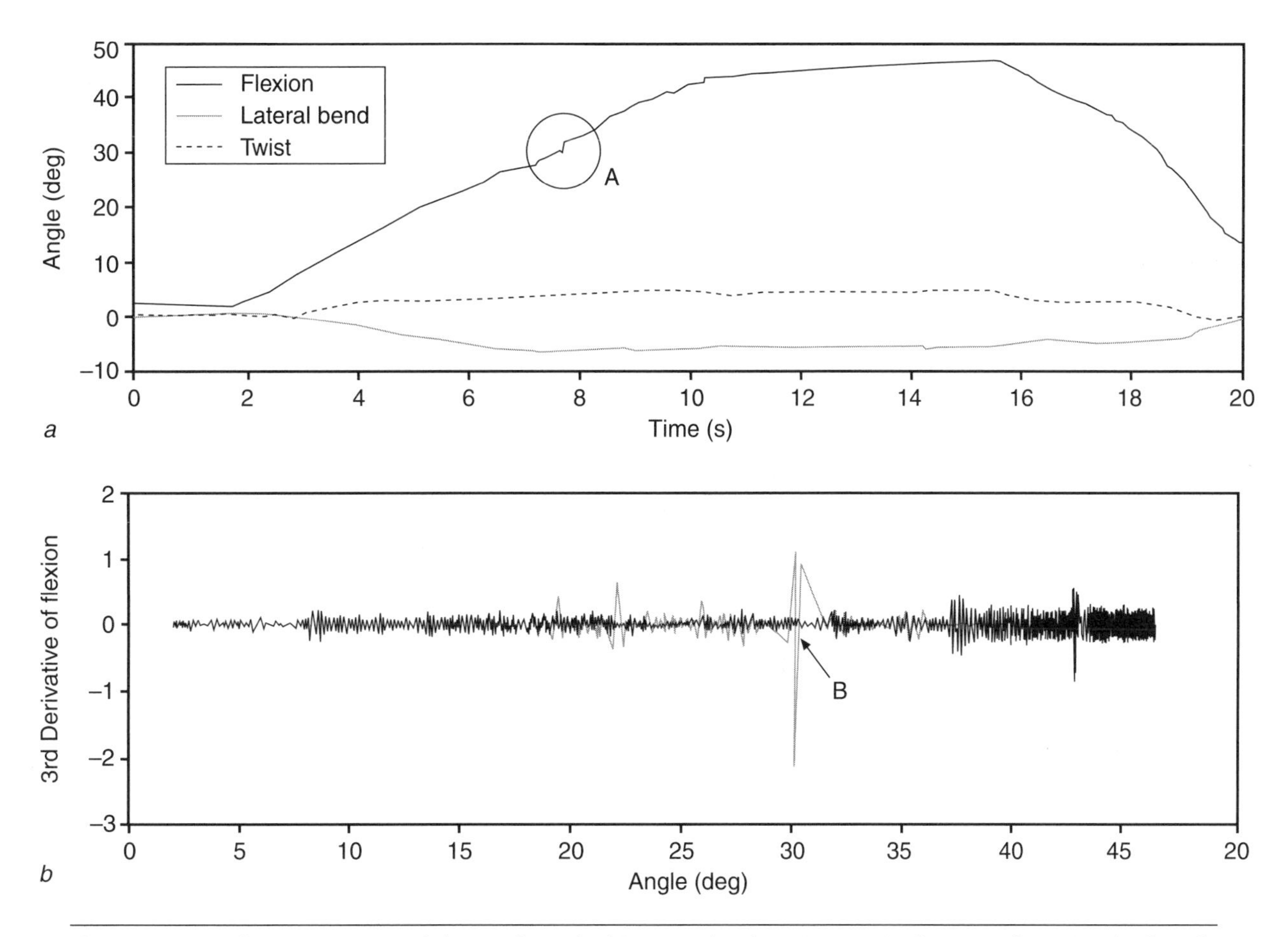

Figure 12.3 In this patient an instability catch or "clunk" occurred in the primary axis of motion (flexion). As he flexed, the clunk was seen in the flexion axis at 30° (event A) *(a)*. This occurred in every trial of forward flexion. The instability is evident (event B) in the third derivative of the flexion axis motion *(b)*. The rehabilitation objective was to stabilize the spine to eliminate the clunk. This was not fully successful over the short term. Symptoms resolved only when the patient developed spine position awareness and was able to avoid flexion to 30°.

the vertebral motion segments. These muscles stiffen against unstable motion in the shear plane. In fact, good stabilization exercise must involve these muscles. A simple test can be performed to see if the patient has pain provoked by shear instability and thus could benefit by reducing the aberrant shear motion. If the test is positive, the patient will do well with stabilization exercises that groove cocontracting motor patterns of the shear musculature.

▶ Manual Testing for Lumbar Joint Shear Stability

The patient lies prone with the body on a table but with the legs over the edge and the feet on the floor. The person must relax the torso musculature. The clinician then applies direct force downward onto each spinous process in turn (a) (starting at the sacrum, then L5, L4, L3, etc.). Unstable segments are identified when either the person reports pain or the clinician feels actual shear displacement, but the patient's reporting of segment-specific pain, in this case, is given more consideration. Then the patient is asked to slightly raise the legs off the floor to contract the back extensors (b). The clinician once again applies force on each spinous process. By virtue of their lines of action, the lumbar

(continued)

Manual Testing for Lumbar Joint Shear Stability *(continued)*

extensors will reduce shearing instability if present. If pain is present in the resting position but then disappears or subsides with the active cocontraction, the test is positive. This test proves conclusively that activating the extensors stabilizes the shear instability and eliminates the pain. Now the trick is to incorporate these extensor motor patterns into exercise prescriptions to carry over to daily activities.

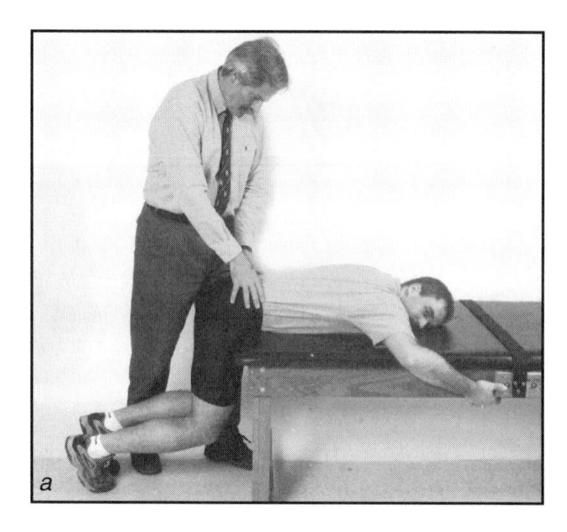

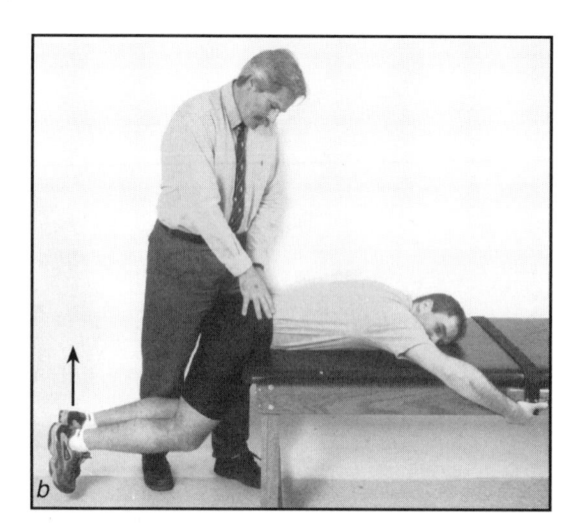

A Note on Motion Palpation

Some clinicians use their hands to feel specific vertebral motion during a whole trunk motion such as flexion or twisting. When they detect a "blocked" or "stiff-feeling" segment, they often declare it pathological and make it the target of mobilizing treatment. Clinicians rarely, however, consider the flexibility in the spinous process and bony neural arch noted in chapter 4. Two of my PhD students questioned whether the asymmetric feel to a specific joint was true pathology or simply asymmetric anatomy—perhaps asymmetric facet joints (Ross, Bereznick, and McGill, 1999). By carefully documenting the motion occurring at the disc and in the spinous process in cadaveric specimens, together with the applied loads (similar to clinical loads), they were able to quantify the effect of anatomy. The results demonstrated that over half of the motion felt by the clinician was actually the spinous process flexing and bending (see figure 12.4) and not vertebral body motion. Those joints that were anatomically asymmetric had asymmetric motion under an applied load from the clinician. Treating such a joint with mobilization therapy would be fruitless.

Testing for Aberrant Motor Patterns During Challenged Breathing

The test for aberrant motor patterns during heavy breathing is rather sophisticated, but it is about the only way to identify a persons' ability to maintain spine stability during tasks that require higher physiological work rates. Candidates for this type of test include athletes and occupational workers such as construction workers, warehouse employees, and so on. We explored the link between breathing and spine stability

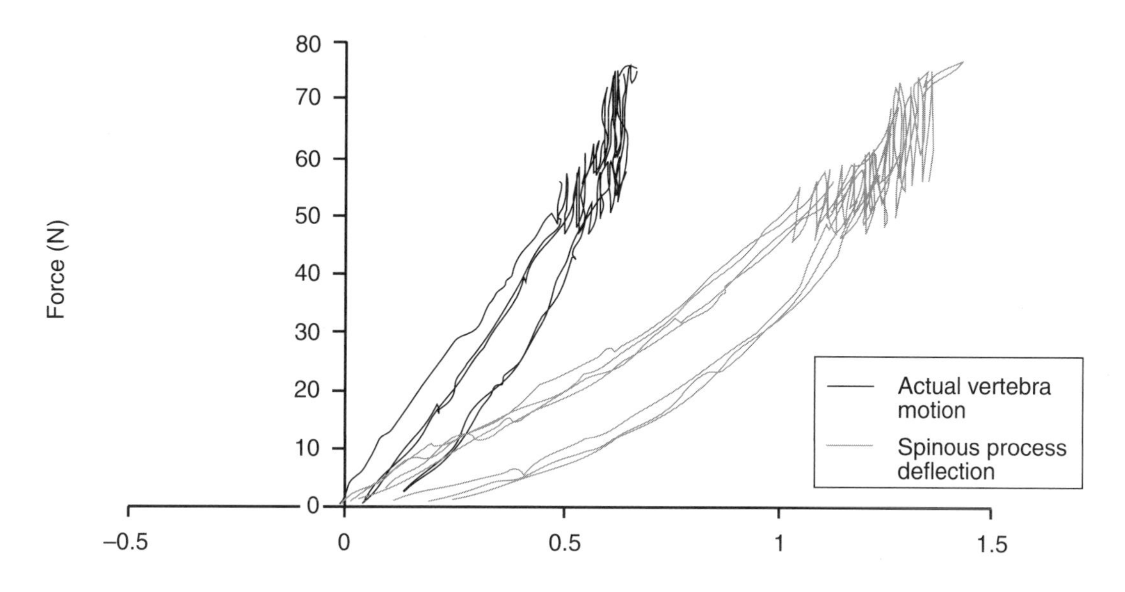

Figure 12.4 Lateral forces were applied to the tip of the spinous process by clinicians to cadaveric motion segments to mimic motion-palpation testing. Most of the motion measured in the spinous process, which is what the clinician feels, took place in the spinous process bending rather than in the disc. Moreover, asymmetric anatomy resulted in asymmetric motion when loaded to mimic the clinician's hand forces, suggesting that this test should not be used solely to determine the site of mobilizing treatment.

previously (pages 145-146). Challenged breathing requires heavy involvement of the abdominal muscles. Because we monitor the muscles with EMG electrodes and because muscle fatigue changes the EMG signal, we avoid creating heavy lung ventilation with exercise for this test. (Note that this consideration is only for testing and that we will intentionally increase the physiological work rate for exercises in chapter 14.)

▶ Challenged Breathing Test

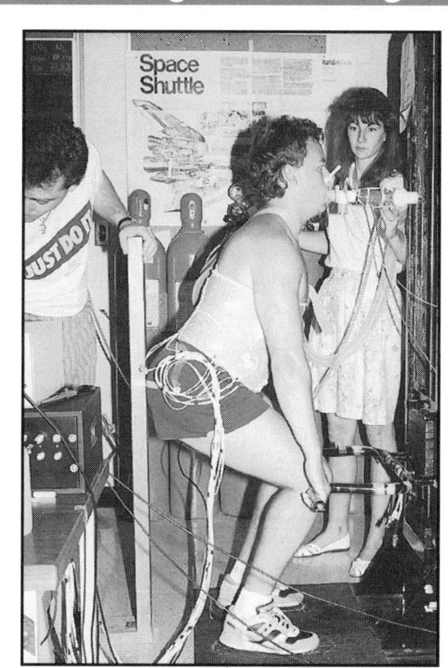

During this test the person breathes a 10% CO_2 gas mixture to elevate the lung ventilation with only a few breaths, but the O_2 is adjusted so that the light-headedness that accompanies hyperventilation does not occur. Then a weight, usually about 15 kg (33 lb) for the average man, is placed in the hands while the torso is flexed forward about the hips to approximately 30°. This test reveals those who have aberrant motor patterns in the abdominals during challenged breathing that compromise lumbar stability.

Good spine stabilization patterns observed in the muscle EMG signal obtained during the challenged breathing test include a constant muscle cocontraction ensuring spine stability. Figure 12.5 shows stable patterns (a) and unstable patterns (b). Our computerized analysis package measures spine stability using the methods explained in chapter 6. Many clinicians will not have this capability but should still look for constant levels of abdominal wall activation. Patients who are poor at stabilization in these situations will demonstrate short, temporary reductions in activation or even loss of activation. These mark the critical instances at which spine instability can occur and classify the person as a candidate for spine stabilization during high physiological work rates. These patients must learn to contract the spine stabilization muscles independently of challenged breathing to ensure spine stability in conditions of heavy work.

Determining Suitability for ROM Training and Stretching

The McKenzie extension approach is well-known for treating acute discogenic troubles. The extension postures can also be used as a test to identify those with posterior disc troubles. Typically, these posterior discogenic patients should never flex the lumbar spine when it is under substantial loading.

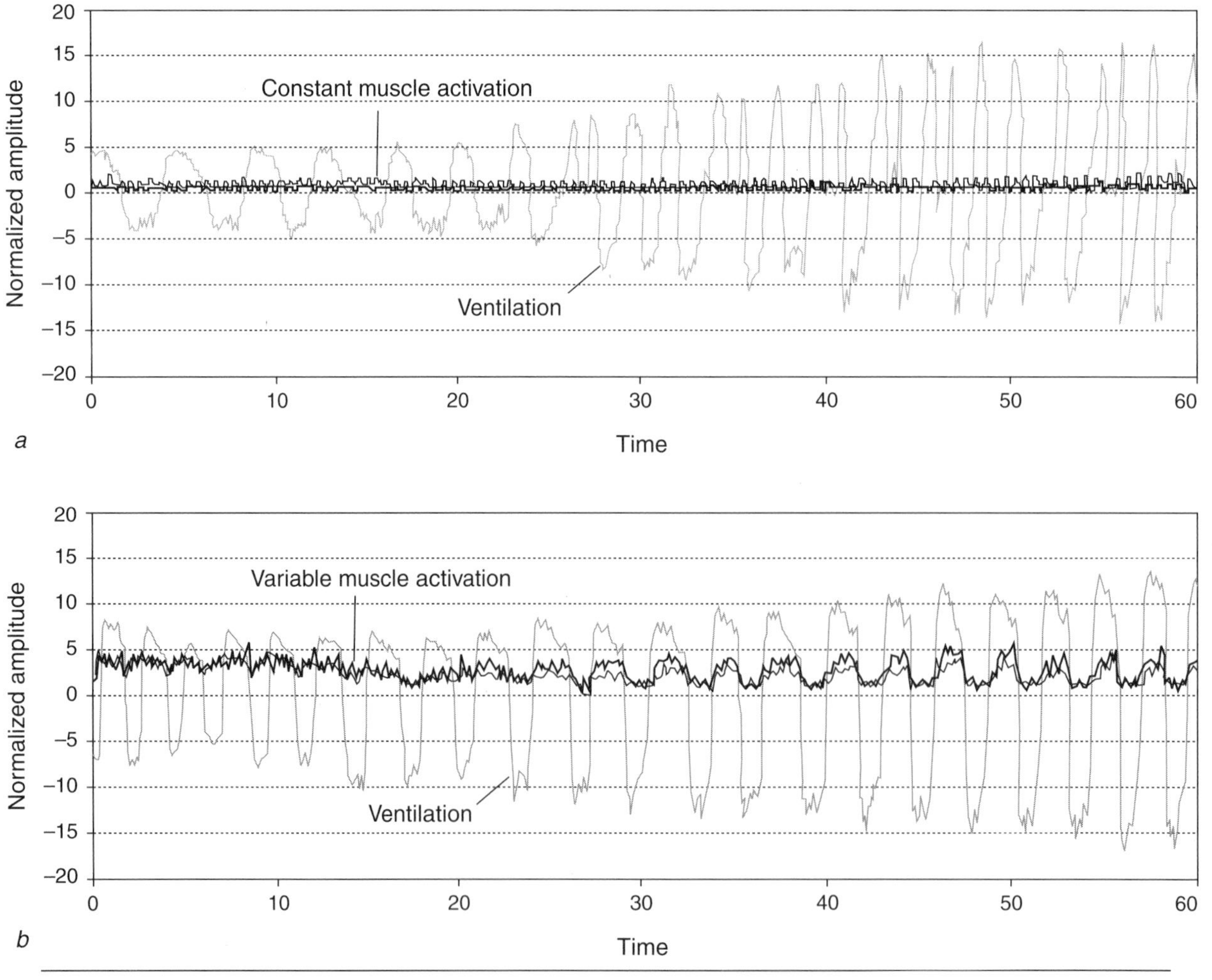

Figure 12.5 Good spine stabilization patterns during the CO_2 breathing test include a constant muscle cocontraction, which ensures spine stability. Both stable *(a)* with constant muscle activation and unstable *(b)* variable abdominal muscle activation patterns are shown.

▶ McKenzie Posture Test

The patient begins in a relaxed standing posture, and the clinician asks him about current pain and the general feeling about the back. Then the patient lies prone, adopting one of the following three levels of the McKenzie posture progression (the clinician decides which is suitable based on the patient's flexibility and symptoms):

- Prone with the arms relaxed (a)
- Prone with the chin resting on the fists (b)
- Prone supported on the elbows (c)

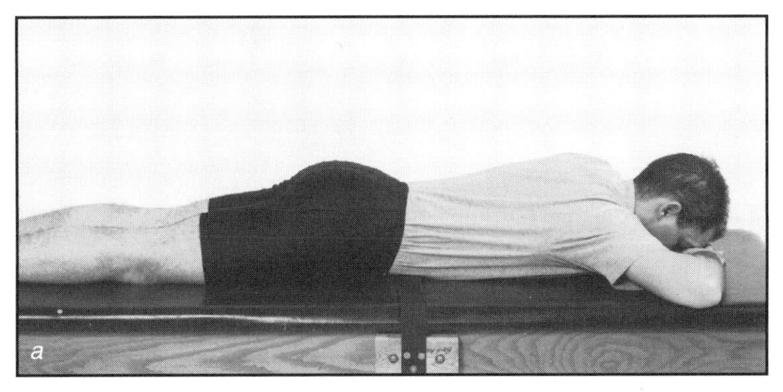

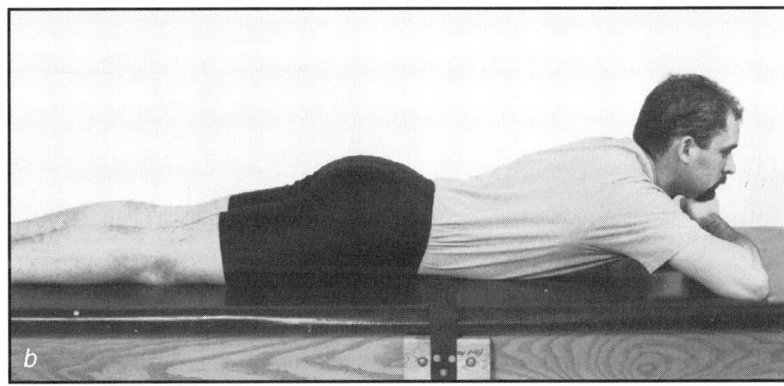

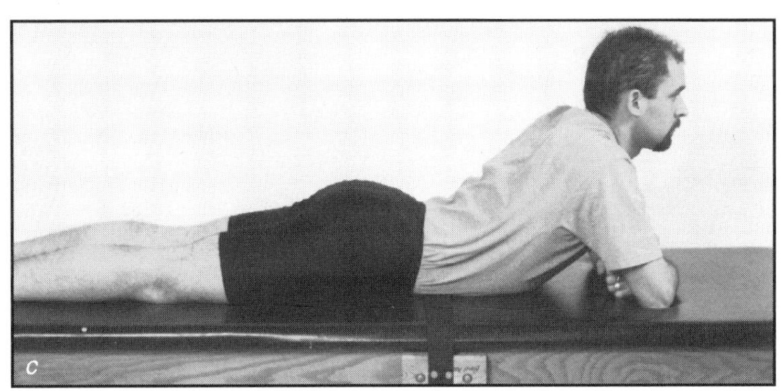

If the patient cannot tolerate one of these postures, the clinician can assume that simple disc disturbances are not the source of troubles or at least not the sole source. Many with uncomplicated posterior disc lesions, or herniations, find relief in these postures. The patient then returns to the standing posture. If he feels more stable and/or has less pain compared to the moments prior to lying prone, he is classified as posterior discogenic.

Patients who are posterior discogenic should avoid flexion stretches and a flexed spine under load when performing exercise therapy. Those who feel worse upon standing could have a host of other possible conditions but will probably experience better progress by adopting a neutral to slightly flexed lumbar posture for exercise therapy.

Distinguishing Between Lumbar and Hip Problems

Pain in the buttock and/or radiating down the leg can have more than one source. For effective treatment, you must discover whether that source is in the lumbar spine or in the sacroiliac (SI) joints, in various muscles, or in tissues around the hip or even the hip joint itself.

▶ Sitting Slump Test

Several forms of the sitting slump test can be used to elicit radiating sciatic symptoms. These are designed to tense the sciatic nerve and irritate the lumbar nerve roots. Typically the patient sits on the table or chair and then slumps or slouches. The intention is to progressively increase the nerve tension. Then

- the leg can then be extended at the knee (a), tensing the nerve from below the lumbar spine;
- foot flexion can be added (b), further tensing the nerve; and finally
- the nerve can be further tensed from above the lumbar region with the addition of cervical spine flexion (c).

If the tension at any of these stages causes symptoms, further progressive testing is stopped, and the patient may be a candidate for nerve flossing (described in chapter 13). Certainly flexion will be contraindicated for these patients during stretching and exercise.

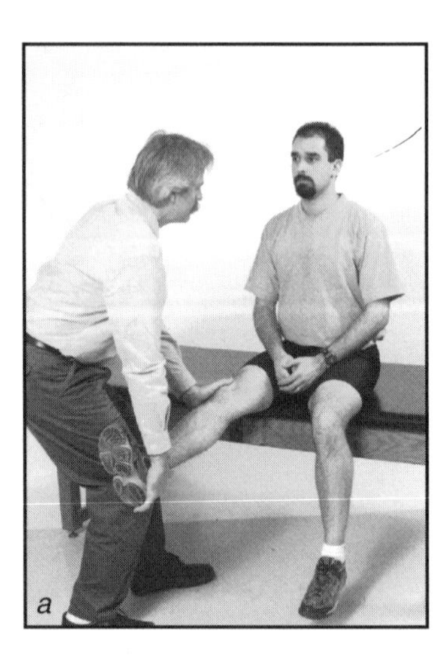
a

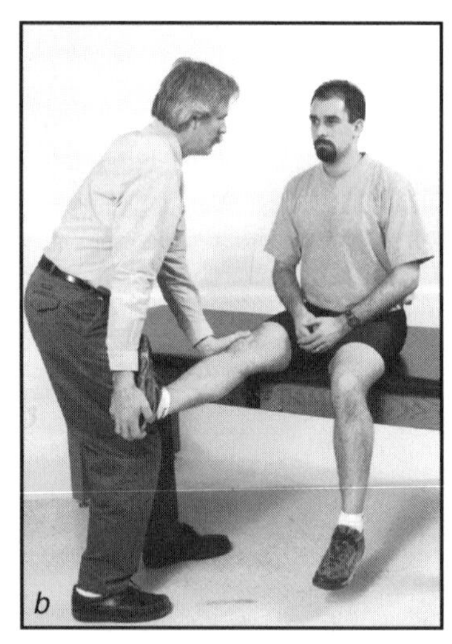
b

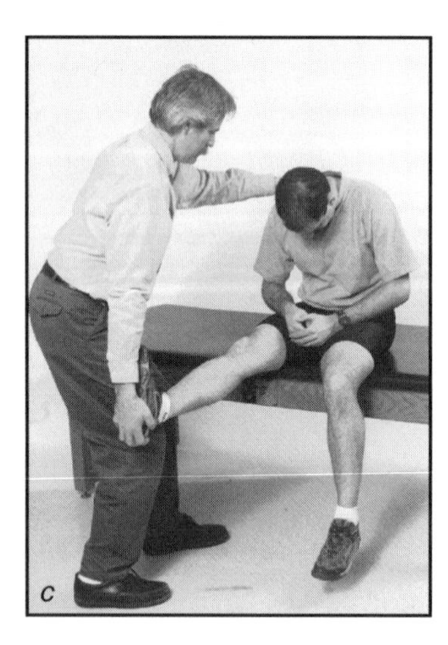
c

If flexing the neck does not increase the pain, then the sciatic nerve is not the tissue of interest and the clinician can begin assessing the SI joints, the hamstrings, piriformis, and other muscle-based syndromes, ending with the hip.

▶ Fajersztajn Test

The Fajersztajn test should always be performed on both legs even when the sciatica is unilateral. Raising the "well leg" is needed to perform the "well leg raising test of Fajersztajn" (DePalma and Rothman, 1970). Raising the well leg tensions the nerve root on the well side together with causing tensions centrally along the midline of

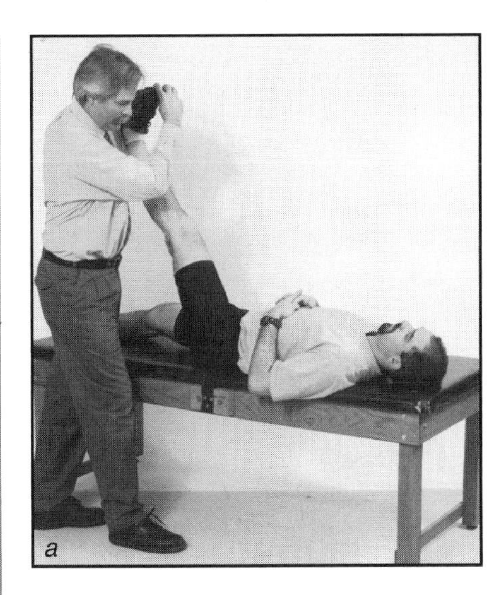
a

the cauda equina and to the nerve roots on the opposite side (a). Sometimes pain is provoked with simultaneous cervical flexion (b). Pain in the symptomatic side (side not raised) is an organic sign of disc lesion, usually a more central lesion. It is not a sign of malingering as some have suggested.

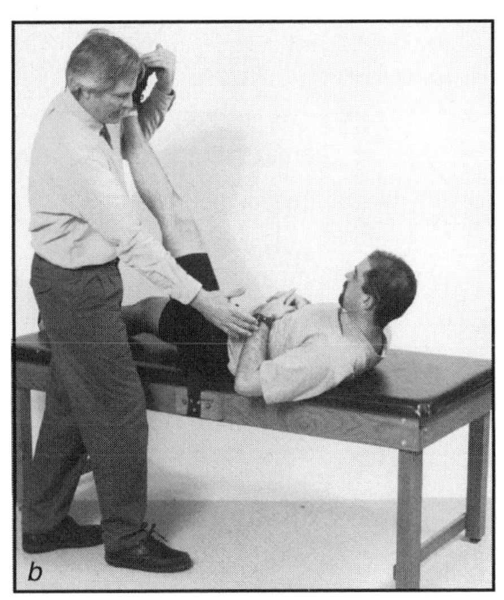
b

I am continually surprised at the number of people with back troubles who also have hip troubles (McGill and colleagues, in press, recently noted a correlation between back troubles and hip pathology in factory workers). It is sometimes difficult to separate hip pain from lumbar trouble since beep buttock pain can be sciatic (lumbar) or originate in the hip joint. Typically, medial and anterior thigh pain are indicators of hip pathology. If hip tests are indicated by the nerve tension tests described in the previous section (i.e., they were negative), flexion and rotation tests should be performed.

► Hip Flexion and Rotation Tests

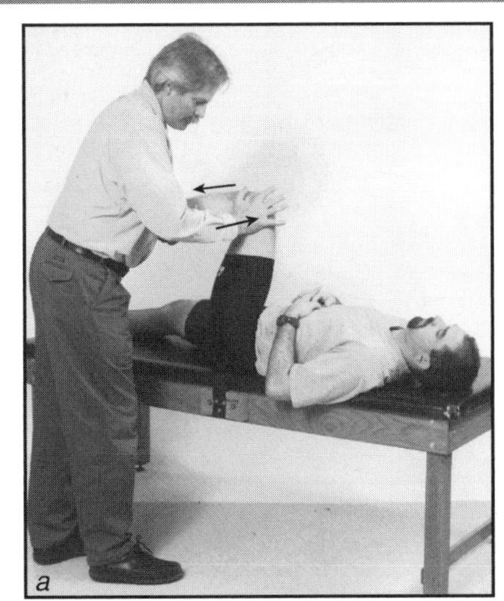
a

Lying supine with the hip flexed, the patient's thigh segment is rotated into hip internal rotation (a). Stiffness or pain is a positive indicator of hip concerns. Hip joint pain is somewhat distinctive with radiation into the groin, and/or along the inguinal crease, and/or medial/anterior thigh. The patient's hip is then flexed (b) while the knee is moved in circular motions, scouring the acetabulum but with the knee bent so as not to confuse any pain with nerve tension related to the back.

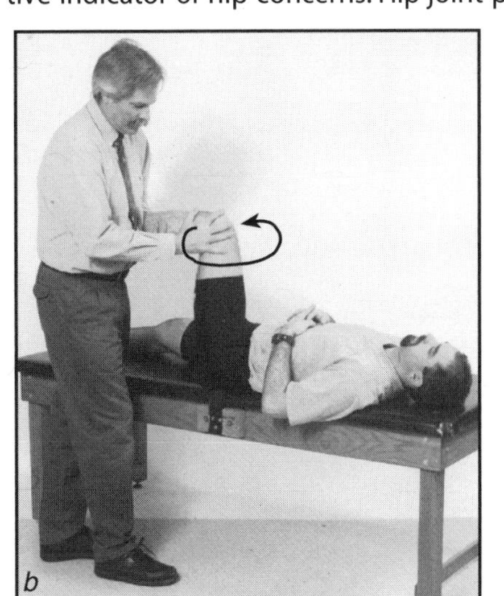
b

References

Biering-Sorensen, F. (1984) Physical measurements as risk indicators for low-back trouble over a one-year period. *Spine,* 9: 106-119.

DePalma, A.F., and Rothman, R.H. (1970) *The intervertebral disc.* Philadelphia: W.B. Saunders.

McGill, S.M., Childs, A., and Liebenson, C. (1999) Endurance times for stabilization exercises: Clinical targets for testing and training from a normal database. *Archives of Physical Medicine and Rehabilitation,* 80: 941-944.

McGill, S.M., Grenier, S., Preuss, R., and Brown, S. (in press) Asymmetries in torso endurance and strength parameters are associated with a history of low back troubles.

McGill, S.M., and Peach, J. (in press) Electromyographic and spine kinematic changes with pain.

Richardson, C., Jull, G., Hodges, P., and Hides, J. (1999) Therapeutic exercise for spinal segmental stabilization in low back pain. Edinburgh, Scotland: Churchill Livingstone.

Ross, J.K., Bereznick, D., and McGill, S.M. (1999) Atlas-axis facet asymmetry: Implications for manual palpation. *Spine,* 24 (12): 1203-1209.

CHAPTER 13

Developing the Exercise Program

The exercises described in this chapter comprise a program founded on the principles discussed in previous chapters. This program is sufficient for many patients and for those interested in optimal back health. For those with performance ambitions, chapter 14 offers more advanced versions. While I claim no expertise beyond the low back, I will broach other issues that have implications for the back. For example, improper technique when performing hip and knee mobilization and stretching can compromise the back; for this reason I have included them here.

Upon completion of this chapter you will be able to design exercise that challenges muscle and establishes stabilizing motor patterns, but spares the spine. The focus here is clearly on the patient and low back health for everyday activity.

Preliminary Matters

Before presenting the exercises, I will review some of the basic ideas discussed in the text thus far, establish how to stretch without stressing the back, clarify issues regarding abdominal hollowing and bracing, and introduce the technique of flossing the nerve roots for those with sciatica.

Philosophy of Low Back Exercise Design

Many traditional notions that exercise professionals consider to be principles for exercise design, particularly when dealing with the low back, may not be as well supported with data as generally thought. As previously noted, many clinicians still prescribe sit-ups—usually with the proviso that the knees remain bent. The resultant spinal loading of well over 3000 N of compression to a fully flexed lumbar spine clearly shows the folly of such a recommendation. Other experts still recommend the posterior pelvic tilt when performing many types of low back exercise. This actually increases the risk of injury by flexing the lumbar joints and loading passive tissues. The recommendation of flattening the lumbar region to the floor when performing abdominal exercise is another version of this ill-founded philosophy. Many continue to believe that having stronger back and abdominal muscles is protective and reduces bad back episodes, even though Luoto and colleagues (1995), among others, showed that muscle *endurance,* not strength, is more protective. This should not be misinterpreted to mean that having stronger muscles is not a good objective; rather, it reflects on many of the strength training approaches currently used that create back patients. Myths still remain regarding the need for greater lumbar mobility, which the evidence suggests leads to more back troubles—not less (e.g., Biering-Sorensen, 1984)! Finally, we are disturbed by the fact that replicating the motion and spine loads that occur during the use of many low back extensor machines used for training and therapy produced disc herniations when applied to spines in our laboratory!

CLINICAL RELEVANCE

Clinical Wisdom

Clearly, some current clinical wisdom needs to be reexamined in the light of the scientific evidence, much of which has been presented in this book. This evidence seems to indicate that the safest and most mechanically justifiable approach to enhancing lumbar stability through exercise is to emphasize endurance over strength. Clinicians using such an approach should encourage patients to maintain a neutral spine posture when under load and use abdominal cocontraction and bracing in a functional way. The bodybuilding approach often used in rehabilitation is not synonymous with health objectives and results in additional risk to create joint loading through a range of motion and the often simple motor patterns that are not optimal for ensuring joint stability.

Sparing the Back While Stretching the Hips and Knees

Given that spasms in the psoas, which produce a shortening and hip extension restriction, are common in those with back troubles (Grenier et al., in press), these people would probably benefit from hip stretching and other range-of-motion exercises. Sometimes, however, poor technique when stretching hip and knee tissues leads to unnecessary loading of the lumbar spine. This can easily be avoided by changing technique. A general guideline for sparing the back is to maintain an upright torso posture while performing hip and knee work. Preservation of a neutral spine ensures minimal loading from passive tissues. An upright torso minimizes the reaction torques, the associated muscle contraction, and spine load. Abdominal bracing is sometimes required for pain control. For example, lunges are a good exercise for challenging strength, endurance, balance, and mobility in the lower extremities. Lower extremity capability is needed to facilitate spine-sparing postures when lifting and when performing a host of other tasks.

CLINICAL RELEVANCE

Sparing the Back

The technique for sparing the back is to maintain an upright torso posture while performing the lunge (see figure 13.1). Some clinicians recommend keeping the back leg straight, which causes patients to flex the torso forward. This is poor form for sparing the back (see figure 13.2). A general principle is to keep the torso upright with a neutral lumbar spine while stretching joints other than the back.

Figure 13.3, for example shows correct postures for stretching the quads (a) and the hip adductors (b). Finally, the lunge with a neutral spine stretches iliacus more than psoas (iliacus is a uniarticular muscle that crosses only the hip). Full stretch on the psoas requires a lateral bend of the torso away from the extended hip, but this is reserved for the more robust backs (see figure 13.4 a-b).

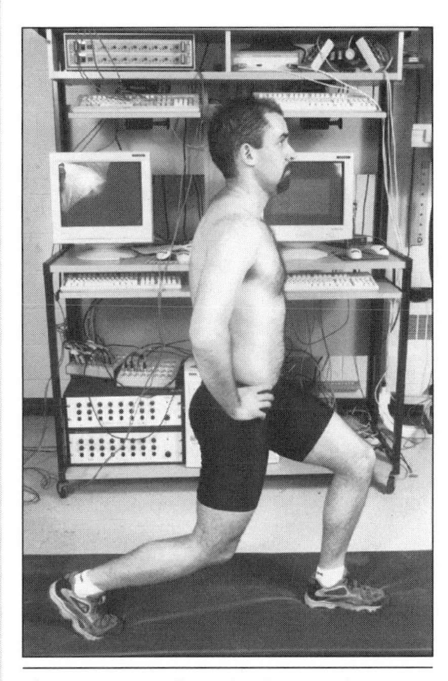

Figure 13.1 The spine is spared performing the lunge by maintaining an upright torso and a neutral spine curvature.

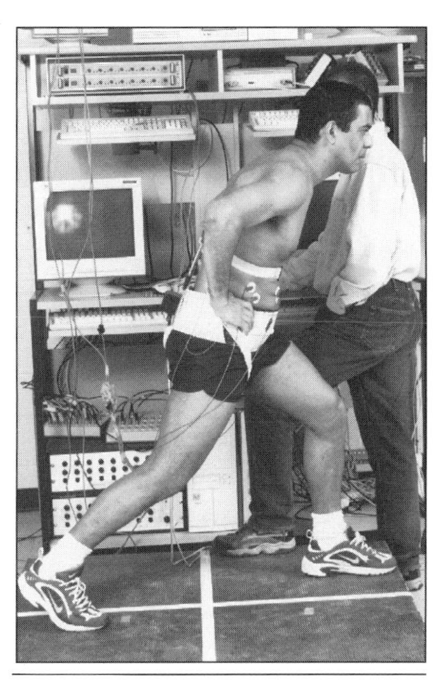

Figure 13.2 Some clinicians recommend keeping the back leg straight while performing lunges, which causes patients to flex the torso forward. This is poor form for sparing the back.

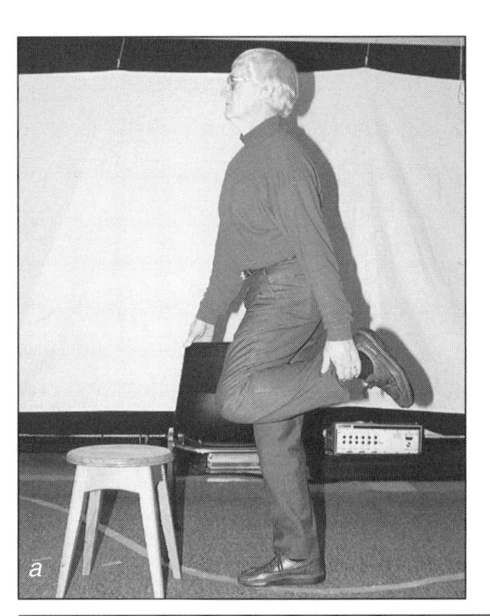

Figure 13.3 During stretching at joints other than the back, the spine is spared by maintaining the upright torso and neutral spine curve. When stretching the quads, holding a chair for balance is a good way to ensure a straight back *(a)*. The person shown in *(b)* has sufficient hip flexion mobility to ensure an upright torso and a neutral spine. Individuals who are unable to maintain this spine posture should forgo this exercise until they can achieve the required hip flexion mobility.

(continued)

Sparing the Back (continued)

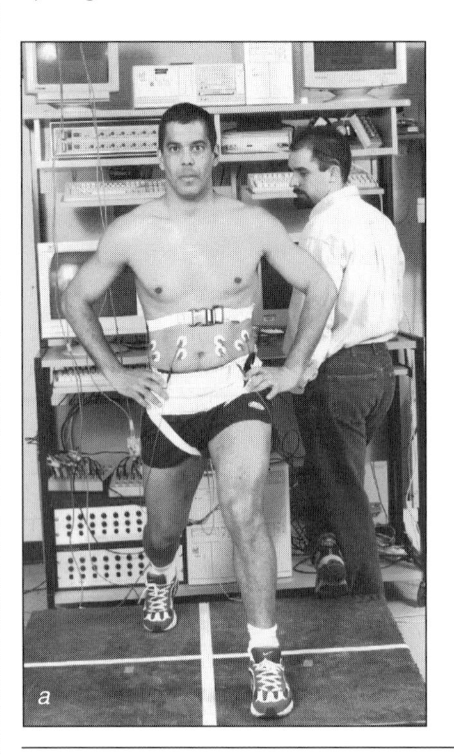

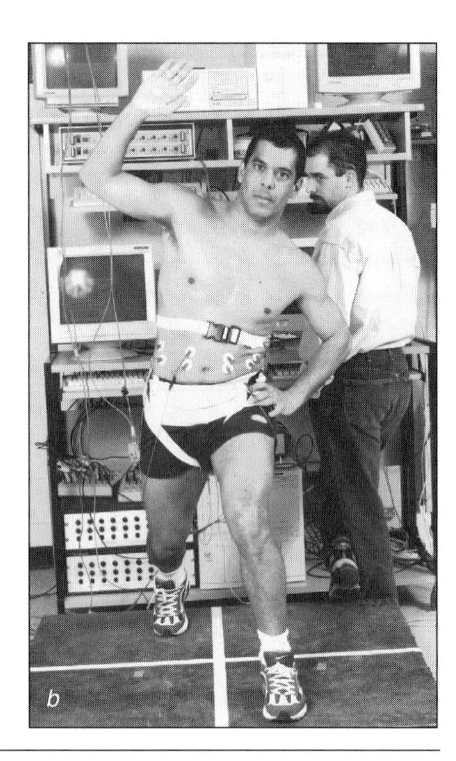

Figure 13.4 Given the uniarticular iliacus and the multiarticular psoas that crosses the hip and entire lumbar spine, a true psoas stretch requires the torso to be laterally bent away from the extended hip; this is an advanced stretch for the more robust backs.

Flossing the Nerve Roots for Those With Accompanying Sciatica

Before presenting the flossing technique for sciatica, some background is required regarding the biomechanics and biochemistry of the spinal cord and nerves.

Excessive tension in the nerve is the cause of sciatica. Normally, the cord nerve roots travel through the foramenae of the neural arch of each vertebra, and their sliding excursions are substantial with spine flexion, hip flexion, and knee extension. According to Louis (1981), the nerve can travel well over a centimeter (over 1/2 in.) during some of these movements. Thus, if a nerve root is impinged and cannot slide, any of these postures that would normally pull that nerve root through the foramen will increase nerve tension. Instead of moving, the nerve is stretched. Further, tension on such a nerve can be increased from the cranial end with simultaneous cervical flexion because the entire spinal cord moves slightly with cervical flexion and thus pulls at the nerve roots all along its length. This is the basis for the slump test for nerve tension presented in chapter 12.

Recently, scientists have suggested that nerves have the ability to create their own pathways as long as they can move. They seem to have some chemically based ability to dissolve—over time—tissues impinging them. The idea of "flossing" is to pull the cord and nerves from one end only while releasing at the other, and then to switch the pull/release direction. In this way the nerve roots are "flossed" through the lateral foramenae and in fact along their entire length. Thus, by working the nerves back and forth in whatever limited range they can manage in spite of impingement, we facilitate the dissolving of the impingement and the gradual release of the nerve to once again move freely (this approach was proposed by Butler, 1999). This flossing action is accomplished with coordinated hip, knee, and cervical motion.

▶ Flossing

The patient, seated and with the legs able to swing freely, flexes the cervical spine (a). This creates a pull on the spinal cord from the cranial end and a release from the caudal end. This should not produce sciatic symptoms (see the discussion of the slump test in chapter 12). Then, the patient extends the cervical spine with simultaneous knee extension on the side with sciatica (b). This pulls the nerve from the caudal end with a corresponding release at the cranial end. The cycle is completed as the knee is flexed with coordinated motion of the cervical spine flexing. This causes the nerves to floss through the vertebral tissues. If the patient experiences minor sciatic symptoms when the cervical spine is flexed or the knee is extended, then he should reduce the range of motion at these joints until no pain is provoked. Some patients who have suffered sciatica pain for years report reductions in their sciatic symptoms within a few days to a couple of weeks; others report increasing symptoms.

A note of caution is needed here: while this can be wonderful in helping chronic sciatica resolution, it can also cause an acute onset. Be *very* conservative in the first session. If the nerve is adhered so that it can't slide, the sciatic symptoms will be exacerbated. However, if the patient reports no change or even relief the next day, then proceed to increase the flossing. We have not been able to uncover a procedure that predicts who will do well or who will be exacerbated from this flossing procedure. Monitor all patients and remove the procedure from the programs of any patients whose symptoms worsen.

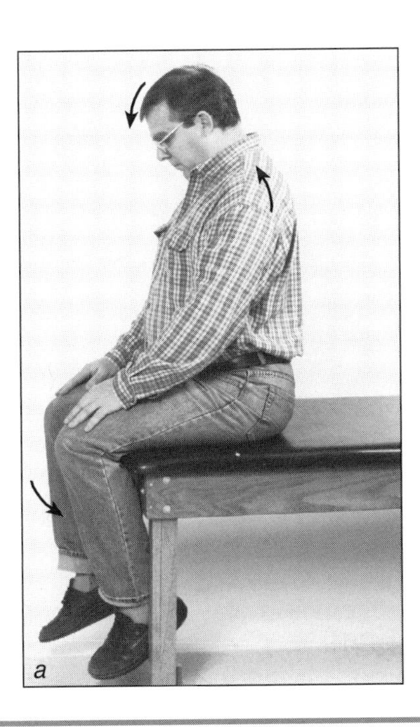
a

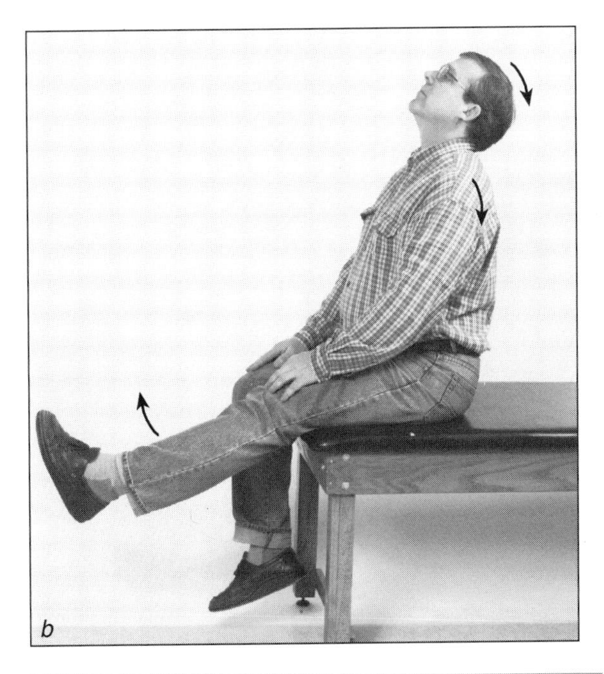
b

Exercises for Training the Stabilizing Muscles

After a brief review of the muscles that stabilize the spine, we will have a look at exercises for training those muscles.

What Are the Stabilizers of the Lumbar Torso?

While many muscles have been regarded as primary spine stabilizers, confirmation of their role requires two levels of analysis. First, engineering—stability analysis must be conducted on anatomically robust spine models to document the ability of each component to stiffen and stabilize. Second, electromyographic recordings of all muscles (even deep muscles requiring intramuscular electrodes) are necessary to confirm the extent to which the motor control system involves each muscle to ensure sufficient stability.

Our intramuscular and surface EMG and modeling studies to quantify spine stability demonstrated that virtually all torso muscles play a role in stabilization. (Some data was presented in chapter 6.) While multifidus, the other extensors, and the abdominal wall all appear to play significant roles, the quadratus lumborum also appears to play an important role in many tasks. The fibers of quadratus lumborum cross-link the vertebrae, they have a large lateral moment arm via the transverse process attachments, and they traverse between the rib cage and iliac crests. In this way, the quadratus buttresses shear instability and appears to be effective in all loading modes, by its architectural design. Typically, the first mode of lumbar buckling is lateral; the quadratus appears to play a significant role in local lateral buttressing. The three layers of the abdominal wall are also important for stability together with the muscles that attach directly to vertebrae—the multisegmented longissimus and iliocostalis and the unisegmental multifidii. Cholewicki and McGill (1996) also presented an argument for the role of the small intertransversarii in producing small but critical stabilizing forces. On the other hand, psoas activation appears to have little relationship with low back demands; the motor control system activates it when hip flexor moment is required (see Andersson et al., 1996, and Juker et al., 1998), compromising its role to stabilize the spine when hip flexion is not involved.

So which are the wisest ways to challenge and train these identified stabilizers? This question will be answered in the following sections.

Training the Stabilizers of the Lumbar Spine: The "Big Three"

Quantitative data have confirmed that no single abdominal exercise challenges all of the abdominal musculature while sparing the back (Axler and McGill, 1997). For this reason, more than one single exercise is required. Unfortunately, many inappropriate exercises have often been prescribed for people with low back difficulties, including the following:

- Sit-ups (both straight-leg and bent-knee) are characterized by higher psoas activation, with consequent high low back compressive loads that exceed NIOSH occupational guidelines.
- Leg raises cause even higher psoas activation and spine compression (actual values were presented in chapter 4).

Upper and Lower Rectus Abdominis?

Myoelectric evidence, normalized and calibrated, suggests that there is no functional distinction between an "upper" and "lower" rectus abdominis in most people; in contrast, the obliques are regionally activated with upper and lower neuromuscular compartments as well as medial and lateral components.

In our investigations we made several relevant observations regarding psoas activation during abdominal exercises. The challenge to the psoas is lowest during curl-ups, followed by higher levels during the horizontal side bridge. Bent-knee sit-ups were characterized by larger psoas activation than straight-leg sit-ups, and the highest psoas activity was observed during leg raises and hand-on-knee flexor isometric exertions. The often-recommended "press-heels" sit-up, which has been hypothesized to activate hamstrings and neurally inhibit psoas, was actually confirmed to increase psoas activation! (See figure 13.5.) (Original data can be found in Juker et al.,1998; some clinicians and coaches who intentionally wish to train psoas will find this data informative.) Once again, the horizontal side support appears to have merit as it challenges the lateral obliques and transverse abdominis without high lumbar compressive loading and ensures a high stability index (a loss of contraction of the involved muscles would cause the patient to fall out of the bridge).

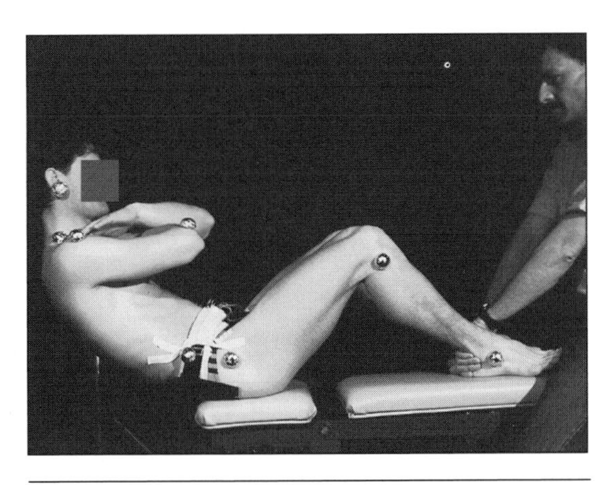

Figure 13.5 The press-heels sit-up was proposed by several clinical groups, on theoretical grounds, to inhibit psoas by activating the hamstrings. In fact, electromyographic assessment (Juker et al., 1998) proved this to be mythical. This original photo from the intramuscular experiments shows the clinicians' hands behind the heels of the subject as they activate hamstrings during the sit-up. Activating the hamstrings creates a hip extensor moment, and sit-ups require hip flexion. During this type of sit-up, the psoas is activated to even higher levels to overcome the extensor moment from the hamstrings and produce a net flexor moment. This type of sit-up produced the highest level of psoas activation of any style of sit-up we quantified!

A wise choice for stabilization exercises in the early stages of training or rehabilitation, and for simple low back health objectives, would be the "big three" that we have quantified to sufficiently challenge muscle, spare the spine of high load, and ensure sufficient stability:

1. Curl-ups for rectus abdominis
2. Several variations of the side bridge for the obliques, transverse abdominis, and quadratus
3. Leg and arm extensions leading to progressions of the "birddog" for the many back extensors

The variation of each of these exercises must be chosen with the patient/athlete's status and goals in mind.

Training Rectus Abdominis

Calibrated intramuscular and surface EMG evidence suggests that the various types of curl-ups challenge mainly rectus abdominis since psoas and abdominal wall (internal and external oblique and transverse abdominis) activity is relatively low (see tables in chapter 4 for relative activation levels in a variety of exercise tasks). Curl-ups performed with poor technique, however, can be counterproductive, either failing to

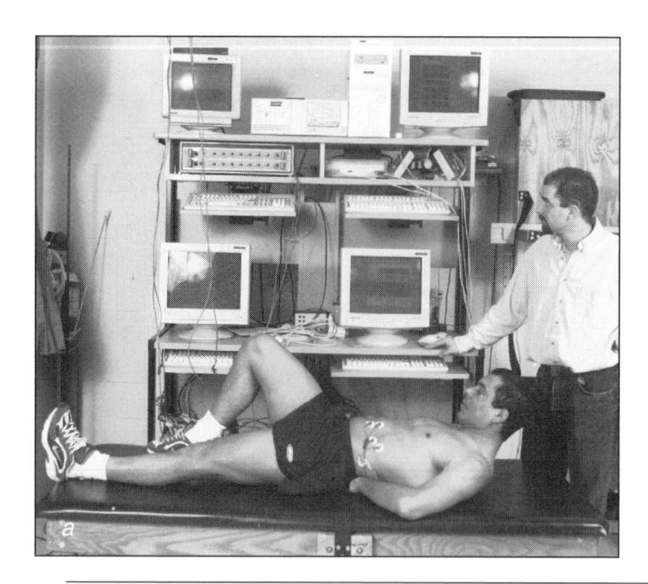

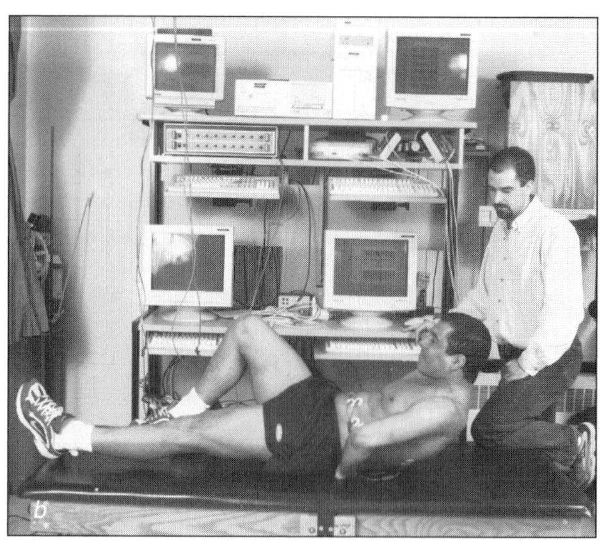

Figure 13.6 Poor form during the curl-up is to flex the cervical spine, loading the neck and not the rectus *(a)*. Correct technique focuses the rotation in the thoracic spine. Another common type of poor form is to elevate the head and shoulders a large distance off the floor *(b)*. This patient is elevating far too much, which is closer to replicating the much higher stresses of the sit-up. The intention is to activate rectus and not to produce lumbar spine motion.

activate the rectus abdominis sufficiently or overstressing the spine (see figure 13.6, a and b). Curl-ups with a twisting motion are expensive in terms of lumbar compression due to the additional oblique challenge. Higher oblique activation with lower spine load is accomplished with the side bridge, which is therefore preferred over twisting curl-ups for training the obliques. This will be presented in the next section. The highest-level curl-up will be presented in chapter 14.

► Curl-Up, Beginner's

The curl-up technique is critical to spare the spine. The basic starting posture is supine with the hands supporting the lumbar region. Do not flatten the back to the floor, which takes the spine out of elastic equilibrium and raises the stresses in the passive tissues. While the position of elastic equilibrium is desired in the lumbar region, the hands can be adjusted to minimize pain if needed. One leg is bent with the knee flexed to 90° while the other leg remains relaxed on the floor. This adds further torque to the pelvis to prevent the lumbar spine from flattening to the floor. The focus of the rotation is in the thoracic spine; many tend to flex the cervical spine, which is poor technique. Rather, picture the head and neck as a rigid block on the thoracic spine. No cervical motion should occur, either chin poking or chin tucking. Individuals who report neck discomfort may try the isometric exercises for the neck that follow. In addition, particularly for patients experiencing neck discomfort, the tongue should be placed on the roof of the mouth behind the front teeth, which helps to promote stabilizing neck muscle patterns. Leave the elbows on the floor while elevating the head and shoulders a short distance off the floor. The rotation is focused in the midthoracic region. The head/neck is locked onto the rib cage. No cervical motion should occur—either chin poking or chin tucking. The intention is to activate rectus and not to produce spine motion.

A very mild form of the curl-up is to just take the weight of the head and shoulders with almost no motion (a), while more challenge is obtained with raising the head and shoulders but focusing the motion to the thoracic spine (b) with no cervical or lumbar flexion.

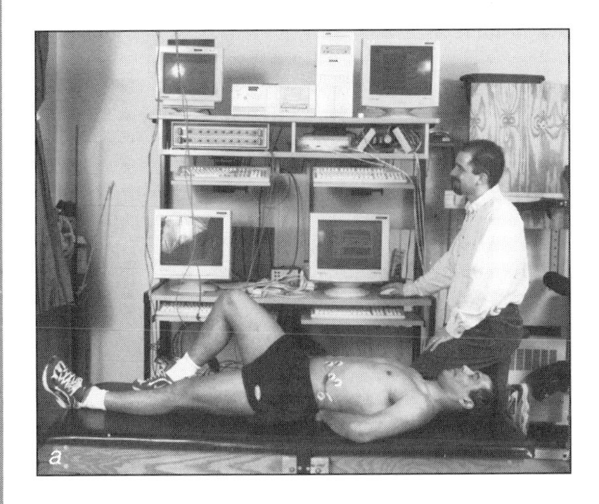
a

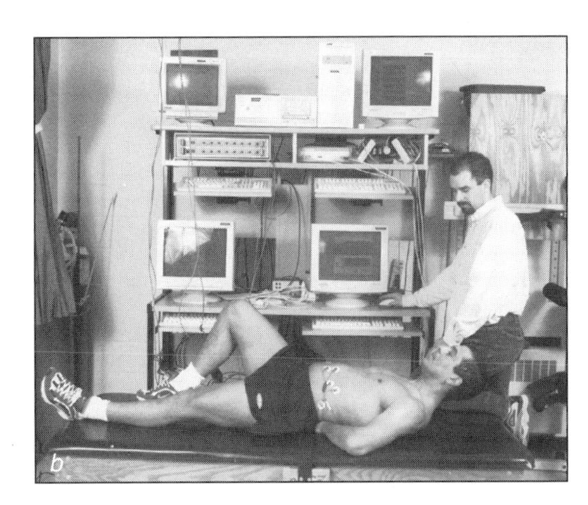
b

▶ Isometric Exercises for the Neck

Those who experience neck symptoms with curl-ups may find relief by building the neck with isometric exercises. In all of these exercises the head/neck does not move, and the tongue is placed on the roof of the mouth behind the front teeth: (a) the hands are placed on the forehead, which resists neck flexion effort; (b) the hand is placed on the side of the head to resist cervical side flexion effort, and then repeated on the other side; (c) the hands are placed on the back of the head to resist cervical extension effort. Hold for several seconds then relax, building up endurance and grooving stabilizing motor patterns by increasing the repetitions of the hold/relax cycles. (Believe it or not, some patients try this while chewing gum, which makes the grooving of stabilizing patterns impossible.)

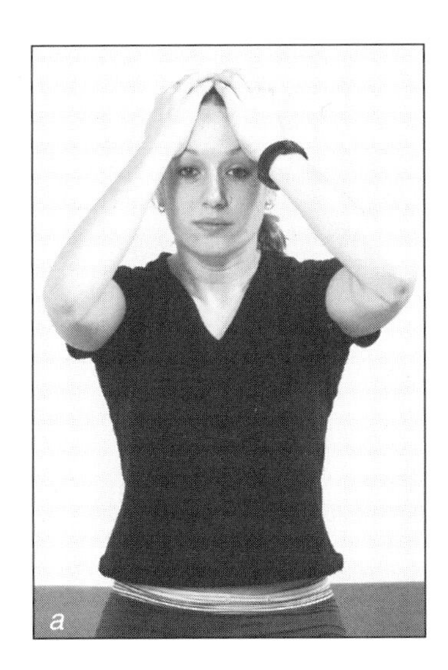
a

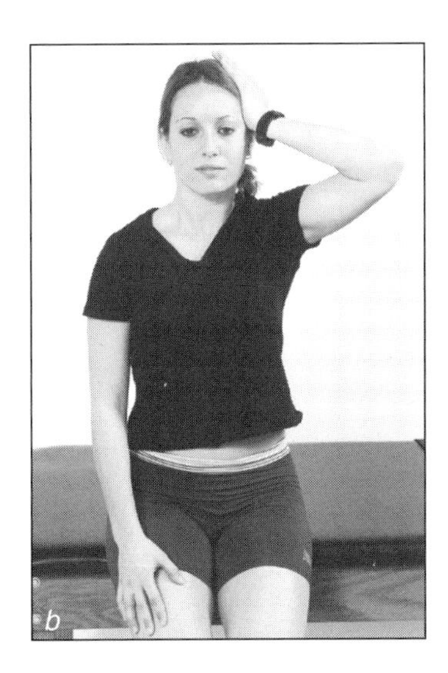
b

c

▶ Curl-Up, Intermediate

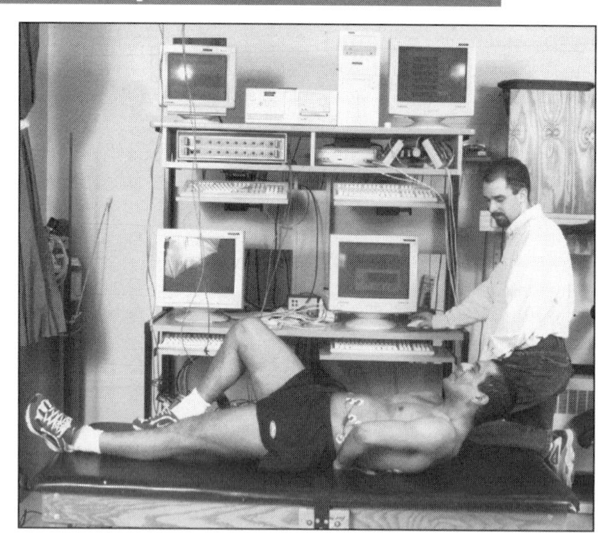

The intermediate progression of the curl-up is to raise the elbows a couple of centimeters so that the arms do not pry the shoulders up, thus shifting more load to the rectus. Do not raise the head/neck any higher than in the beginner's curl-up.

▶ Curl-Up, Advanced

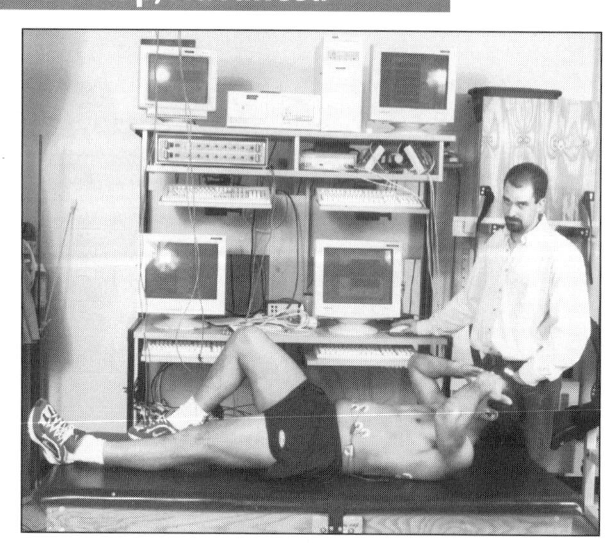

An advanced curl-up requires placing the fingers lightly on the forehead (never behind the head, which causes the tendency to pull and flex the cervical spine), as long as lumbar posture is controlled and no pressure is applied to the head by the hands. Note: The average patient need not go to this level if spine health is the objective. The head and neck must move as a unit, maintaining their rigid-block position on the thoracic spine.

Training the Quadratus Lumborum, Lateral Obliques, and Transverse Abdominis

Given the architectural and electromyographic evidence for the quadratus lumborum, transverse abdominis, and abdominal obliques as spine stabilizers, the optimal technique to maximize activation but minimize the spine load appears to be the side bridge. Abdominal bracing is emphasized in all forms of this exercise. Maintaining the bridge ensures constant muscle activation while the brace introduces new combinations of muscle recruitment to ensure stability. It is almost impossible for the spine to become unstable while performing a side bride with a neutral spine.

▶ Side Bridge, Remedial

Many special patient cases deserve consideration—for example, the chronic patient who is quite deconditioned. These patients are sometimes unable to perform the side bridge even from the knees. Start these people with a side bridge while standing against a wall. Instruct them to move smoothly from the beginning position (a) through the intermediate positions (b-d) to the final position (e), in one flowing motion. The patient should pivot over the toes, as if performing a fluid dance move.

Once patients graduate to the floor, they can spare the spine by beginning with the knees and hips quite flexed (f) and moving into the side-bridge posture with an accordion-like unfolding of the legs, all the time keeping a braced neutral spine (g).

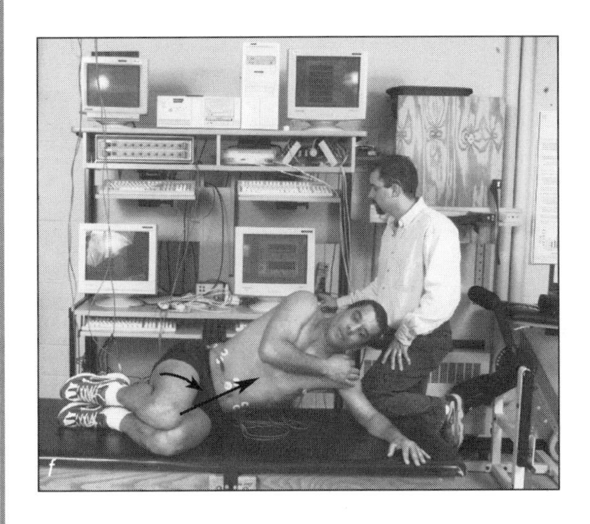

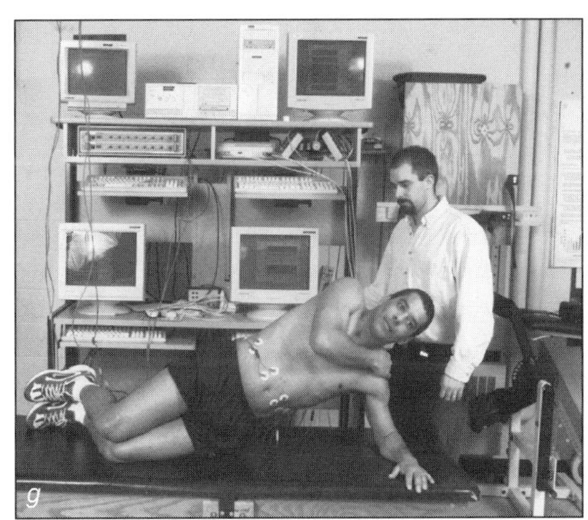

Another type of patient is the football player whose shoulders are so painful he can not tolerate the shoulder load (some elderly women also fall into this patient group). These patients can perform a modified side bridge by lying on the floor and attempting to raise the legs laterally (h and i) or simply attempting to take the weight off the legs.

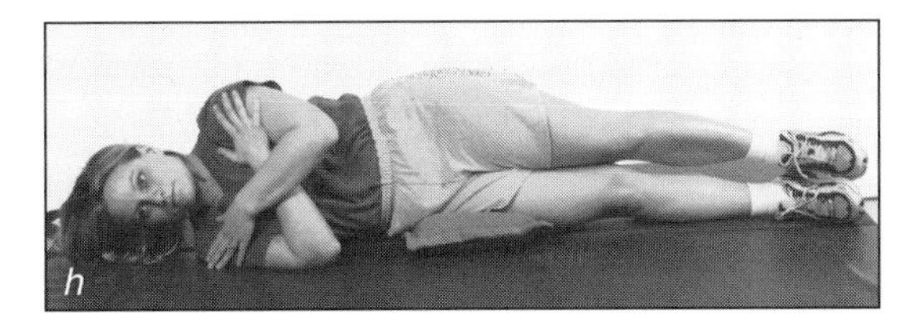

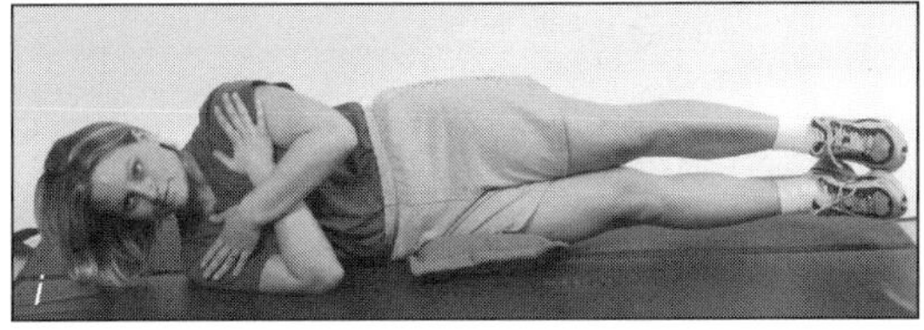

(continued)

Side Bridge, Remedial *(continued)*

Another option for patients with shoulders that cannot tolerate load is to stand on a 45° bench with the feet anchored (j), which spares the shoulders.

Using other devices, the torso can either be elevated from the pad (k and l) or bridged with the shoulder (m).

j

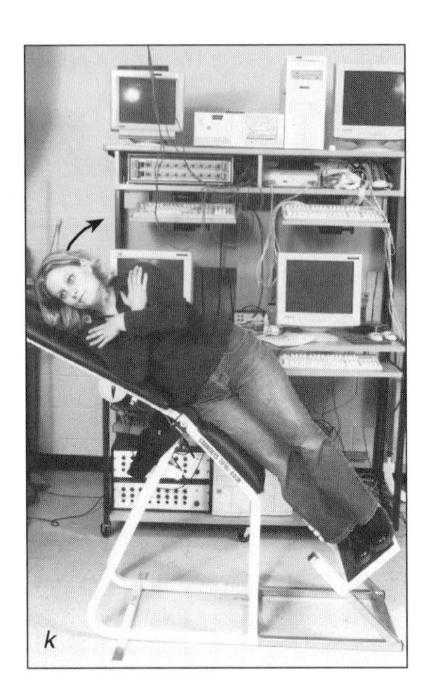
k

l

m

▶ Side Bridge, Beginner's

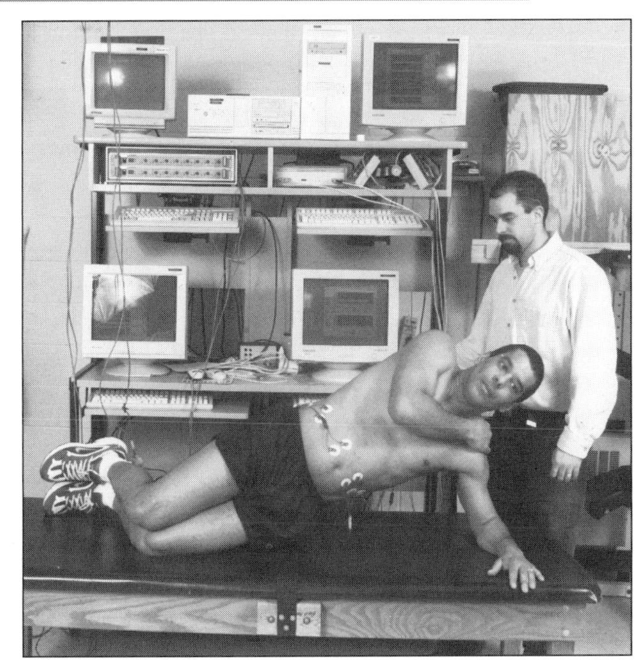

Beginners bridge from the knees. In the beginning position, the exerciser is on his side, supported by his elbow and hip. The knees are bent to 90°. Placing the free hand on the opposite shoulder and pulling down on it will help stabilize the shoulder. The torso is straightened until the body is supported on the elbow and the knee, with some input from the lower leg. The beginner's side bridge can be slightly advanced by placing the free arm along the side of the torso—effectively placing more load on the bridge.

▶ Side Bridge, Intermediate

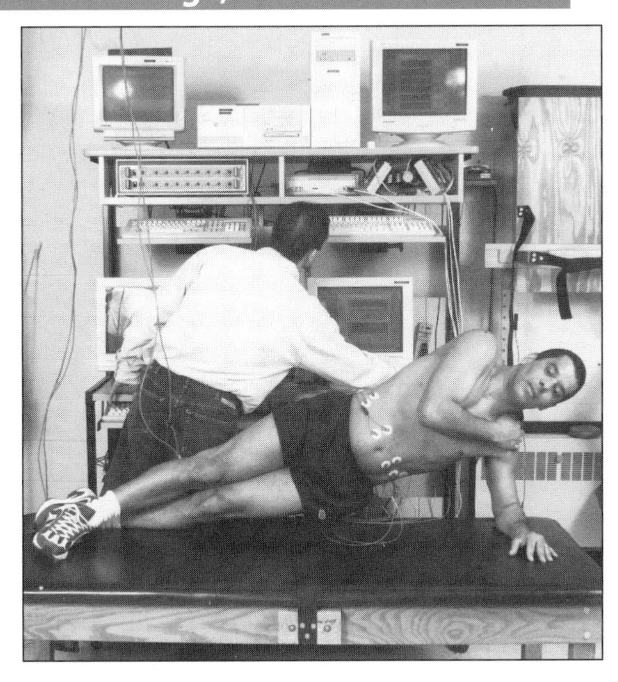

The beginning position for the intermediate side bridge is like that for the beginner, except that the legs are straight. The torso is straightened until the body is supported on the elbow and feet. (See figure.) When supported in this way, the lumbar compression is a modest 2500 N, but the quadratus closest to the floor appears to be active up to 50% of MVC (this is a preferred exercise for the obliques since they experience similar levels of activation).

▶ Side Bridge, Intermediate Variation

Placing the upper leg/foot in front of the lower leg/foot will enable longitudinal "rolling" of the torso to challenge both anterior (a) and posterior (b) portions of the abdominal wall.

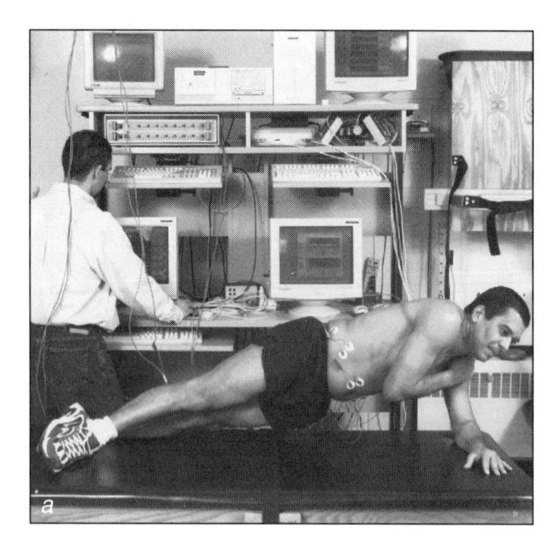

a

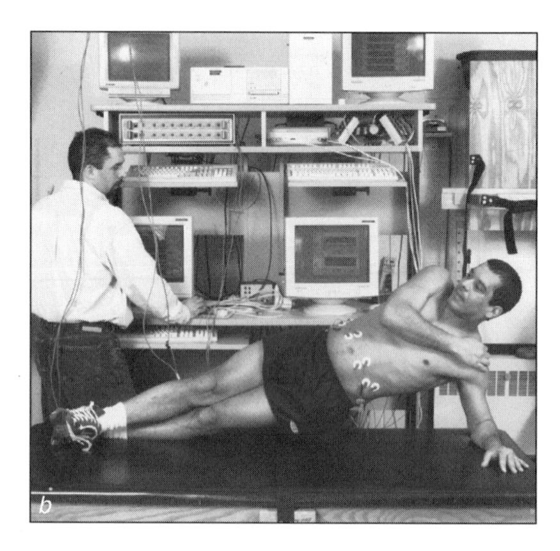

b

▶ Side Bridge, Advanced

Advanced technique to enhance the motor challenge of the side bridge is to transfer from one elbow to the other while abdominally bracing (a, b, and c) rather than repeatedly hiking the hips off the floor into the bridge position. Ensure that the rib cage is braced to the pelvis and that this rigidity is maintained through the full roll from one side to the other. Still higher levels of activation would be reached with the feet on a labile surface (Vera-Garcia, Grenier, and McGill, 2000).

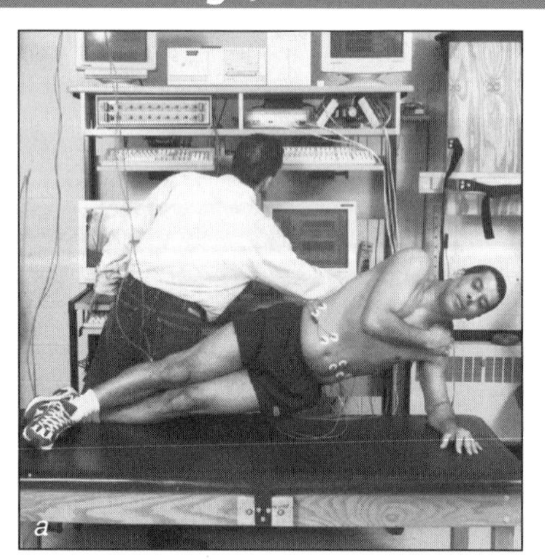

a

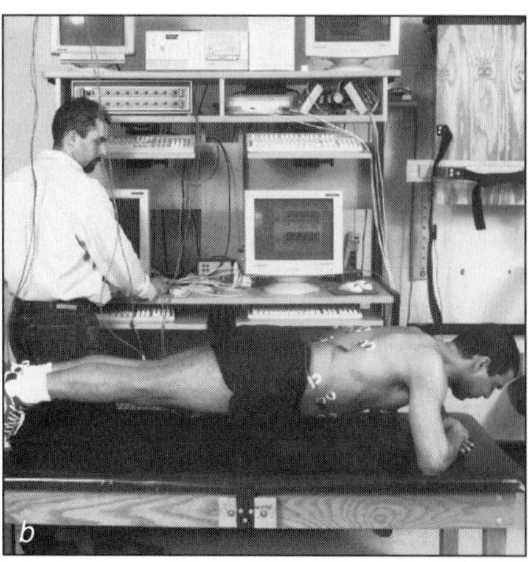

b

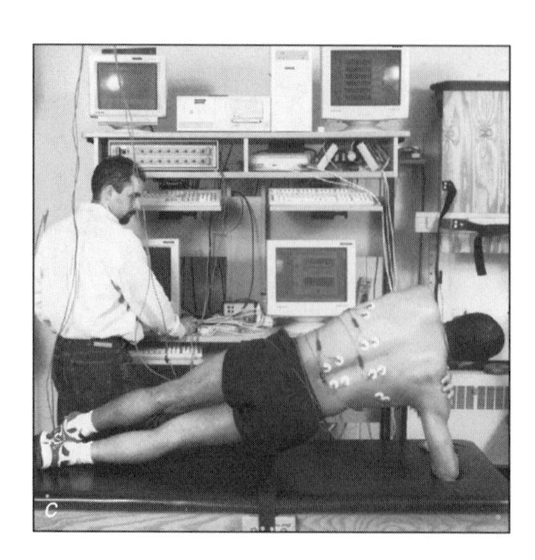

c

Training the Back Extensors

Most traditional extensor exercises are characterized by very high spine loads, which result from externally applied compressive and shear forces (either from free weights or resistance machines). A commonly prescribed spine extensor muscle challenge involves lying prone while extending the arms and legs (see figure 13.7). This results in over 6000 N of compression to a hyperextended spine, transfers load to the facets, and crushes the interspinous ligament. Needless to say, this exercise is contraindicated for anyone at risk of low back injury—or re-injury! Although some may believe that putting the hands on either side of the head rather than extending them may make this exercise safe, that is not true. It should be not be done in any form. Further, recall that the mechanism for disc herniation is reproduced by back machines that take the lumbar spine from full flexion and through the range of motion under load from muscle contraction.

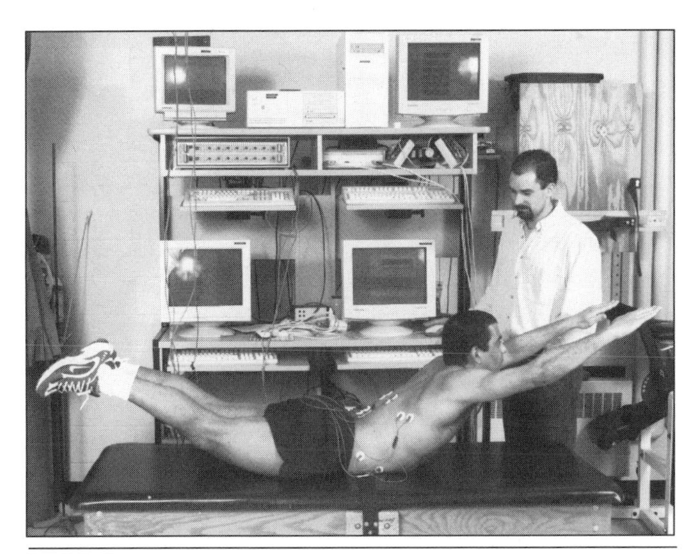

Figure 13.7 A commonly prescribed spine extensor muscle challenge involves lying prone while extending the arms and legs. This results in over 6000 N of compression to a hyperextended spine. It is a poorly designed exercise.

In our search for methods to activate the extensors (including longissimus, iliocostalis, and multifidii) with minimal spine loading, we have found that the single leg extension hold results in tolerable spine loading (<2500 N) for many and activates one side of the lumbar extensors to approximately 18% of MVC. Simultaneous leg extension with contralateral arm raise (the "birddog") increases the unilateral extensor muscle challenge (approximately 27% MVC in one side of the lumbar extensors and 45% MVC in the other side of the thoracic extensors) but also increases lumbar compression to well over 3000 N (Callaghan, Gunning, and McGill, 1998). This exercise can be enhanced with abdominal bracing and deliberate mental imaging of activation of each level of the local extensors. Once again, technique for challenging the extensors should be guided by the patient's status and goals. This is an individual clinical decision.

Clinicians should remember to emphasize abdominal bracing and a neutral spine throughout all versions of this exercise. Poor form includes "hip hiking" or any other configuration that causes deviation (twist, flexion, or lateral bending) to the spine (see figure 13.8, a-b).

Figure 13.8 Common mistakes made in the birddog are "hiking" the hip, which twists the spine *(a)* or not achieving a neutral spine—a flexed posture is seen in *(b)*.

▶ Birddog, Remedial

The starting position is on the hands and knees with the hands under the shoulders and the knees directly under the hips (a). For the patient with a very deconditioned back, this exercise involves simply lifting a hand or knee about an inch off the floor (b). After the patient is able to raise a hand or knee without pain, he can progress to raising the opposite hand and knee simultaneously.

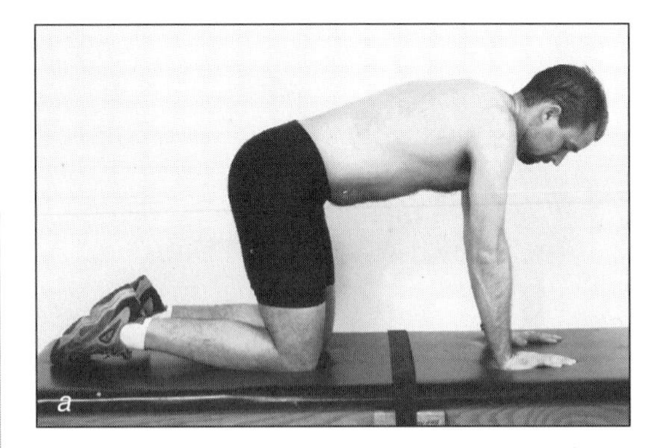
a

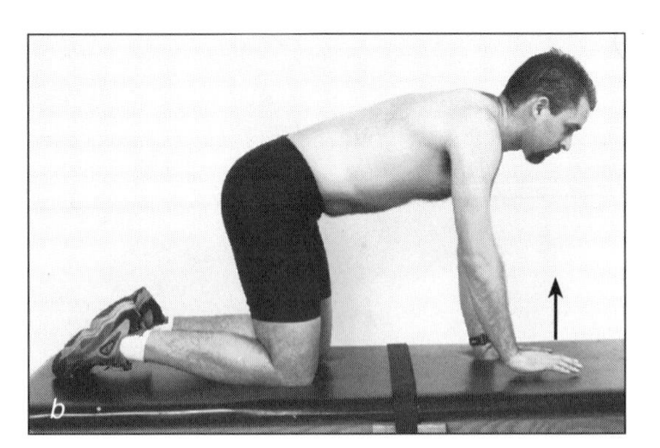
b

▶ Birddog, Beginner's

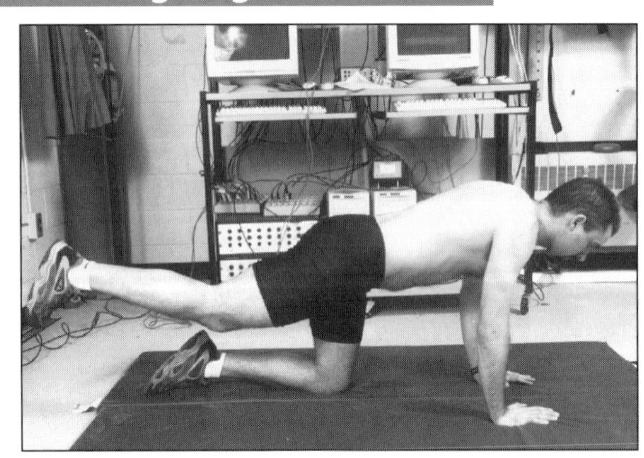

The progression continues with raising one leg or one arm at a time.

▶ Birddog, Intermediate

The intermediate birddog is achieved when the patient is able to raise the opposite arm and leg simultaneously. He should avoid raising either the arm or the leg past horizontal. The objective is to be able to hold the limbs parallel to the floor for about six to eight seconds. Good form includes a neutral spine and abdominal bracing.

▶ Birddog, Advanced

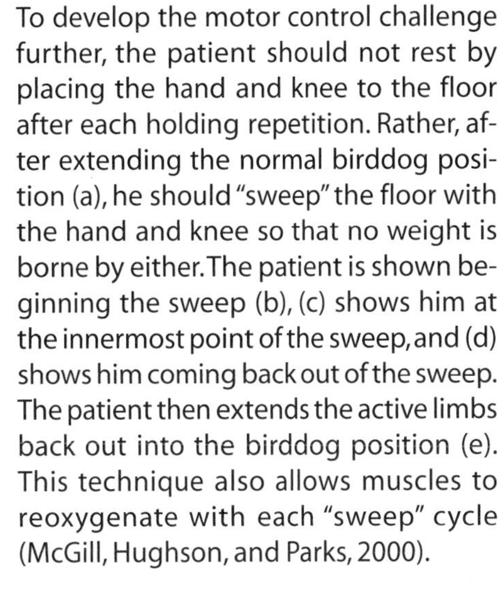
To develop the motor control challenge further, the patient should not rest by placing the hand and knee to the floor after each holding repetition. Rather, after extending the normal birddog position (a), he should "sweep" the floor with the hand and knee so that no weight is borne by either. The patient is shown beginning the sweep (b), (c) shows him at the innermost point of the sweep, and (d) shows him coming back out of the sweep. The patient then extends the active limbs back out into the birddog position (e). This technique also allows muscles to reoxygenate with each "sweep" cycle (McGill, Hughson, and Parks, 2000).

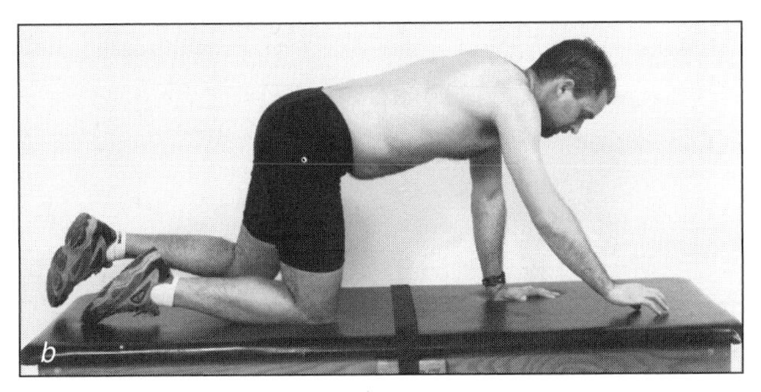

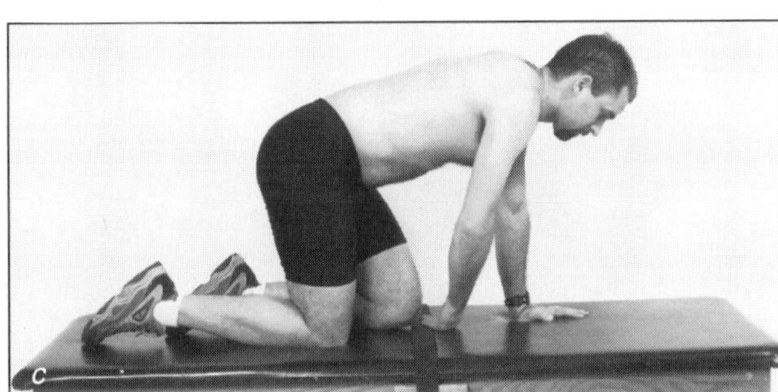

Beginner's Program for Stabilization

Now that you are familiar with some specific recommended low back exercises, how do you put these together to create the best possible program for your patient? We recommend considering the following sequence:

1. Begin with the flexion/extension cycles, also called the cat/camel motion (see figure 13.9, a and b) to reduce spine viscosity. Note that the cat/camel is intended as a motion exercise, not a stretch, so the emphasis is on motion rather than "pushing" at the end ranges of flexion and extension. We have found that five or six cycles are often sufficient to reduce most viscous stresses—additional cycles rarely reduce viscous friction further. Those with sciatica may find increased symptoms during the flexion phase. Use pain to guide the suitable pain-free range of motion.
2. Slow lunges, with an upright and braced torso, satisfy the objectives of sparing the spine while achieving hip and knee endurance and mobility (see figure 13.1).
3. Sciatics may try the nerve-flossing technique.
4. These motions are followed by anterior abdominal exercises, namely, appropriate curl-ups.
5. Lateral musculature exercises follow, namely, the side bridge for quadratus lumborum and the muscles of the abdominal wall for optimal stability.
6. The extensor program consists of leg extensions and the birddog.

In general, we recommend that the isometric holds performed in the curl-ups, bridges, and birddogs be no longer than seven or eight seconds, given recent evidence from near infrared spectroscopy, indicating rapid loss of available oxygen in torso muscles contracting at these levels. Short relaxation of the muscle restores oxygen (McGill, Hughson, and Parks, 2000). The endurance objectives are achieved by building up repetitions of the exertions rather than by increasing the duration of each hold.

Motivated by the evidence for the superiority of extensor endurance over strength as a benchmark for good back health, we recently documented normal ratios of endurance times for the torso flexors relative to the extensors and lateral musculature (see chapter 11). Use these values to identify endurance deficits—both absolute values

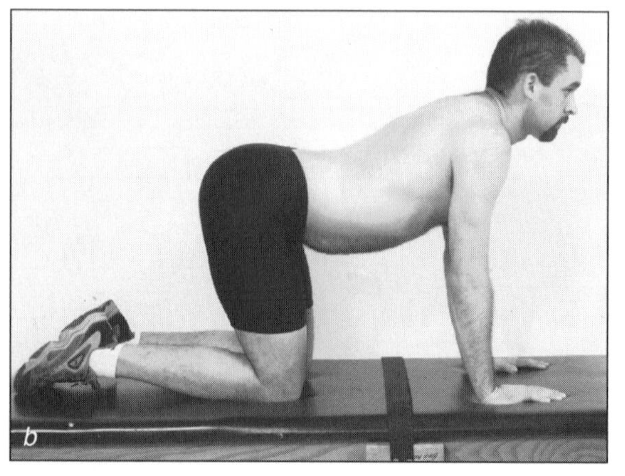

Figure 13.9 Cat/camel exercise. Note that this is a motion exercise and not a stretch; do not push at the end range of motion. Viscosity is measurably reduced after just a few cycles.

and for one muscle group relative to another—and to establish reasonable endurance goals for your patients.

References

Andersson, E.A., Oddsson, L., Grundstrom, O.M., Nilsson, J., and Thorstensson, A. (1996) EMG activities of the quadratus lumborum and erector spinae muscles during flexion-relaxation and other motor tasks. *Clinical Biomechanics*, 11: 392-400.

Axler, C., and McGill, S.M. (1997). Low back loads over a variety of abdominal exercises: Searching for the safest abdominal challenge. *Medicine and Science in Sports and Exercise,* 29 (6): 804-811.

Biering-Sorensen, F. (1984) Physical measurements as risk indicators for low-back trouble over a one-year period. *Spine,* 9: 106-119.

Butler, D.S. (1999) Mobilization of the nervous system. Edinburgh, Scotland: Churchill Livingstone.

Callaghan, J.P., Gunning, J.L., and McGill, S.M. (1998) Relationship between lumbar spine load and muscle activity during extensor exercises. *Physical Therapy*, 78 (1): 8-18.

Cholewicki, J., and McGill, S.M. (1996) Mechanical stability of the in vivo lumbar spine: Implications for injury and chronic low back pain. *Clinical Biomechanics*, 11 (1): 1-15.

Grenier, S., Preuss, R., Russell, C., and McGill, S. (in press) Deficits in a wide variety of fitness variables linger after back trouble episodes.

Juker, D., McGill, S.M., Kropf, P., and Steffen, T. (1998). Quantitative intramuscular myoelectric activity of lumbar portions of psoas and the abdominal wall during a wide variety of tasks. *Medicine and Science in Sports and Exercise,* 30 (2): 301-310.

Louis, R. (1981) Vertebroradicular and vertebromedullar dynamics. *Anatomica Clinica*, 3: 1-11.

Luoto, S., Heliovaara, M., Hurri, H., and Alaranta, M. (1995) Static back endurance and the risk of low back pain. *Clinical Biomechanics*, 10: 323-324.

McGill, S.M., Hughson, R., and Parks, K. (2000) Lumbar erector spinae oxygenation during prolonged contractions: Implications for prolonged work. *Ergonomics,* 43: 486-493.

Vera-Garcia, F.J., Grenier, S.G., and McGill, S.M. (2000) Abdominal response during curl-ups on both stable and labile surfaces. *Physical Therapy*, 80 (6): 564-569.

CHAPTER

Advanced Exercises 14

The beginner's program described in chapter 13 should be sufficient for daily spine health. Several situations, however, may call for further training:

- Certain occupational tasks with specific demands that require unique preparation and training
- Some athletic endeavors that demand higher challenges of low back training (although this is achieved with much higher risk of tissue damage from overload)
- Some patients who, as they progress with the isometric stabilization exercises described in chapter 13, want to continue increasing the challenge

This chapter will discuss how to increase low back challenges safely, how to address specific worker and athletic concerns regarding advanced training, and some ideas about the next steps to be taken in the study of low back stability.

Upon completion of this chapter you will be familiar with some guidelines to prescribe exercise that will enhance performance in a way that spares the spine. A more complete book on this topic is forthcoming from the author.

Safely Increasing Challenges

To optimize patient safety, the clinician must be aware of issues associated with training on labile surfaces and with machines, and how to progress the "big three" exercises described in chapter 13 (curl-ups, side bridges, and leg and arm extensions) to their highest level. Some guidance is provided here.

Labile Surfaces and Resistance Training Machines

Not all clinicians are aware of the factors they should take into account when deciding whether to prescribe exercise using labile surfaces or machines. Following are guidelines about both of these issues:

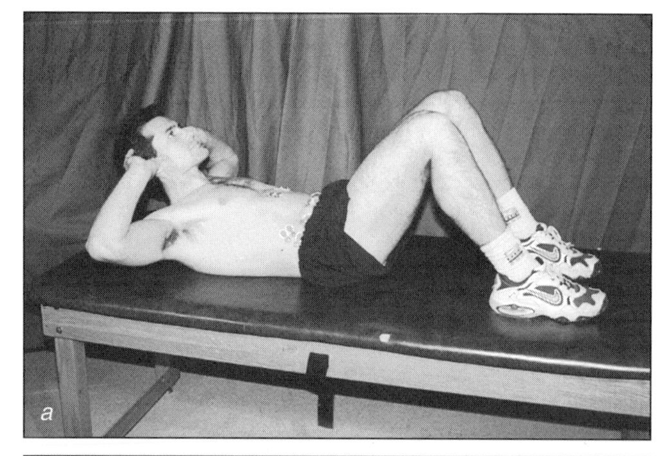

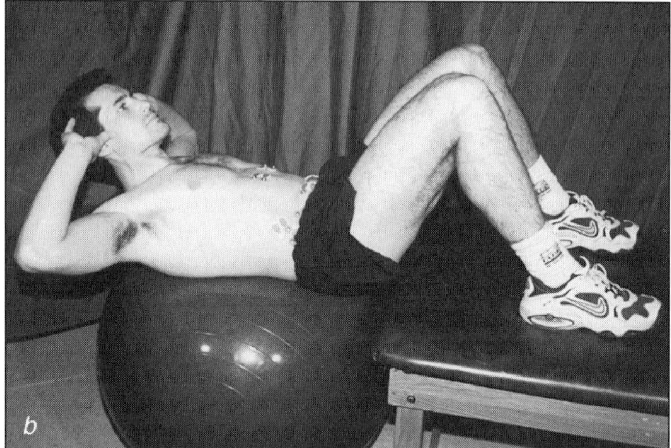

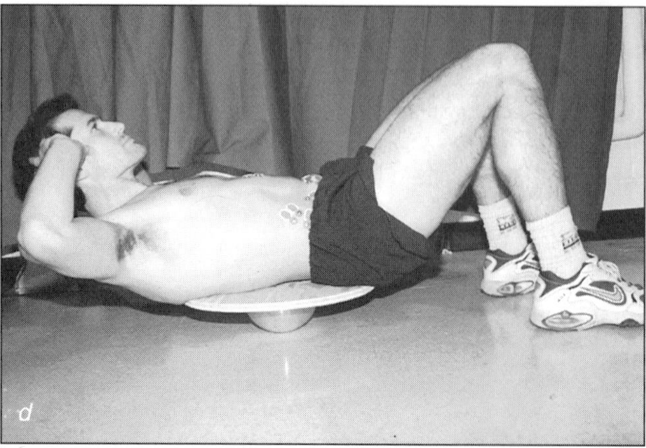

Figure 14.1 The simple curl-up was assessed for the effect of a labile surface on muscle activation patterns. The percent of maximal voluntary contraction (% MVC) in the rectus abdominis and external and internal obliques caused by simply moving from a stable surface *(a)* to varying types of labile surfaces *(b)*, *(c)*, and *(d)* is shown in figure 14.2.

• **Training with labile surfaces.** Challenges to the spine during daily activity include maintaining stability during static, steady-state postures; unexpected loading events; and planned dynamic or ballistic movement. This has motivated some clinicians to recommend exercising on labile surfaces such as gym balls. Certainly, these labile surfaces challenge the motor system to meet the dynamic tasks of daily living or specific athletic activities and can be very helpful for advanced training. But might this type of training be of concern for some patients? Our recent quantification of elevated spine loads and muscle coactivation when performing a curl-up on labile surfaces (Vera-Garcia, Grenier, and McGill, 2000) suggests that the rehabilitation program should begin on stable surfaces. In this case, we assessed the simple curl-up for the effect of a labile surface on muscle activation patterns (see figure 14.1, a-d). Simply moving from a stable surface to a labile surface caused much more cocontraction, which in many cases virtually doubled the spine load (see figure 14.2). The practice of placing patients on labile surfaces early in the rehabilitative program can delay improvement by causing exacerbating spine loads. We therefore suggest beginning exercises on a stable surface and establishing a positive slope to improvement. Introduce labile surfaces judiciously only once the patient has achieved spine stability and sufficiently restored load-bearing capacity, and can tolerate additional compression. This same principle can be extended to sitting. Sitting on a gym ball greatly elevates spine load through increased muscle coactivation. For this reason, nonpatients should avoid prolonged sitting on gym balls, and patients should use them only once they have achieved spine stability and increased load-bearing capacity. There is a time and a place for labile surfaces.

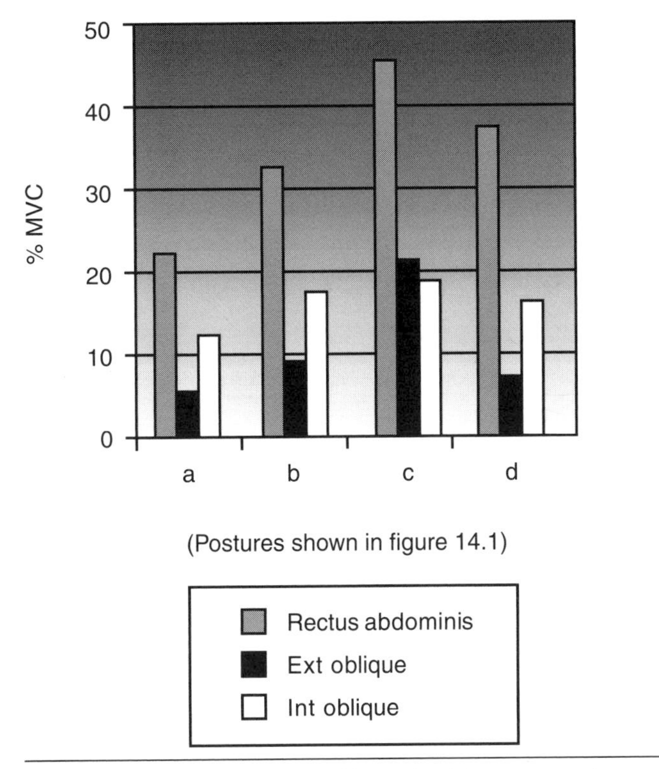

Figure 14.2 The % MVC caused by each of the postures shown in figure 14.1. A curl-up with the body over a ball and the feet on the floor (figure 14.1c) virtually doubles the abdominal muscle activation seen in a curl-up on a stable surface (figure 14.1a) and, correspondingly, the spine load. Note that the % MVC required of the three muscles studied is also much higher in curl-ups with the body over a ball and the feet on a bench (figure 14.1b) and with the body on a wobble board (figure 14.1d) than on a stable surface. Clearly, a gym ball can be wonderful for advanced training but is contraindicated for many patients.

Figure 14.3 The typical back extensor devices will impose generally twice the muscle loading of the birddog since both right and left sides of the lumbar and the thoracic extensors are activated. These may be necessary for some types of performance training, but they are ill-advised for most patients.

- **Training with machines and equipment.** Generally, the goal of establishing stabilizing motor patterns requires the individual to support body weight and coordinate the stabilization of all joints involved in the task. In other words, it is a whole-body, even whole-person, endeavor. Many training machines, on the other hand, are made to isolate a specific joint. Since neither workers nor athletes perform their tasks in this artificially stabilized manner, these types of motor patterns may not be transferable and, worse, may cause inappropriate grooving of motor/motion patterns. The only time machines that isolate joints can be helpful is when an injury to a specific body part requires its protection during rehabilitative training. Some incorrectly believe that isolating a joint somehow reduces the loading and thus reduces the risk. Consider, for example, the typical extensor benches or Roman chair, which generally imposes twice the muscle loading than results from performing the birddog since both right and left sides of the lumbar and the thoracic extensors are activated (see figure 14.3). Patients who must use machines should consider certain back-sparing techniques. The leg press rack, for example, sometimes causes the pelvis to rotate away from the back rest when the weight is lowered (see figure 14.4). The resultant lumbar flexion produces herniating conditions for the disc! If patients must use this exercise, we suggest they use just one leg at a time to ensure that the pelvis remains in contact with the back pad and that they preserve a neutral spine (see figure 14.5). Having stated this, however, we would never attempt to groove these motor patterns in any of our programs; they are nonfunctional except for leg-pushing a weight up a ramp!

For the purposes of this discussion, cables to weight stacks are not considered machines since they do not isolate joint motion and add resistance to whole-body motion patterns. One particular exercise that enhances the ability to stabilize the spine is the cable pulldown. This

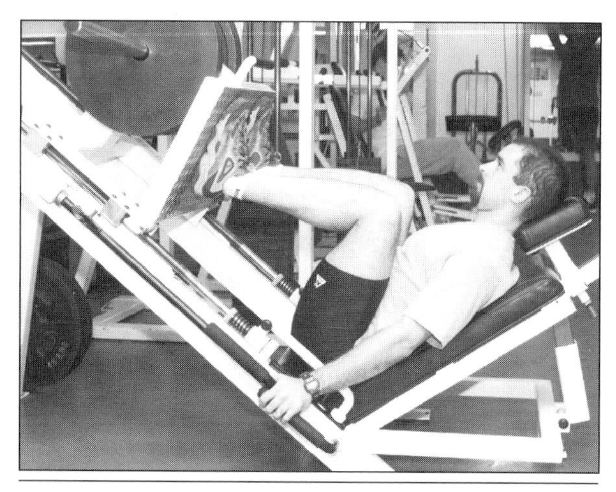

Figure 14.4 For back patients who have compromised hip flexion, lowering the weight sometimes causes the pelvis to rotate away from the back rest. This also can occur if the weight is lowered too far. This produces herniating conditions for the disc!

Figure 14.5 For patients who must use the leg press exercise, we suggest they use just one leg at a time to ensure that the pelvis remains in contact with the back pad and that they preserve a neutral spine. (Note that we still consider this a nonfunctional motor/motion pattern).

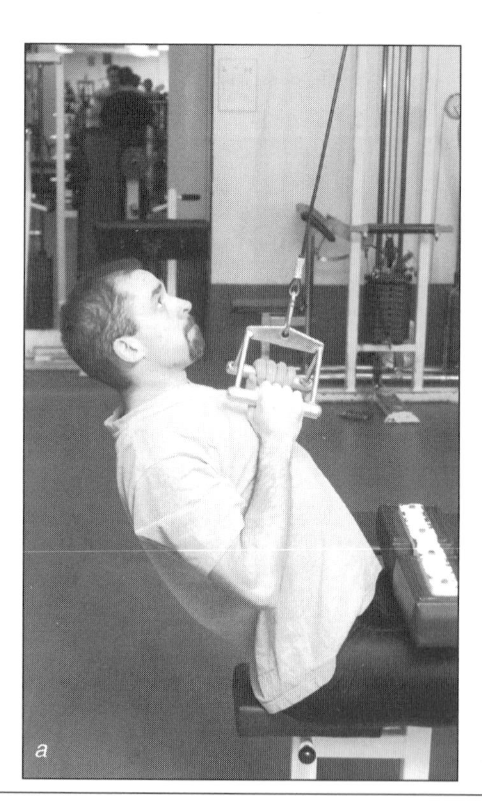

Figure 14.6 Two variations, *(a)* and *(b)*, of the cable pulldown exercise should be performed with the handlebar lowered to the chest rather than to the back. This enhances the role of several spine extensors and in particular the latissimus dorsi for lumbar stabilization.

exercise, when the handlebar is brought down to the chest (rather than the back, which is traditional technique) challenges the latissimus dorsi and other stabilizers (see figure 14.6, a-b). As we will see in a subsequent section, these types of exercises form a part of a performance-based program.

Safely Progressing Back Exercises

The curl-ups, side bridges, and birddogs pictured in chapter 13 can be made more challenging and yet still incorporate the safety features noted. After explaining these advanced forms of the big three exercises, I will present an additional advanced back exercise.

▶ Curl-Up, Highest Level

Once the patient has mastered the advanced curl-up described in chapter 13, she may be instructed to try the following:

1. Brace the abdomen (if necessary, review the instructions on pages 210-213).
2. Curl up against the brace, but do not curl up any higher than in other forms of the exercise.
3. Perform deep breathing while in the "up" curl-up position and while maintaining bracing. Remain in the up position long enough to take a few deep breaths.

This level of curl-up will be challenging for the toughest NFL linebacker!

▶ Side Bridge, Highest Level

The side bridge can be more challenging by incorporating the roll described in chapter 13 (page 252), together with dynamically contracting and relaxing the abdominal wall while sustaining position.

▶ Birddog, Highest Level

The birddog is more demanding when the patient consciously stretches the hand out forward and the foot out behind—all the while ensuring that spine motion (particularly bending) does not take place. If performance objectives are desirable and if the patient has no exacerbation of symptoms, small wrist and ankle weights may be allowed.

▶ Advanced Exercises for Back Strength and Endurance

When searching for ways to obtain maximum myoelectric activity from the back extensor muscles, we discovered that isometric back extensor exercises do not recruit the full pool of motor units. Many more motor units fire with some extensor motion. As a result of this discovery, this back extensor exercise, which builds strength and endurance and helps to hypertrophy the muscles, was developed.

While the athlete is lying prone on a bench with the torso supported on a movable stiff pad, place weight in one hand. The edge of the pad is placed under the mid ribcage. The cantilevered portion of the spine is slightly flexed and then extended, combined

(continued)

Advanced Exercises for Back Strength and Endurance *(continued)*

with some slight twist back to neutral (a). The object is not to actually twist but to focus activation on one side of the extensors—to use mental imagery (chapter 11) to assist in activating the maximum number of available motor units. The spine never extends past neutral. After a set, place the weight in the athlete's other hand and have the athlete repeat the exercise. Then move the pad downward (perhaps about the level of the navel) so that a greater portion of the torso is cantilevered (b). Have the athlete repeat the sets. Move the pad farther down the athlete's body (perhaps to about the pelvis), leaving more of the torso cantilevered, and have the athlete repeat the entire process (c).

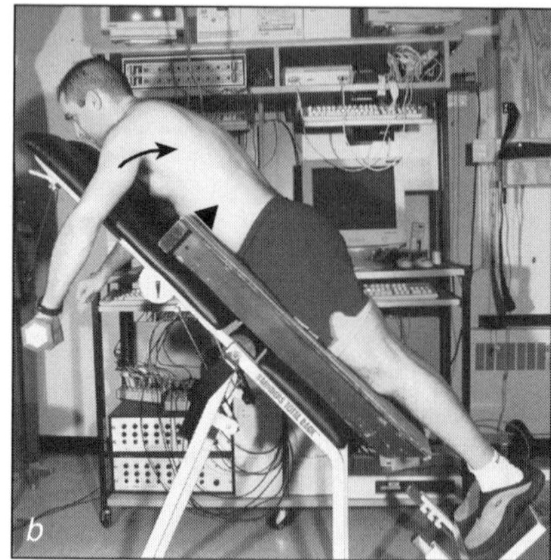

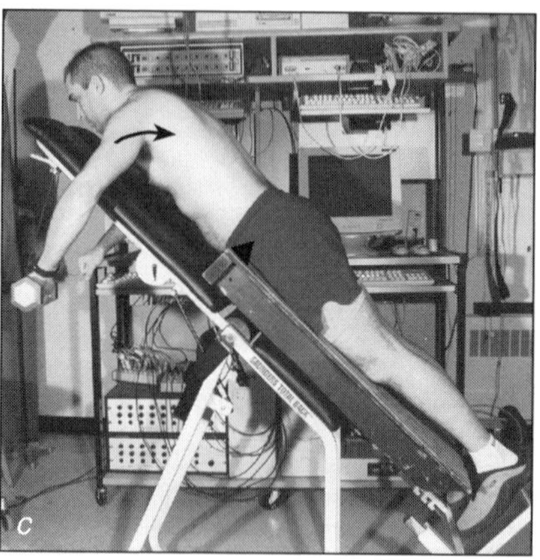

Another variation of this exercise is to perform the extension motion over a gym ball where the feet are secured (d) and the ball is moved progressively toward the pelvis with each set (e). (See Goldenberg and Twist, 2002) or the legs are cantilevered (f) and the ball moved progressively toward the chest (g).

While some strength athletes have successfully used this exercise to enhance performance, we are still quantifying the muscle activation and joint loads. Full data will be presented on this and other "strength" exercises in my upcoming book for athletic performance.

f

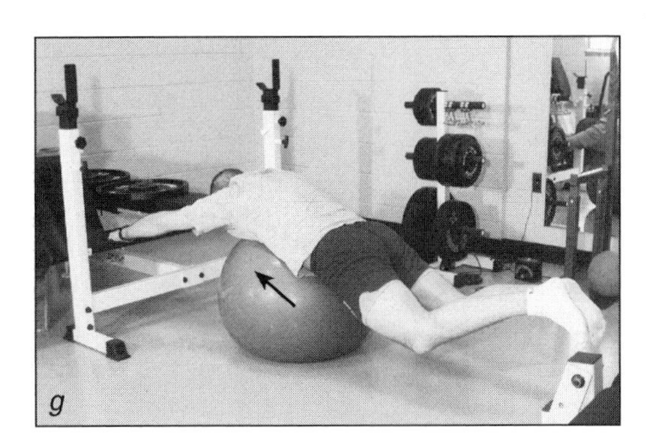

g

Occupational and Athletic Work Hardening

Certain workers and athletes encounter demands that require training that we would never recommend to patients unless they were absolutely necessary. Anyone considering going beyond the exercises described previously should be fully aware of the increased risks of back injury that these exercises will pose. These exercises aimed at workers or occupational athletes should be restricted to that group because they present a level of risk that is unacceptable for any but those who are willing to assume them for the sake of athletic or work performance.

Once again, choosing the appropriate challenge involves a blend of clinical art and science. Too little load will not produce the training adaptation, and too much will break tissue down. Listen to the body, and give adequate rest.

Low Back Exercises for High-Performance Workers or Athletes

I will be writing a book in another year or so on evidence-based high-performance training for the back. Given the emphasis of this book on disorders, this section is brief and introductory.

In general, we stage training of the athlete with the same steps used in our rehabilitation approach, only two more steps are added. We begin in stage 1 by identifying the essential motions and grooving appropriate motion/motor patterns. Stage 2 is directed toward ensuring joint and whole-body stabilizing patterns. Stage 3 is to develop muscle endurance around these patterns. Stage 4 is directed to enhancing strength. And, finally, stage 5 is to establish power. Unfortunately, too many exceptional athletes are trained, or given a rehabilitation program based on strength and power, without an adequate foundation of stabilizing motion/motor patterns. They end up with back problems and are referred to us.

In this section we will briefly consider exercises that may be helpful to either workers with demanding jobs or athletes. The requirements of the activities are not excessive for those who have adequately prepared by mastering all of the exercises previously discussed in the text. Specifically, these individuals should have mastered spine position awareness and be able to produce low back stabilizing patterns but may need more work to successfully move through the ranges of motion specific to their task. In addition, they may need more strength and endurance.

Qualifying the Worker/Athlete

The clinical decisions involved in staging a worker/athlete through the progression of challenge is an art that can be assisted with scientific data. Those who are looking for a set recipe, however, will fail at creating a successful program for an individual. Generally we blend the worker/athlete's current exercise status and history of injury with our own knowledge of spine tissue loads that result from various activities and our knowledge of injury mechanisms to qualify an individual for a specific exercise progression. We then put our educated guess to the test and monitor patient progress to ensure the maintenance of a positive slope of improvement in symptoms and function.

Lumbar Stability With Elevated Simultaneous Physiological Work Rates

While some individuals can maintain spine stability over all sorts of activities, including those that require the stabilizing musculature to assist with physiological challenges such as elevated breathing, others cannot (McGill, Sharratt, and Seguin, 1995). We have noted a compromise in the ability to stabilize more often postinjury (McGill et al., in press), although we have also detected that compromise in "virgin backs." Interestingly, tall athletes tend to be poorer in their ability to cocontract the abdominals to ensure sufficient stability during high work rates and highly challenged breathing than their shorter counterparts. This perception appears to be supported by the most recent evidence showing that taller workers have a greater likelihood of having perturbed motor patterns while breathing heavily and holding spine loads (McGill et al., in press).

Consider the warehouse worker, firefighter, or football player who must work at a high physiological rate that results in deep and elevated lung ventilation. The inability to maintain constant cocontraction in the abdominal wall (i.e., the muscles tend to relax during deep inhalation) is an indicator of compromised spine stability, particularly when heavy external loads that demand a stable spine are required. These individuals must develop stabilizing motor patterns that will transfer to all activities.

► Establishing Spine Stability When Breathing Is Elevated

One training exercise we have developed is to have the individual

1. ride an exercise bike at an intensity to elevate ventilation, and then
2. immediately dismount and adopt the side-bridge posture on the floor.

In this position the stabilizing musculature must remain isometrically contracted; otherwise, the bridge posture is lost. Heavy ventilation is also required, however, acting to groove the motor patterns that coordinate diaphragm contraction and other thoracic muscles involved in efficient lung function. The curl-up and birddog postures are also used after vigorous stationary biking to groove stable patterning in all of the supporting muscles. In this way we establish spine stabilization patterns in workers and athletes alike.

Torsional Capability

Some activities that athletes or workers must perform require twisting and the creation of substantial torsional moments. The difference between the mechanics involved in twisting and the generation of torsional moments was discussed in chapter 5. The question is, how can we maximize stability and minimize the risk of injury during training for trunk torsion? Because, as previously noted, generating torque about the twist axis imposes approximately four times the compression on the spine as an equal torque about the flexion/extension axis, it is unwise to train for torsion generation until the back is quite healthy.

▶ Training for Torsional Capacity

The technique we have found for producing low spine loads while challenging the torsional moment generators is to raise a handheld weight while supporting the upper body with the other arm and abdominally bracing (see a and b) to resist the torsional torque with an isometrically contracted and neutral spine. Dynamic challenged twisting is reserved for the most robust of athletes. We would never recommend training on the torsional machines (such as those where the athlete is seated with the pelvis belted down and the upper body twists against resistance) unless specific athletic performances were the training objective, and then the individual must be made aware of the elevated risk of pursuing this approach.

Some very useful and transferable motion patterns are produced with the various cable exercises (e.g., high pulls, low pulls, and "chopping" exercises), which produce high torsional challenges. Because of the lumbar compressive loads associated with these torsional torque challenges, we suggest discipline in bracing the torso in a neutral posture. It would also be wise to consider the transmissible vector (the perpendicular distance between the cable force and the lumbar spine) when choosing the technique to control the torsional torque and resultant spine load.

Antalgic Posture From Sitting

At our university clinic we notice a rash of back troubles at exam time. The students are sitting too long studying. Typically, they sit for a long spell and then when they stand they have the typical antalgic posture, leaning forward at the hips and unable to immediately stand upright.

► Special Suggestion for Those Whose Back Is Worse With Sitting and Then Have Trouble Standing

For these patients I may suggest

1. putting on a backpack weighing about 5-10 kg (11-22 lb) and
2. going for a walk over rough ground for an hour or two.

Amazingly (to some), many patients report that this works wonders. The typical forward-flexed antalgic posture requires the back extensors to be active (which imposes a substantial load penalty on a flexed spine). The backpack acts as a counterweight to extend the spine and bring the torso upright. The back extensors are no longer needed to contract, effectively removing their contribution to spine load. Wearing the backpack reduced spine loading!

Low Back Exercises Only for Athletes

Specific athletic objectives require specific training techniques. As noted earlier, however, too many patients make the mistake of looking to athletes for training exercises under the misconception that the same approach will help their own backs. This critically important notion must be emphasized again: Athletic exercises are not for enhancing back health in patients. In addition, too many athletes use bodybuilding principles for building mass. These must not be emulated by patients. The following exercises are reserved for athletes only.

Training-Specific Athletic Maneuvers

The various trunk torsional machines or resisted twisting motions are reserved for those who want to excel at very special tasks such as Olympic discus throwing. (I must add that some track athletes with whom I have been associated clinically do not perform these exercises because it exacerbates their back symptoms.) Otherwise, the torsional moment capabilities are developed with the spine in a neutral position, with a fully braced torso musculature, but with a torsional moment challenge. The neutral position is the most robust posture to withstand the elevated spine loads; it is also the spine posture that is transferable to other activities that require torsional moments with the least risk of spine damage.

Sprinting and other power running events such those performed by football players should be addressed here. Top-end speed for most sprinters is usually limited by the recovery of the leg in flexion (hip flexion), not a lack of hip extension power. Thus, many train the psoas according to the power philosophy—and end up with a bad back. It is critical for these athletes to maintain spine stability and be conscious of the braced and neutral spine during rigorous hip flexor training required for sprinting events.

All sorts of exercises and special equipment have been developed for performance training. But very few performance approaches have been documented with substantial scientific foundation to warrant discussion in a book such as this. However, one exception is the book *Supertraining* by Dr. Mel Siff, where an impressive number of studies have been synthesized and reviewed. For example, Dr. Siff makes a case for avoiding the development of motions that are guided predominantly by machines, and instead advocates cables and free weights with data-based argument. Further, substantiated principles of functional anatomy, physiology, biomechanics, and motor aspects such as facilitation are used to justify the approaches documented in *Supertraining* to enhance strength, power, speed, and muscle endurance. These performance objectives cannot exist without the foundation of stability, mobility, and the additional principles I have documented in this book. While there is no question that these principles and exercises are excellent for performance, there is a high resultant spine load. Some backs simply will not tolerate some of these higher-demand exercises—but not everyone can tolerate the rigors of training to be a champion, regardless of the soundness of the scientific technique. The exercise variations are endless and are outside of the scope of this book.

Finally, when designing workouts, the "big three" can be performed at the beginning of the workout to help groove and establish the stable motor patterns for the rest of the training session. Some trainers also like to finish the session with these exercises if they desire to take the exertions to fatigue.

▶ Star Exercises

One method growing in popularity is known as the star exercises, which are designed to promote stability throughout a range of motion. A plus sign (+) is drawn on the floor. The patient begins with the feet on the center of the + and performs various practice movement patterns, moving into each quadrant. What is performed in one quadrant is repeated in another quadrant for motion/motor symmetry. Various pairs of positions may be combined with different postures and may be done with or without resistance, which may be provided by either dumbbells or hand grips on cables. See, for example, the pairs (a) and (b), and (c) and (d). Star exercises are ideal for grooving patterns such as a tennis serve. It is crucial that you emphasize abdominal cocontraction and limiting lumbar motion to the patient's safety zone.

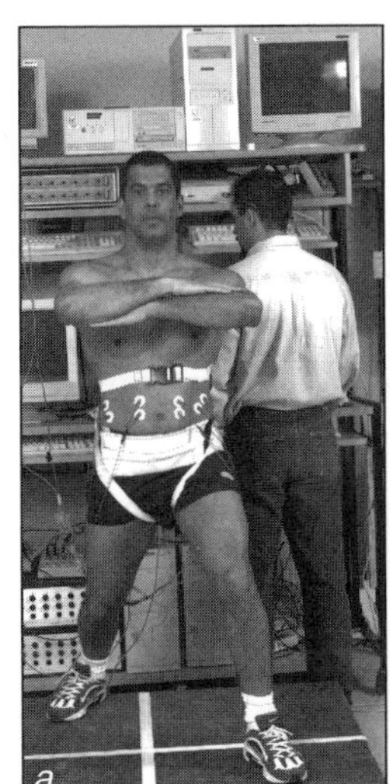

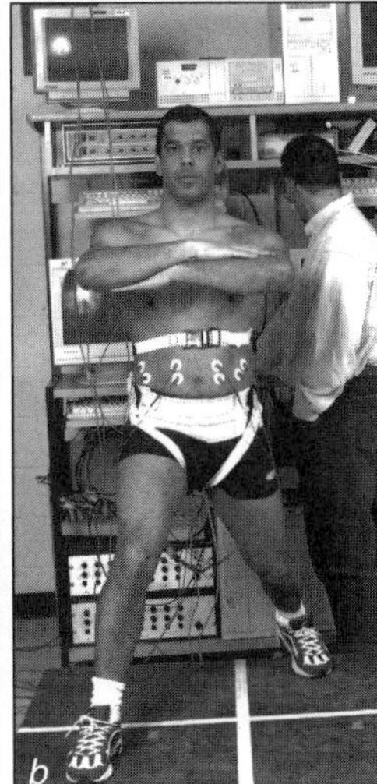

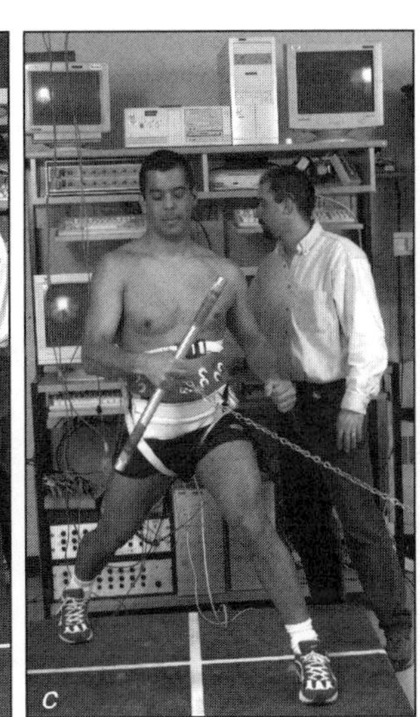

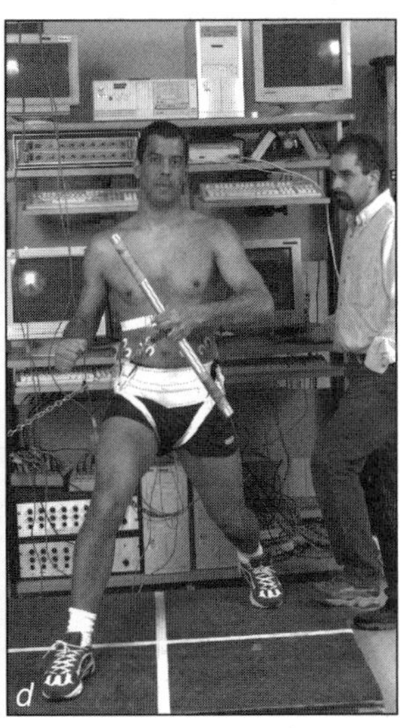

Training to Squat and Power Clean

Squats and power cleans are good exercises for developing power for athletic performance. The instructions from our perspective are very simple: Never sacrifice form for lifting more weight. Year after year I give this advice to young athletes, and year after year a substantial proportion will ruin their backs by not heeding this simple guideline. As the old saying goes, "It's amazing how much your parents learned as you grew older."

Very few people in North America can perform squats and power cleans well and with a low risk of injury. The Eastern Europeans are technical masters, and it is beyond the scope of this book to delve into the nuances of the technique. But it is interesting to see the emphasis placed on grooving the correct motion/motor patterns long before substantial weight is attempted. Young Eastern European athletes spend years developing the form by lifting broomsticks. Only when the form is perfect is strength increased and weight added to the bar. Generally, preserving the neutral lumbar spine will solve many of the safety issues, but this depends on the ability to take the shoulders and hips to extremes in the range of motion. Some have found that placing the weight bar on blocks to raise the starting height improves the utility of this exercise for athletes other than competitive weightlifters. This way a lifter can accomplish a fast lift without the larger back loads associated with the initial crouched posture needed to pull from the floor. In this power exercise speed is important; participants are advised to "train slow to be slow, train fast to be fast." For those interested in these exercises, the actual weight kinematics for the full power clean and the coordination of the leg muscles, torso muscles, and those involved in the shoulder pull can be found elsewhere.

Power athletes who are not weightlifters are given power lifting exercises with things like medicine balls (see figure 14.7, a-c) in our programs. We rarely recommend power cleans with a bar despite their widespread use.

Figure 14.7 Medicine ball lift *(a)* and *(b)* and toss *(c)* for power.

Looking Forward

Rehabilitation endeavors are continuing to embrace techniques that consider notions of torso stability and various components involving posture, motor patterning, and appropriate progressive challenge. Many groups continue to work

- to understand the contributions to stability of various components of the anatomy at particular joints—and the ideal ways to enhance their contribution;
- to understand what magnitudes and patterns of muscle activation are required to achieve sufficient stability while sparing the joints;

- to identify the best methods to reeducate faulty motor control systems to both achieve sufficient stability and reduce the risk of inappropriate motor patterns occurring in the future; and
- to develop motor patterns for optimal performance in athletes.

Collegial efforts between scientists and clinicians continue to develop the scientific foundation to justify better low back injury prevention and rehabilitation approaches. Much remains to be done.

References

Goldenberg, L., and Twist, P. (2002) *Strength ball training*. Champaign, IL: Human Kinetics.

McGill, S.M., Sharratt, M.T., and Seguin, J.P. (1995) Loads on spinal tissues during simultaneous lifting and ventilatory challenge. *Ergonomics*, 38: 1772-1792.

McGill, S.M., Grenier, S., Preuss, R., and Brown, S. (in press) Motor performance characteristics of the low back: Links to having a history of low back troubles.

Vera-Garcia, F.J., Grenier, S.G., and McGill, S.M. (2000) Abdominal response during curl-ups on both stable and labile surfaces. *Physical Therapy*, 80 (6): 564-569.

Epilogue

Now you have another perspective to employ in your efforts to prevent and rehabilitate bad backs. I hope that from now on, when you read that bad backs just happen, or that they cannot be diagnosed to guide therapy, or that they are solely the function of psychological factors or the compensation system, you will pause and reflect that you are reading the musings of individuals who have reached the end of their expertise.

Not that anyone has all the answers. With each experiment that we perform we may obtain some new insight, but generally we are confronted with many more questions. So, with each experiment we become relatively more ignorant. And after hundreds of experiments we are so ignorant that we are tempted to think that we are no longer qualified to make recommendations about the low back to anyone!

Consider the viewpoint described in this text and blend it with your own clinical wisdom and experiences. The very best clinicians, with whom I have had the pleasure of working, had wonderful clinical skills and insights but made use of a scientific foundation.

To the clinicians—I wish the confidence to continue with what you know works and the inspiration and leadership to try new approaches when things are problematic. To the students of clinical disciplines and those involved in the continuing development of the scientific foundation—there are so many more exciting experiments to perform on the wonderful fascination that we share, the low back. May we all enjoy the continuing journey.

Appendix

A.1 Raw Cross-Sectional Areas (mm²) (Standard Deviation in Parentheses) Measured Directly From MRI Scans

Muscles	Vertebral level					
	L5	L4	L3	L2	L1	T12
R rectus abdominis	787(250)	750(207)	670(133)	712(239)	576(151)	
L rectus abdominis	802(247)	746(181)	693(177)	748(240)	514(99)	
R external oblique		915(199)	1276(171)	1158(222)		
L external oblique		992(278)	1335(213)	1351(282)		
R internal oblique		903(83)	1515(317)	1055(173)		
L internal oblique		900(115)	1424(310)	1027(342)		
R trans. abdominis	119(22)	237(82)	356(110)	596(50)		
L trans. abdominis	175(57)	224(48)	376(115)	646(183)		
R abdominal wall*	1104(393)	2412(418)	3269(422)	3051(463)		
L abdominal wall*	1146(377)	2420(475)	3329(468)	3111(556)		
R longissimus thor.			747(162)	1175(370)	1248(228)	1095(222)
L longissimus thor.			782(129)	1089(251)	1180(184)	1258(347)
R iliocostalis lumb.			1368(341)	1104(181)	1181(316)	921(339)
L iliocostalis lumb.			1395(223)	1150(198)	1158(247)	835(400)
R multifidus			447(271)	343(178)	290(96)	289(66)
L multifidus			472(269)	366(157)	324(95)	312(76)
R latissimus dorsi			232(192)	429(202)	717(260)	1014(264)
L latissimus dorsi			256(217)	372(161)	682(260)	960(310)
R erector mass**	905(331)	2151(539)	2831(458)	2854(547)	2615(405)	2614(584)
L erector mass**	986(338)	2234(476)	2933(382)	2833(456)	2723(428)	2601(559)
R psoas	1606(198)	1861(347)	1594(369)	1177(285)	513(329)	330(210)
L psoas	1590(244)	1820(272)	1593(291)	1211(298)	488(250)	462(190)
R quadratus lumb.		725(209)	701(212)	552(192)	392(249)	320(197)
L quadratus lumb.		625(249)	746(167)	614(189)	404(220)	326(5)
Disc area	1360(276)	1459(270)	1415(249)	1332(294)	1334(285)	1241(166)
Total area	52912(9123)	51813(9845)	54286(8702)	55834(8112)	59091(6899)	63287(9153)

*Abdominal wall includes external and internal oblique and transverse abdominis.

** Erector mass includes longissimus thoracis, iliocostalis lumborum, and multifidus.

Vertebral level						
T11	**T10**	**T9**	**T8**	**T7**	**T6**	**T5**
938(49)						
938(21)						
556(234)						
551(170)						
331(89)	351(90)	312(97)				
327(80)	353(53)	355(73)				
1254(281)	1368(330)	1458(269)	1581(159)	1764(289)	1876(432)	2477(246)
1102(316)	1239(257)	1417(293)	1582(281)	1697(189)	2013(422)	2596(721)
1832(282)	1690(210)	1413(304)	1049(201)	842(165)	777(189)	743(70)
2041(285)	1722(279)	1471(351)	1129(100)	879(114)	779(95)	675(76)
1133(124)	1015(125)	933(112)	798(91)	797(104)	741(80)	671(82)
59249(7272)	61051(7570)	61732(6960)	65794(5254)	67782(3982)	66410(2372)	69337(2233)

A.2 Raw Lateral Distances (mm) Between Muscle Centroids and Intervertebral Disc Centroid (Standard Deviation in Parentheses)

Muscles	Vertebral level					
	L5	L4	L3	L2	L1	T12
R rectus abdominis	32(5)	38(7)	43(7)	46(8)	37(8)	
L rectus abdominis	−33(6)	−36(7)	−38(8)	−43(7)	−35(17)	
R external oblique		125(13)	130(10)	140(5)		
L external oblique		−120(9)	−125(9)	−133(7)		
R internal oblique		109(11)	116(8)	123(9)		
L internal oblique		−103(9)	−112(8)	−121(11)		
R trans. abdominis	99(1)	108(11)	122(9)	117(9)		
L trans. abdominis	−101(1)	−101(9)	−107(7)	−109(9)		
R abdominal wall*	102(8)	113(12)	119(8)	123(9)		
L abdominal wall*	−102(9)	−115(14)	−114(7)	−120(9)		
R longissimus thor.			22(4)	32(2)	32(6)	30(2)
L longissimus thor.			−19(5)	−30(6)	−37(12)	−34(4)
R iliocostalis lumb.			52(4)	58(4)	68(10)	65(7)
L iliocostalis lumb.			−48(6)	−60(10)	−65(9)	−67(7)
R multifidus			11(1)	13(4)	13(3)	10(3)
L multifidus			−14(7)	−12(3)	−11(3)	−11(2)
R latissimus dorsi			102(8)	108(8)	122(12)	129(10)
L latissimus dorsi			−104(15)	−107(9)	−117(11)	−128(7)
R erector mass**	22(6)	34(7)	40(4)	42(4)	44(5)	42(3)
L erector mass**	−21(5)	−33(6)	−38(5)	−41(6)	−41(7)	−40(4)
R psoas	54(4)	50(3)	44(3)	39(2)	32(3)	32(3)
L psoas	−54(4)	−48(4)	−42(3)	−38(3)	−31(3)	−32(2)
R quadratus lumb.		81(5)	75(6)	63(5)	46(6)	46(11)
L quadratus lumb.		−78(12)	−73(4)	−64(5)	−50(6)	−47(5)
Total area	0(2)	1(3)	−2(4)	−1(3)	−1(4)	0(3)

*Abdominal wall includes external and internal oblique and transverse abdominis.

** Erector mass includes longissimus thoracis, iliocostalis lumborum, and multifidus.

Vertebral level						
T11	**T10**	**T9**	**T8**	**T7**	**T6**	**T5**
29(1)						
–36(7)						
61(4)						
–67(11)						
8(2)	11(2)	12(2)				
–12(2)	–12(2)	–15(10)				
129(9)	140(9)	141(8)	145(7)	146(7)	153(7)	153(4)
–129(10)	–137(9)	–139(8)	–143(6)	–147(10)	–153(5)	–151(5)
34(4)	34(4)	32(4)	31(7)	30(4)	25(5)	27(2)
–40(3)	–36(3)	–35(4)	–33(6)	–31(2)	–29(3)	–27(6)
1(3)	0(2)	0(2)	2(1)	2(1)	1(3)	2(3)

A.3 Raw Anterior/Posterior Distances (mm) Between Muscle Centroids and Intervertebral Disc Centroid (Standard Deviation in Parentheses)

Muscles	Vertebral level					
	L5	L4	L3	L2	L1	T12
R rectus abdominis	81(16)	73(14)	79(13)	91(14)	109(8)	
L rectus abdominis	80(15)	73(14)	81(14)	92(14)	112(6)	
R external oblique		35(10)	20(14)	28(12)		
L external oblique		32(18)	19(11)	28(11)		
R internal oblique		41(12)	25(9)	36(17)		
L internal oblique		41(17)	26(12)	40(16)		
R trans. abdominis	55(0)	28(11)	22(11)	36(6)		
L trans. abdominis	50(5)	30(14)	23(10)	44(5)		
R abdominal wall*	58(16)	31(12)	17(12)	30(15)		
L abdominal wall*	59(17)	32(13)	20(12)	31(11)		
R longissimus thor.			–61(6)	–62(7)	–60(7)	–60(6)
L longissimus thor.			–61(5)	–63(6)	–60(7)	–59(8)
R iliocostalis lumb.			–57(7)	–61(7)	–62(5)	–59(7)
L iliocostalis lumb.			–57(7)	–61(6)	–61(4)	–58(6)
R multifidus			–55(7)	–55(6)	–52(6)	–51(3)
L multifidus			–53(7)	–56(5)	–51(5)	–50(3)
R latissimus dorsi			–45(16)	–47(12)	–47(10)	–39(8)
L latissimus dorsi			–43(17)	–46(10)	–46(7)	–37(8)
R erector mass**	–64(6)	–61(5)	–61(5)	–61(5)	–59(5)	–56(5)
L erector mass**	–63(5)	–61(5)	–61(5)	–62(5)	–60(4)	–57(5)
R psoas	18(9)	1(5)	–7(5)	–9(5)	–11(6)	–14(2)
L psoas	19(8)	2(4)	–6(4)	–8(2)	–11(4)	–11(1)
R quadratus lumb.		–36(9)	–37(6)	–37(6)	–35(4)	–31(6)
L quadratus lumb.		–31(5)	–34(6)	–36(5)	–34(4)	–32(8)
Total area	1(10)	–2(9)	1(8)	9(8)	18(5)	24(7)

*Abdominal wall includes external and internal oblique and transverse abdominis.

** Erector mass includes longissimus thoracis, iliocostalis lumborum, and multifidus.

Vertebral level						
T11	**T10**	**T9**	**T8**	**T7**	**T6**	**T5**
−56(4)						
−52(4)						
−57(1)						
−56(2)						
−47(5)	−49(4)	−48(2)				
−47(5)	−47(3)	−47(2)				
−32(7)	−24(7)	−22(7)	−18(9)	−17(6)	−12(3)	−17(5)
−28(9)	−23(7)	−19(7)	−17(7)	−15(8)	−11(7)	−19(3)
−54(4)	−54(4)	−52(4)	−52(3)	−52(4)	−47(4)	−50(3)
−52(4)	−52(4)	−51(4)	−51(3)	−51(4)	−46(5)	−50(3)
29(8)	30(9)	32(7)	36(7)	37(6)	32(10)	34(5)

Photo Credits

Steve Brooks: Figure 9.4

Stuart McGill: Figures 1.3a; 2.2, a and b; 2.3; 4.7; 4.9; 4.12; 4.13; 4.14, a and b; 4.17; 4.21; 4.22; 4.24; 4.25; 4.26, a through d; 4.32, a through d; 4.37; 4.38; 5.12; 5.13, a and b; 5.14; 5.15; 6.1, a and b; 6.5, a and b; 8.2; 8.3, a and b; 8.6, a and b; 8.7; 8.10, a and b; 8.11; 8.13, a and b; 8.14; 8.15, a through f; 8.17; 8.18; 8.19, a and b; 8.20, a and b; 8.24; 8.25, a and b; 10.1, a through c; 11.2, a and b; 11.3a; 11.4, a through d; photo on page 233; figures 13.3a; 13.5; 14.1, a through d; photos on page 264; photos on page 265; figure 14.7, a through c.

William Marras: Figure 7.1

Glossary

anisotropic—Behavior of an anisotropic material is dependent on the direction of the applied load. For example, adult bone is stronger in compression than in tension.

antagonist—A muscle that opposes the function of another in terms of torque generation but that may be necessary to ensure a stable joint.

breaking strain—Strain at which tissue fails.

buckle—The spine may experience several modes of unstable behavior. The column may buckle under compressive loads if sufficient stiffness is not present in the rotational degrees of freedom. The buckle is abnormally large rotation at a single joint, resulting in very large passive tissue forces.

dermatome—An area of skin supplied with afferent nerve fibers by single spinal nerve root.

eigenvalue—The element of the Hessian matrix that, if less than zero, indicates the potential for possible unstable behavior.

eigenvector—Describes the mode and site of unstable behavior, and what deficit "allowed" the instability to develop.

electromyography—The measurement of the electromyogram, or electrical event associated with muscle contraction.

etiology—Analysis and description of the cause of injury or disease.

failure tolerance—Load at which a tissue fails.

lordosis—The curvature of the lumbar spine where neutral lordosis is the natural curve of the spine, hypolordosis occurs with a flexed lumbar spine, and hyperlordosis occurs with an extended lumbar spine.

moment—Moment of force; occurs when a force is applied through a distance about an axis to create rotational effort.

moment arm—The perpendicular distance between the axis of rotation and a force, which creates a moment.

myoelectric—Of or relating to an electric signal obtained from contracting muscle.

myotome—Electric signal obtained from contracting muscle.

neutral lordosis—When the lumbar spine is neither flexed nor extended.

NIOSH—National Institute for Occupational Safety and Health.

nociception—Sensation of pain with a nociceptor (pain-sensing organ).

odds ratio—The increase in odds of an event occurring. (Full explanation via an example is provided on page 30.)

osteophytes—Bony spurs that grow on the stress points of a vertebra usually indicative of an unstable joint.

palpation—Using the hands to feel for spinal motion or pathology.

pars interarticularis—Part of the vertebra that connects the facet articulating surface to the neural arch and is the site of fracture.

pathomechanics—Mechanics associated with pathology.

Poisson's effect—A material property of an elastic sheet in which strain longitudinally produces proportional strain in the lateral axis.

provocative/functional diagnosis—Diagnosis based on tasks that exacerbate symptoms.

provocative testing—Provoking tissues with movement or direct load to elicit pain.

psychophysical—Variables that are associated with psychological or social aspects of an individual.

psychosocial—These variables influence behavior and include psychological traits such as perception of control over a task and social traits such as how individuals are viewed by others.

spondylolisthesis—A fracture of the pars that causes an anterior shear displacement of the superior vertebrae (usually L4 or L5).

strain—A change in length normalized to rest length (dimensionless ratio) and expressed as a percentage.

subfailure—An applied load that is less than the load required to cause damage if applied once.

synergist—A muscle that assists another in the same functional objective.

tolerance—A load value above which injury occurs. It is modulated by repetition, duration, rest, and so on.

transmissible vector—The transmissible property of any force vector is due to it having an effect along the line on which it is directed. A perpendicular line can be drawn from the force vector to any body joint to determine the moment arm, or the potential of that force to cause a moment.

viscoelastic—The property of being both elastic and viscous that absorbs the shock of loads.

viscosity – Friction as a function of velocity; it always moves in the direction of motion or impending motion.

yield strength—The load at which a tissue begins to experience damage.

Index

Note: Tables are indicated by an italicized *t* following the page number; figures by an italicized *f*.

About the Author

Stuart McGill, PhD, is a professor at the University of Waterloo in Ontario, Canada and internationally recognized lecturer and expert in spine function and injury prevention and rehabilitation. He has written more than 200 scientific publications that address lumbar function, low back injury mechanisms, investigation of tissue loading during rehabilitation programs, and the formulation of work-related injury avoidance strategies. He has received several awards for his work including the Volvo Bioengineering Award for Low Back Pain Research from Sweden. Dr. McGill has been an invited lecturer at many universities and delivered more than 200 invited addresses to societies around the world. As a consultant, he has provided expertise on assessment and reduction of the risk of low back injury, and rehabilitation exercise design, to government agencies, corporations, professional athletes and teams, and legal firms. He is one of the few scientists who consults and to whom patients are regularly referred.